Voice/Data Telecommunications for Business

STAN SCHATT
Analog Intelligence Corporation

STEVEN FOX
AT&T
Technical Senior Consultant

PRENTICE HALL
Englewood Cliffs, N.J. 07632

Library of Congress Cataloging-in-Publication Data

Schatt, Stanley.
  Voice/data telecommunications for business / Stan Schatt, Steven Fox.
    p.   cm.
  ISBN 0-13-107889-5
  1. Telecommunication. 2. Business communication. I. Fox, Steven [date]. II. Title.
  TK5102.5.S353 1989
  651.7—dc20   89-35267
                                                              CIP

The cover photo depicts the power and potential of today's integrated voice/data communications—unifying a diversity of information into a single, coherent resource.
© Copyright Fujitsu America, Inc.

Editorial/production supervision and
  interior design: Diane Delaney and Fred Dahl
Cover design: Lundren Graphics
Manufacturing buyer: Michael Woerner

 ©1990 by Prentice-Hall, Inc.
A Division of Simon & Schuster
Englewood Cliffs, New Jersey 07632

All rights reserved. No part of this book may be
reproduced, in any form or by any means,
without permission in writing from the publisher.

Printed in the United States of America

10 9 8 7 6 5 4 3 2 1

ISBN 0-13-107889-5

Prentice-Hall International (UK) Limited, *London*
Prentice-Hall of Australia Pty. Limited, *Sydney*
Prentice-Hall Canada Inc., *Toronto*
Prentice-Hall Hispanoamericana, S.A., *Mexico*
Prentice-Hall of India Private Limited, *New Delhi*
Prentice-Hall of Japan, Inc., *Tokyo*
Simon & Schuster Asia Pte. Ltd., *Singapore*
Editora Prentice-Hall do Brasil, Ltda., *Rio de Janeiro*

To **Jane**:
This book is for you.
I can't imagine anyone
more supportive or more wonderful.

To **Ellen**:
With love and gratitude.
Your support and devotion
made this achievement possible.

Contents

Introduction, vii

CHAPTER **1** *Voice Transmission, 1*

CHAPTER **2** *The PBX, 49*

CHAPTER **3** *Data Communications and Computers, 79*

CHAPTER **4** *Local Area Networks, 110*

CHAPTER **5** *Bridges, Gateways, and Micro-Mainframe Communications, 148*

CHAPTER **6** *The OSI Model and Data Communications Protocols, 171*

CHAPTER **7** *Packet-Switched Networks, 199*

CHAPTER **8** *ISDN, 217*

CHAPTER **9** *AT&T PBX Systems, 235*

CHAPTER **10** *Northern Telecom, 261*

CHAPTER	11	*IMB/Rolm/Siemens and Other Major PBX Vendors, 283*
CHAPTER	12	*Designing a Voice/Data Network, 304*
CHAPTER	13	*Network Management and Security, 337*
APPENDIX	A	*Voice and Data Periodicals for the Telecommunications Manager, 354*
APPENDIX	B	*The Telecommunications Manager's Bookshelf, 358*
APPENDIX	C	*On-Line Information Services and Databases, 360*
APPENDIX	D	*Manufacturers of Local Area Network Vendors, 362*

Glossary, 363

Index, 380

Introduction

A very subtle revolution is taking place right before our eyes. Instead of telephone companies and computer companies, we now have communications companies. In some cases, data processing departments have been renamed telecommunications departments to reflect this growing trend toward the integration of voice and data. *Voice/Data Telecommunications for Business* is designed not only to provide you with a solid foundation in the basics of voice transmission and data communications, but also to show you how these principles have been incorporated in integrated voice/data products found in many Fortune 500 offices.

The first section of this book focuses on voice communications. Chapter 1 explains analog wave theory and basic switching techniques. These are the foundations behind our public switched network, a topic we also examine. Chapter 2 provides an in-depth discussion of the Private Branch Exchange (PBX). We trace it through its various generations and then focus on its basic architecture and significant features. In Chapters 9–11 we will return to this very important subject and apply what we've learned about the generic PBX to specific models offered by AT&T, Northern Telecom, and IBM/Rolm/Siemens.

The second section of this book concentrates on providing a solid foundation in data communications. Chapter 3 begins with the bit, the fundamental building block computers use to describe data. We survey a number of major data coding schemes and methods of data transmission as well as discuss some of the more popular error checking schemes. We then examine the mainframe world of terminals and computers as well as the current state of microcomputers.

Chapter 4 provides an in-depth look at local area networks including network components, media, and architecture. We discuss the major IEEE 802 standards, including the latest trends toward the use of twisted pair wire on high-speed Ethernet networks, and the development of the FDDI standard for fiber optic networks.

In keeping with our desire to blend theory with actual practice, we focus on such major products as IBM'S Token Ring Network and Novell NetWare, as well as look at Macintosh networks and Macintosh communications with IBM PCs and compatibles. We also discuss the significance of OS/2 and LAN Manager, a topic that is bound to come up in any data processing or telecommunications department.

Chapter 5 is devoted to bridges, gateways, and micro-mainframe communications. We look at how a local area network can communicate with another LAN as well as establish a gateway to the IBM world of System Network Architecture (SNA). We provide a tutorial on the major features and architecture of SNA and then examine the implications of LU 6.2 on the future of peer-to-peer communications. We also discuss how Macintosh networks can communicate with this IBM mainframe environment. While we can't claim to give DEC equal time, we do look at its network architecture as well as its links to the microcomputer world.

Chapter 6 marks our final chapter on data communications; it focuses on the OSI model and some popular computer protocols such as TCP/IP and XNS. We also look at the structure of the Government OSI Profile (GOSIP) and the effect it will have on subcontractors in the very near future.

The third section of this book focuses on how major PBX vendors such as AT&T, Northern Telecom, and IBM integrate their voice and data communications as well as use their PBXs to provide gateways to packet switched networks, local area networks, and mainframe computers. Chapter 7 explains the basics of packet switched networks and then looks at public, private, and hybrid networks. Chapter 8 focuses on ISDN and provides the latest information available on what is no longer a dream but a reality—at least a reality with beta test status. Chapters 9, 10, and 11 focus on AT&T, Northern Telecom, and IBM/Rolm/Siemens respectively. While actual products change rapidly, these companies have definite PBX architectures that have proven remarkably stable and warrant careful examination. We hope these chapters will give students a glimpse of real telecommunications in practice.

The final section of our book covers voice/data network design and management. We lead the reader through the step-by-step process required to design a voice/data network using a case study as the basis for discussion. We look at network optimization as well as network management and control issues, and discuss how companies such as IBM and AT&T provide these services.

While we assume no prior voice or data background for our readers, we do cover a lot of material and have made a real effort to discuss today's major telecommunications issues with some degree of depth. We hope you enjoy using this text and find it provides a solid basis for exploring the exciting world of telecommunications.

Voice/Data Telecommunications for Business

Voice Transmission

CHAPTER 1

OBJECTIVES

Upon completion of this chapter, you should:

- be able to describe how a telephone operates
- know the difference between *analog* and *digital*
- understand basic analog wave theory
- be able to describe how digital transmission differs from analog transmission
- be able to list and explain different switching techniques
- be able to list different types of networks
- understand the design of the Public Switched Network

INTRODUCTION

Voice communications is nothing more and nothing less than methods for allowing people to talk to one another over long distances. One of the most universal methods is the telephone, an instrument that allows people to converse beyond the distance they can normally speak and hear. Extending the distance over which a conversation may be held is the telephone's chief benefit. But what a benefit it is. A person can use a telephone to call home whether home is two blocks away or 2,000 miles away.

We take for granted the convenience a telephone provides. For instance, picking up a telephone to call a store to be sure it is open or calling a theater to check show times. If we are late for an appointment, it is relatively easy to call ahead and say we will be delayed. No other means of long-distance voice communication can be used as easily or as effectively.

On a larger scale, the growth of cities and businesses owes quite a bit to the telephone. Cities and businesses both must be managed. Managing them requires a flow of information between areas of activity and decision-making points. City services departments that otherwise might have to be located at city hall can be dispersed throughout the community because they can communicate with centrally located decision-makers by telephone. Businesses can expand into new areas because branch offices can quickly and easily call top management for decisions and direction.

The telephone has come to have other benefits as well. The telephone network is now used not only for conversations, but for nonvoice applications, too. Television networks rely on it to send programs to affiliates throughout the country. Facsimile transmissions, which allow copies of documents to be sent from location to location, are carried through the telephone network. Computer-generated information is sent via telephone line wherever it is needed.

Various technologies provide the means to accomplish these remarkable things. The telephone started as a single idea, but imagination and changing needs boosted it to become a system of diverse communications capabilities.

A discussion of the telephone must necessarily include the telephone network. While the telephone converts speech to electric current, the network is the key to making it useful. The interconnection of locations and the ability to call any number of places give the telephone its appeal and broad utility. This utility has found application far beyond original expectations. The balance of this chapter looks at the technologies that make the telephone and the telephone network function.

VOICE BASICS

How a Telephone Operates

When a person speaks, vibrations are created that pass through the air as sound waves (see Figure 1.1). The waves reach the telephone's mouthpiece (a microphone) and cause its diaphragm to move. A part of the diaphragm is embedded in carbon granules. The granules compress and decompress with the movement of the diaphragm. When compressed, they emit an electric current that varies in strength, according to the degree of compression. Lower frequencies and louder sounds generate stronger currents, while higher frequencies and softer sounds generate weaker currents. The current is amplified and passed to the telephone line.

At the receiving end the electric current is turned back into sound (see Figure 1.2). In the telephone earpiece (a speaker) is another diaphragm. Permanent magnets create an electromagnetic field around the diaphragm. When electric current flows, a second electromagnetic field is induced by the electromagnet. The second

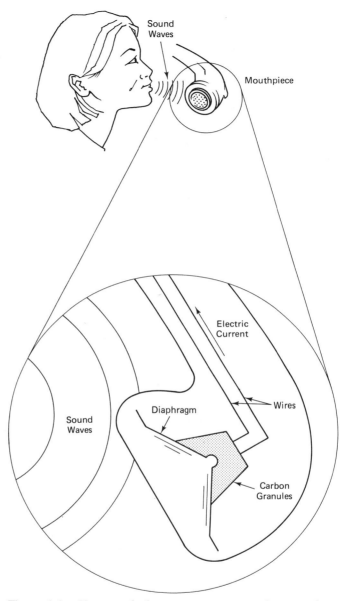

Figure 1.1: How a telephone operates converting sound to electric current. Sound waves strike the telephone's diaphragm, causing it to vibrate. The vibrations induce an electric current from the carbon granules.

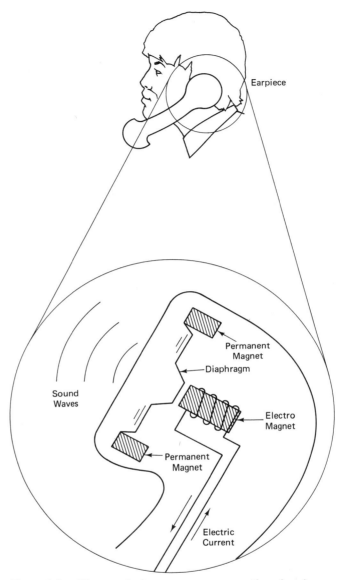

Figure 1.2: How a telephone operates converting electric current to sound. Electric current causes fluctuations in the electromagnetic field around the diaphragm. The diaphragm resonates according to the fluctuations, creating sound waves.

field disturbs the first one, which causes the diaphragm to vibrate. The vibrations produce sound waves that are heard as the caller's voice. The quality achieved is quite acceptable. The distinctive characteristics of a person's voice are transmitted, making it easy to recognize who is calling.

The mouthpiece in a telephone is a transmitter and the earpiece a receiver. They operate independently of one another, but it is interesting to note that a bit of the transmitted conversation is crossfed to the receiver. This is called sidetone and allows a person to hear his or her own words. The amount fed back is small, but it has an important purpose. When two people face each other and talk, they hear a portion of their own words because the sound waves created by their mouths are received not only by the other person's ears, but by their own ears as well. Early telephones did not provide any kind of feedback, and people felt like they were talking into a vacuum. Thus they tended to speak very loudly, thinking it necessary to be properly heard. Sidetone was introduced to correct this "human" problem. With it, a telephone conversation more closely approximated a face-to-face conversation and abated the human inclination to raise the voice.

Dialing

Dialing is a telephone's way of signaling the network. The telephone can only generate signals and relies on equipment within the network to interpret them. Two methods of dialing are in use today: rotary dialing and tone dialing.

Figure 1.3 shows a picture of a rotary dial telephone. This telephone is used by rotating the dial and then releasing it. When released, the telephone generates electrical pulses corresponding to the number dialed. The number 1 generates one pulse, 2 generates two pulses, 3 generates three pulses, and so on up to 0, which generates 10 pulses. Table 1.1 lists the pulses generated by each number.

Figure 1.3: The rotary-dial telephone. (Courtesy of AT&T)

TABLE 1.1
Pulses Generated by Dialed Numbers

1	One pulse	∼
2	Two pulses	∼ ∼
3	Three pulses	∼ ∼ ∼
4	Four pulses	∼ ∼ ∼ ∼
5	Five pulses	∼ ∼ ∼ ∼ ∼
6	Six pulses	∼ ∼ ∼ ∼ ∼ ∼
7	Seven pulses	∼ ∼ ∼ ∼ ∼ ∼ ∼
8	Eight pulses	∼ ∼ ∼ ∼ ∼ ∼ ∼ ∼
9	Nine pulses	∼ ∼ ∼ ∼ ∼ ∼ ∼ ∼ ∼
0	Ten pulses	∼ ∼ ∼ ∼ ∼ ∼ ∼ ∼ ∼ ∼

When the pulses are received at the other end by central office equipment, they are interpreted and the telephone is connected to the correct line based on the numbers dialed (see SWITCHING TECHNIQUES, below).

The terms *rotary dialing* and *pulse dialing* are often used interchangeably. Both refer to dialing methods that generate electrical pulses. While rotary dial telephones are rarely seen any more, there are many telephones on the market that have a pushbutton dialpad and use the pulse-dial method. Some telephones even have a switch allowing users to choose between pulse and tone dialing. That way the telephones can be used on all types of circuits, whether the circuits support tone and rotary dialing or rotary dialing only.

In the early 1960s, a second method of dialing was introduced: tone dialing. Instead of the electrical pulses used by rotary telephones, tones are generated. A dialpad replaces the rotary dial, and each number has a button. Two additional buttons are also part of the dialpad, and these are labeled * ("star" or "asterisk") and # ("pound" or "octothorp"). These additional buttons increase the flexibility of the telephone by providing more types of signaling than the digits 0 to 9 allow.

Figure 1.4: The tone-dial (DTMF) telephone. (Courtesy of Northern Telecom)

The * and # are used by PBXs and in private networks to access special features. As public networks become more intelligent, the use of these buttons for special functions will become increasingly common.

Each tone that is generated is the combination of two frequencies. There are seven single frequencies: 697 hertz, 770 hertz, 852 hertz, 941 hertz, 1,209 hertz, 1,336 hertz, and 1,477 hertz. Figure 1.5 shows how they are combined in pairs to produce a unique tone for every button. Each column on the dialpad has one frequency and each row has one frequency. The combination of frequencies for each button results from the button's position on the keypad.

A more proper name for tone dialing is **dual-tone multi-frequency (DTMF)**. This is the name used within the telephone industry. Among the public at large, commercial names, such as TouchTone, an AT&T tradename, are more common.

ANALOG TRANSMISSION

The sounds created by our voice and the electric currents generated by the telephone are waves that carry information through the telephone network. They can vary by subtle amounts, which is why a wide range of sound qualities and voice inflections can be transmitted. In fact, they can theoretically vary by infinitely small amounts. In more common parlance, they vary *continuously*. Signals of this type are called **analog** signals. They follow the principles of analog wave theory and can be thought of as analog waves.

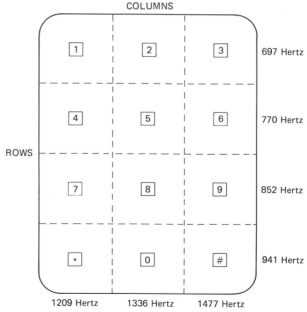

Figure 1.5: The tone (DTMF) dialpad.

Analog Wave Theory

Figure 1.6 shows a point on a circle. Imagine that the point moves around the circle in the direction shown at a steady rate. As it moves around and around it is said to **oscillate.** Oscillation, however, is more commonly thought of as a back-and-forth motion. If the circle were viewed from the edge, this back-and-forth motion would be more easily seen—the circle would appear as a vertical line segment with the point oscillating from end to end.

Every trip around the circle is called a *cycle*. The speed that the point travels can be measured in terms of how many cycles it makes in a given amount of time. This is called its **frequency.** Most often, frequency is measured as cycles per second, or **hertz.** If the point moves around the circle once a second, it has a frequency of 1 hertz. If the speed is increased to two cycles per second, the frequency is 2 hertz, three cycles per second is 3 hertz, and so on. 1,000 hertz is often expressed as 1 kilohertz and 1,000,000 hertz is often expressed as 1 megahertz. In sound waves traveling through the air, the higher the frequency; the higher the **pitch** of the sound. Normal human hearing can detect frequencies between 20 and 28,000 hertz. Standard voice conversations carried over the telephone are in the 300-to-3,300-hertz range (see Figure 1.7).

Another attribute of analog waves is **amplitude.** Amplitude corresponds to the intensity, or loudness, of the sound. When a person speaks softly, waves of small amplitude are created. When a person speaks loudly, waves of large amplitude are created. Returning to the example of the circle, the size of the circle indicates the amplitude. Soft sounds are represented by small circles and loud sounds by large circles. As the point moves around a smaller circle, it does not oscillate as far as in a larger circle (see Figure 1.8). The amplitude is half the total distance that it oscillates.

Sine Waves

Wave motion is continuous. Just as ripples in water undulate across the unbroken surface, analog waves undulate through the electrons in a wire. To better conceptualize this, think of the circle in our example as moving from left to right (see Figure 1.9). Tracing the path of the point as it goes around the circle leaves a trail

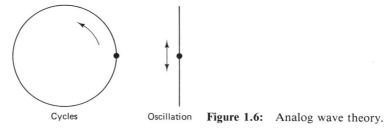

Cycles   Oscillation   **Figure 1.6:** Analog wave theory.

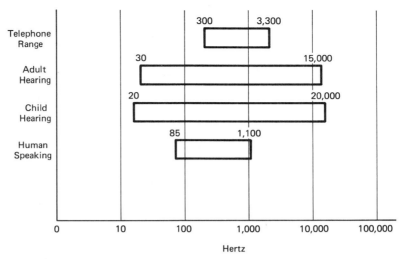

Figure 1.7: Frequency ranges.

that arcs upward, then returns to the midpoint, dips down and returns again to the midpoint. This pattern is called a **sine wave.**

Sine waves alternate between positive and negative, always beginning and ending at zero (the midpoint). Each time a wave travels in a positive direction, returns to zero, then travels in a negative direction and returns to zero, it completes one cycle. The frequency of a wave in hertz is the number of cycles it completes in 1 second (see Figure 1.10).

Sound Waves

The distance from the **crest** to the midline or the **trough** to the midline of a wave is its amplitude (see Figure 1.11). The basic unit of measurement is *watts per square centimeter* (W/cm^2), a measure of the power (intensity) of the wave. This is useful for determining the absolute power of a wave. However, with sound waves, it is

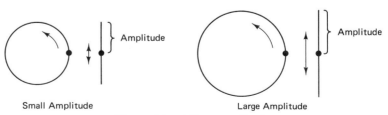

Figure 1.8: Wave amplitude.

Analog Transmission 9

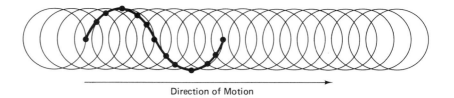

Direction of Motion

Imagine the circle in motion from left to right.

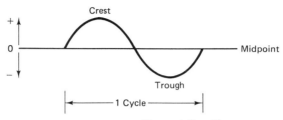

Figure 1.9: Sine waves.

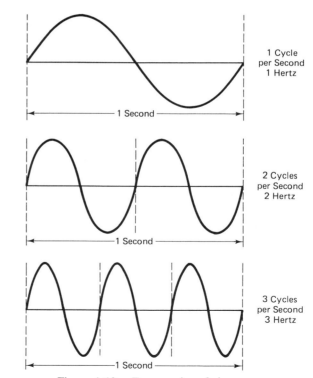

Figure 1.10: Frequencies of sine waves.

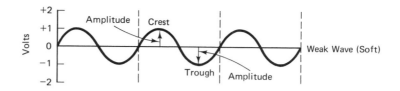

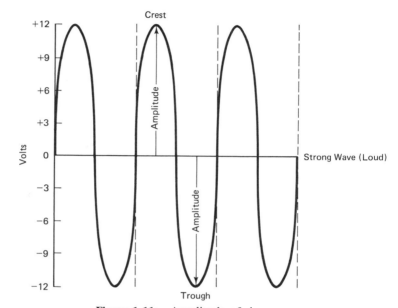

Figure 1.11: Amplitude of sine waves.

common to guage sounds in comparison to the threshold of human hearing. When two sounds are compared, a relative reference is used instead of the absolute reference of watts per square centimeter. The relative reference is called the **decibel (dB).** Decibels are calculated from the ratio of the signal strength to a reference value. The reference is 10^{-16} W/cm², a sound barely audible to an average adult. It has a decibel level of 0. The level of other sounds are computed using the formula

$$\text{decibels} = \log \left(\frac{\text{Input Signal Power}}{10^{-16} \text{ W/cm}^2} \right) \times 10$$

The decibel level is equal to the logarithm of the input signal power divided by the reference value, multiplied by 10. A conversation between two people with an input signal power of 10^{-11} W/cm² has a decibel level of

Analog Transmission 11

$$\text{decibels} = \log\left(\frac{10^{-11} \text{ W/cm}^2}{10^{-16} \text{ W/cm}^2}\right) \times 10$$

$$= \log\left(\frac{1}{10^{-5}}\right) \times 10$$

$$= \log 10^5 \times 10$$

$$= 5 \times 10$$

$$= 50$$

The result is decibel level of 50.

Note that decibels are a logarithmic measurement. This means that a level of 10 dB is 10 times more intense than a dB of 0, 20 dB is 100 times more intense, 30 dB is 1,000 times more intense, and so on. Table 1.2 shows a comparison of the levels of some common sounds.

Another way to think of it is that doubling the strength of a signal increases its decibel level by about 3. For example, if a quiet conversation has a decibel level of 45 dB, doubling the intensity (loudness) of the conversation increases the decibel level to 48 dB.

Note also the number 10 in the formula. It is included for the following reason: originally, the relative intensity of a sound was measured in **bels,** named for Alexander Graham Bell. The value 1 bel is too large a unit for many applications, so it has become common to measure in tenths of a bel, or decibels. The constant 10 in the formula converts the value from bels to decibels.

TABLE 1.2
Decibel Levels of Common Sounds

Sound Level (dB)	Intensity (W/cm^2)	Sound
0	10^{-16}	Threshold of hearing
10	10^{-15}	Rustle of leaves
20	10^{-14}	Whisper
30	10^{-13}	Quiet home
40	10^{-12}	Average home. Quiet office
50	10^{-11}	Average office
60	10^{-10}	Normal conversation
70	10^{-9}	Noisy office
80	10^{-8}	Busy traffic
90	10^{-7}	Inside subway train
100	10^{-6}	Machine shop
120	10^{-4}	Threshold of pain
140	10^{-2}	Jet plane

Reprinted by permission of the publisher from Alan H. Cromer, *Physics for the Life Sciences* (New York: McGraw-Hill, 1977), p. 287.

Electric Signals

While the basic unit of measure for the amplitude of sound waves is watts per square centimeter, once the waves are converted to electrical signals passing through a wire, the unit of measure becomes **volts**. The greater the voltage, the greater the amplitude and the stronger the signal. Voltage alternates between positive and negative values (see Figure 1.11). Signal strength depends upon the the absolute value of the voltage, not whether it is positive or negative.

Just as sound waves can be rated in decibels, so can electrical signals. Remember that decibels measure the relative strength of two waves. A reference value is used to provide a common basis of comparison. When dealing with electrical signals, the reference value becomes 1 milliwatt (mW) of power. It is given a decibel level of 0. The formula for computing the level of other signals is

$$\text{decibels} = \log\left(\frac{\text{Input Signal Power}}{1 \text{ mW}}\right) \times 10$$

An input signal of 2,000 mW has a decibel level of 33 dB, as shown below.

$$\begin{aligned}\text{decibels} &= \log\left(\frac{2000 \text{ mW}}{1 \text{ mW}}\right) \times 10 \\ &= \log(2000) \times 10 \\ &= 3.3 \times 10 \\ &= 33\end{aligned}$$

Comparing the output level of an electrical signal to its input level is also very common. The need for it comes up when a signal must be carried through a substantial length of wire. For instance, on a college campus, telephones may be located in buildings that are 15,000 feet or more from the central switching equipment. The signal may be strong at the originating end, but as it travels through the cable, it gradually weakens because of resistance from the wire's atoms. This weakening of the signal is known as attenuation. At the receiving end, the attenuated signal may be too weak to be heard. To check for problems of this type, decibel level readings are made at the originating and receiving ends. The amount of signal lost, called the **decibel loss** (or **dB loss**), is the difference between them. It indicates the amount of resistance in the wires and is used to determine if the signal needs boosting with amplifiers.

The formula for calculating decibel loss is similar to the formula for electrical signal strength, except the reference value of 1 mW is replaced by the output signal strength, and the output signal is compared to the input signal (not vice versa). The formula is

$$\text{decibels} = \log\left(\frac{\text{Output Signal Power}}{\text{Input Signal Power}}\right) \times 10$$

Analog Transmission

A signal with an input strength of 5,000 mW and an output strength of 50 mW would have a loss of

$$\text{decibels} = \log\left(\frac{50 \text{ mW}}{5000 \text{ mW}}\right) \times 10$$

$$= \log\left(\frac{1}{100}\right) \times 10$$

$$= -2 \times 10$$

$$= -20$$

A loss of -20 dB means that output is 100 times weaker than the input.

MODULATION

The sine waves discussed to this point are regular and unchanging in shape. They are most representative of pure tones. Natural sounds, however, such as the human voice, produce complex tone patterns. They change frequency constantly as the pitch goes up and down, and they change amplitude as the sounds get louder and softer. Their wave patterns are correspondingly more complex and are characteristic of those shown in Figures 1.12 and 1.13.

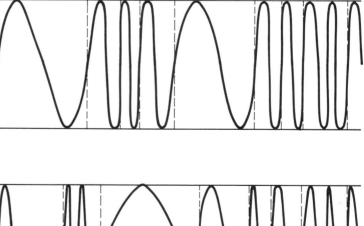

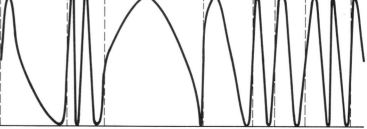

Figure 1.12: Sine waves of varying frequency.

These complex waves patterns are found elsewhere, too. For instance, commercial radio depends on them. A pure sine wave (carrier wave) is **modulated,** or changed, by the voice or music that is being transmitted (see Figure 1.14). The voice or music create signal waves that are superimposed on the carrier. The result is a wave embedded with voice or music information. At the receiving end, the wave is demodulated; the carrier signal is removed leaving the original voice or music.

Two techniques are commonly used: frequency modulation (FM) and amplitude modulation (AM). These are the techniques used by FM radio and AM radio, respectively. As their names imply, FM modulates the frequency of the carrier wave and AM modulates the amplitude.

A fair question would be to ask: "Why modulate a carrier and then demodulate it? Why not simply transmit the signal waves, since they carry the information?" The answer is that the carrier waves "shift" the range of the signal waves to a different part of the radiowave spectrum.

We have seen that the bulk of human speech falls within the 300-to-3,300-hertz range. Music extends the range to 20,000 hertz, near the upper limit of human hearing. This range from roughly 0 to 20,000 hertz is the frequency band of signal waves. Without a carrier wave to shift the range, all radio stations would transmit over the same band. They would interfere with one another and none would be intelligible.

Carrier waves are employed to solve the problem. They are pure tones set at

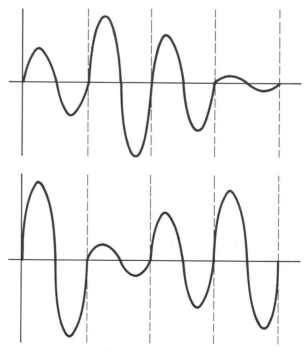

Figure 1.13: Sine waves of varying amplitude.

Frequency-Division Multiplexing 15

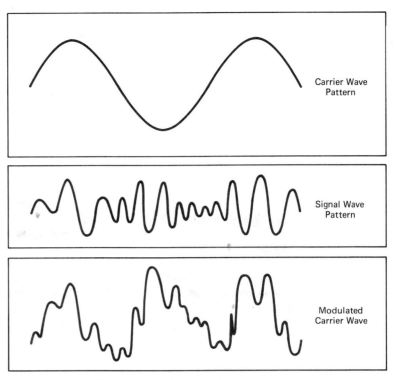

Figure 1.14: A modulated carrier wave.

particular frequencies and amplitudes. Radio stations are assigned a carrier frequency by the Federal Communications Commission (FCC). Assignments are made such that one station's signals will not interfere with another. Figure 1.15 shows some examples. For instance, a radio station is assigned a carrier wave frequency of 640 kilohertz. The signal waves require a range, or **bandwidth,** of 20,000 hertz. This means the station will transmit over a bandwidth from 640 to 660 kilohertz. In addition, guard bands will be added above and below this range. Guard bands are buffer zones where no information will be transmitted. They are included because analog waves have a tendency to drift from their assigned frequency. The guard bands allow for moderate drift without affecting the transmission of adjacent carriers. Total bandwidth for the radio station, then, is about 40,000 hertz, from 630 to 670 kilohertz. No other local radio stations will be assigned in this range.

FREQUENCY-DIVISION MULTIPLEXING

The technique of moving signal waves to different frequencies is used in the telephone industry, too. It allows telephone companies to make more efficient use of their cable and wire facilities.

The usual transmission medium of telephone calls is a pair of copper wires,

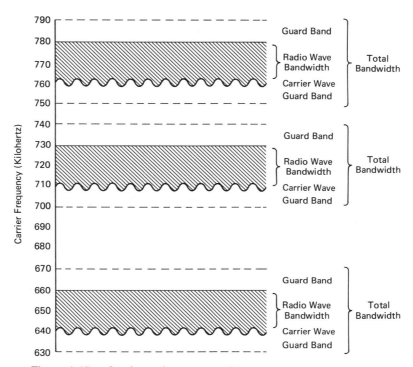

Figure 1.15: Carrier assignments and bandwidths of radio waves.

which has a bandwidth of about 1 megahertz. Since the voice range over telephone networks is about 300 to 3,300 hertz, a voice telephone call uses only a small portion of the wire pair's bandwidth. Adding guard bands above and below the voice range gives a total bandwidth for a voice call of 4,000 hertz (see Figure 1.16). Based on this, a single pair of wires can carry 250 calls. The first call would be in the 0-to-4,000-hertz range, the second call in the 4,000-to-8,000- hertz range, the third call in the 8,000-to-12,000-hertz range, and so on.

The technique of dividing a large bandwidth into smaller bandwidths and then spacing them by frequency so that a single **facility** contains multiple channels is called **frequency-division multiplexing.** Figure 1.17 shows an example of what it would look like to place 250 voice channels over a single pair of wires.

In practice, voice channels are not put on wires this way. Figure 1.18 depicts the how it is done in the public switched telephone network. Voice channels are grouped in a tiered, or hierarchical, manner. This allows them to be manipulated as a group. First, 12 voice channels are combined into a group. Frequency-division multiplexing is used to boost the channels to the 60-to-108- kilohertz range. Then, five groups are joined into a supergroup of 60 voice channels. Again, the frequency range is modulated to the 312-to-552-kilohertz range. Ten supergroups are combined to form a single mastergroup. The mastergroup contains 600 voice channels and is modulated to a frequency range of 564 to 3,084 kilohertz. At this point, there are

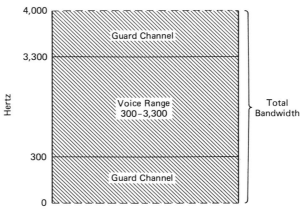

Figure 1.16: Voice channel bandwidth.

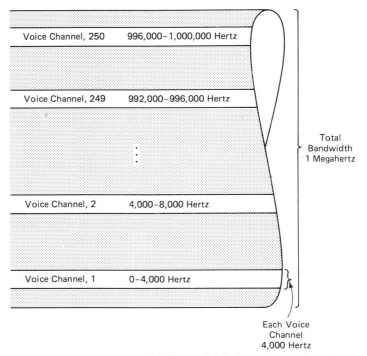

Figure 1.17: Frequency-division multiplexing. Shown is a hypothetical example of dividing 1-megahertz wire into 250 voice channels.

too many voice channels to be carried over a wire pair and, typically, coaxial cable is used.

It is possible to multiplex the channels further still. For instance, six mastergroups can be combined into a jumbogroup of 3,600 voice channels. Other combinations can be made as well. As you can see, very large numbers of channels can be involved. Grouping serves to bundle them so that all within the group can be handled as a unit. For instance, a supergroup in Los Angeles that is connected to a mastergroup in Denver may be moved to a mastergroup in San Francisco. Instead of moving 60 channels one at a time, all 60 are moved in one transaction. This provides a fast, efficient, and flexible way of making changes.

TIME-DIVISION MULTIPLEXING

Frequency-division multiplexing divides bandwidth by frequency into multiple channels. Bandwidth can also be divided by time. This is referred to as **time-division-multiplexing (TDM)**.

With TDM, bandwidth is partitioned into segments. Each segment is called a **timeslot** and represents a short interval of time. There are many timeslots and each is equivalent to a voice channel.

Figure 1.19 depicts timeslots in a wire. Four timeslots are shown indicating four channels are available and four calls can be supported. A more detailed discussion of time-division multiplexing can be found in Chapter 6.

One of the major advantages of TDM is that it can handle data transmission. Streams of data can be broken up to fit into timeslots without much difficulty. Furthermore, TDM can transmit both voice and data through the same wire, if the voice transmission is digitized.

Digitized voice is used quite commonly in the public switched network and amounts to converting an analog voice signal to a digital code. The digital code is passed through the network and converted back to an analog voice signal at the receiving end. Digitizing is used extensively for two reasons: (1) the sound quality of a call can be controlled better than on an analog circuit and (2) network circuits are utilized more efficiently when time-division rather than space division multiplexing is used.

SWITCHING TECHNIQUES

Up to this point, we have looked at calls and how they pass through facilities. This is one-half of the telephone transmission story. To complete the picture, we must look at **switching**.

Switching is the ability to connect any telephone to any other telephone in a network. It is switching that provides the flexibility and usefulness to the network. Without it, a telephone would be permanently connected to one other telephone only! To talk to two locations, two telephones would have to be installed.

This is, in fact, precisely what happened in the early days of the telephone

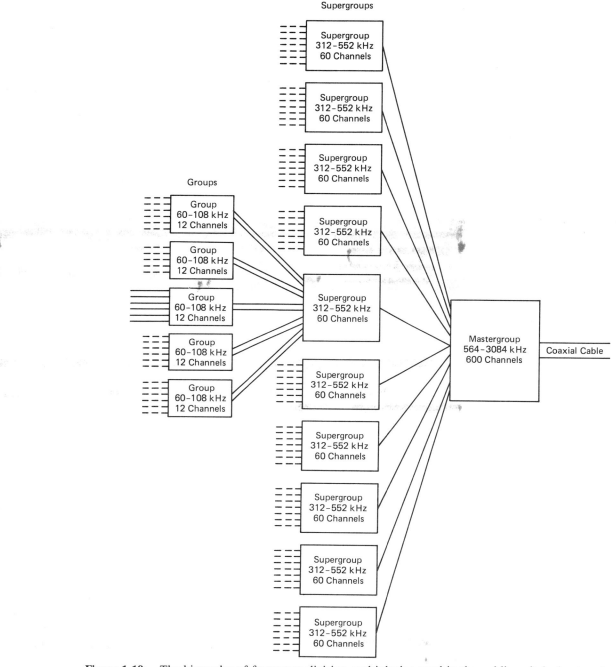

Figure 1.18: The hierarchy of frequency-division multiplexing used in the public switched network.

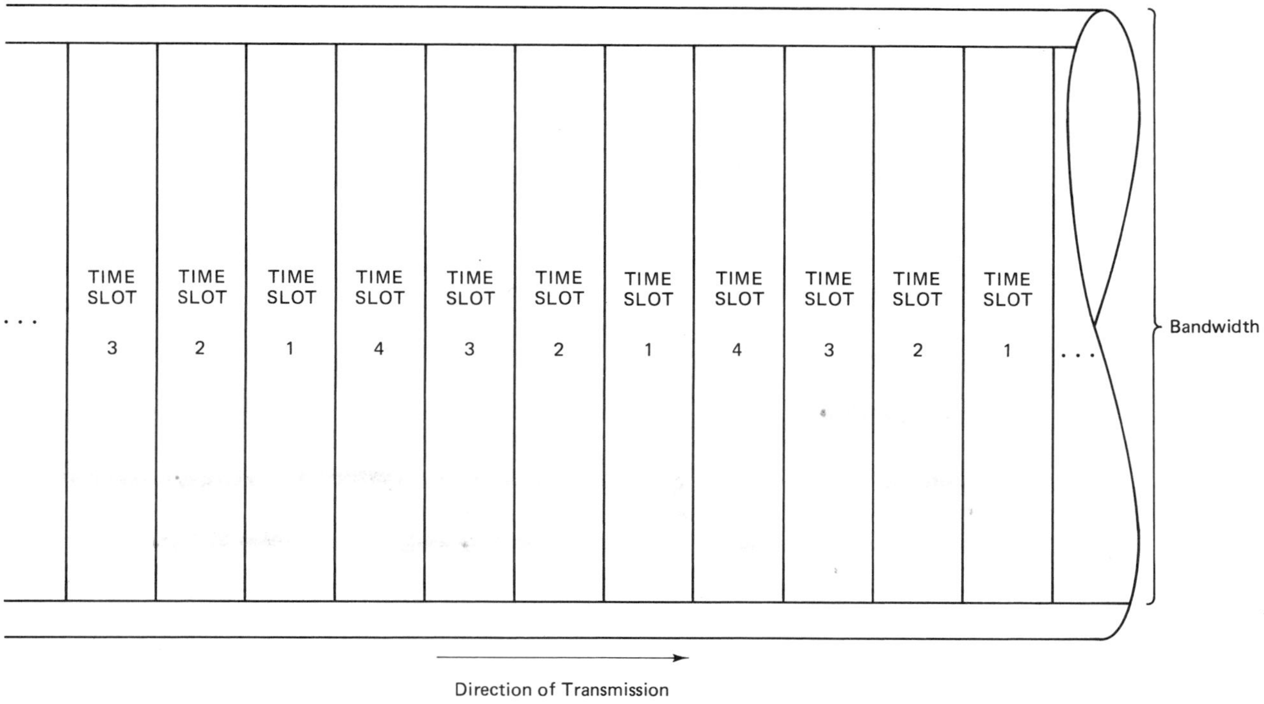

Figure 1.19: Time-division multiplexing. Four timeslots are shown, although an actual system would have many more.

industry. There was no switching equipment or operators. When a telephone was installed, it was directly connected by wire to another telephone. Within a short time, the need for switching was recognized and developments toward that end were implemented.

The first "switch" was an operator placing calls through a switchboard. Telephones, instead of being directly connected to one another, were connected to a switchboard. The operator would connect one telephone to another according to a subscriber's request. The operator and switchboard came to be known as an *exchange*.

Names were initially used to tell an operator where to place a call. Gradually, as the number of telephones increased, this was replaced by the use of numbers. Today telephone numbers in the United States and many other countries are seven digits long and are written in the form 555-1234. The first three digits ("555" in this case) are called the A, B, and C digits. They are called the *prefix* or *exchange number* and identify the central office serving the telephone number. All telephones with the same prefix are connected to the same central office. Central offices have replaced the operator and switchboard in being known as an exchange.

In our hypothetical telephone number, the numbers 1, 2, 3, and 4, are called the *thousands digit,* the *hundreds digit,* the *tens digit,* and the *ones digit,* respectively. Once a call reaches the serving central office, these digits are used to "steer" the call to the correct telephone. As such, these digits are sometimes called steering digits.

From human-operated switchboards came automatic switching equipment. Several types evolved over a period of years. They are described below.

ELECTROMECHANICAL SWITCHES

The earliest class of automatic switching equipment was electromechanical switches. These combined the use of electricity and mechanical movement to switch calls.

Step-by-Step

Step-by-step switches were the first type to be put into service. They were first used in a public exchange in 1919.

These switches operate by a technique called **progressive control.** At the time they were introduced, all telephones were rotary dial (pulse). Pulses from the caller's telephone directly controlled the movement of the switching equipment. Each pulse caused the equipment to move one step, hence the name *step-by-step*.

Figure 1.20 shows the operating elements. Wipers move vertically and horizontally to connect with banks of contacts. Note that the wipers rotate in an arc when they move horizontally. To conform with this motion, equipment is shaped like a cylinder, sometimes called a *can*.

The equipment is used to perform three functions, and each function is carried out by a different component of the switching system. The components are line finders, selectors, and connectors.

Figure 1.21 is a diagram depicting the connections among the components. Line finders connect to selectors, and selectors connect to connectors. The banks of the line finders are wired to incoming telephone lines from subscribers. Line-finder wipers are wired to the banks of selectors. Selector wipers are wired to connector wipers, and connector banks are wired to outgoing telephone lines.

When a subscriber picks up the telephone and goes off-hook, the wiper of an idle line finder moves first vertically and then horizontally to seize the contact from the subscriber's line. A selector, wired to the wiper's cords, returns a dial tone to the subscriber. When the subscriber dials the first digit, the selector's wiper steps vertically one to ten times, according to the number dialed. This selects the hundreds group of the number being called. It then steps horizontally looking for an idle connector. When one is found, the second dialed digit can be accepted. The connector receives the second digit and steps vertically in accordance with the number of pulses. This selects the tens bank of the number being called. When the third digit is dialed, the wiper steps horizontally the appropriate number of times. It then rests on the contact of the telephone number being dialed and the call is completed.

If you reflect for a moment, you will realize that the components of a step-by-step switch can switch three digits only. In order to handle larger telephone numbers,

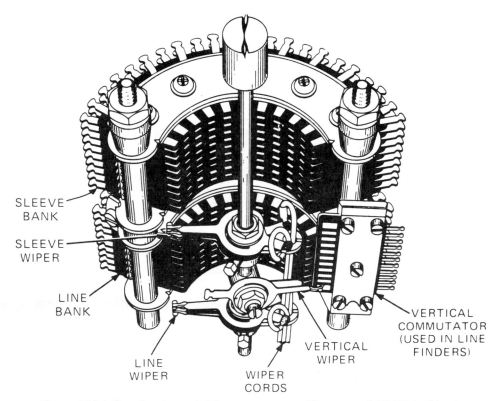

Figure 1.20: Step-by-step switching components. (Courtesy of AT&T Archives)

additional selectors must be put in place so that some selectors are tied to other selectors rather than directly to connectors.

The original design of the switches is attributed to Almon B. Strowger, an undertaker who suspected a local telephone operator of directing calls intended for him to a competitor. To eliminate the need for the operator, Strowger invented the switch. It proved effective and reliable and was widely used. Many are still in service today. Sometimes step-by-step switches are referred to as Strowger switches in his honor.

As important as the design is, there are some inherent problems making it less desirable than more recent designs. First, the pulses of a rotary dial telephone must be used to make it work. In today's world, DTMF dialing is common, and special equipment is required to convert tones to pulses. Second, adding capabilities such as alternate routing is expensive and difficult. Additional equipment must be added and a great deal of rewiring must be done. Third, some features, such as call forwarding, cannot be added at all. This means the switch cannot be used to offer such services as Centrex.

Electromechanical Switches

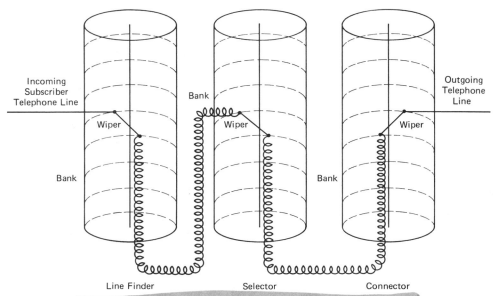
Figure 1.21: Making a connection through a step-by-step switch.

Panel

Another type of progressive controlled switch is the panel switch. Its name derives from the banks of contacts that are arranged in large, flat panels as opposed to the cylindrical cans of the step-by-step switch (see Figure 1.22). Contact brushes are used instead of wipers. They are located on rods and move vertically to touch the contacts on the panels.

Call progress through the switch is as follows (see Figure 1.23). When a telephone subscriber goes off-hook, a line finder connects to the line. At the same time, a sender is brought in and supplies dial tone. The dialed digits are received by the sender, which passes them to a decoder. The decoder determines the central office of the number being called and identifies the trunk group over which the call should be routed. The decoder then disconnects from the call, leaving it free to handle others. The sender routes the call according to the decoder's instructions from the local central office to the appropriate remote central office. It then controls the progress of the call at the remote office until a connection is made to the called telephone number. The sender drops off the call to be available for others while the subscribers begin their conversation over the newly made connection.

Panel switches were first put in use in 1921. Designed for large, metropolitan areas, their panels gave them greater capacity than step-by-step switches. However, they suffered from many of the step-by-step's shortcomings. For instance, adding features required expensive equipment, as did adding tone-dialing capability. Some features could not be added at all.

Figure 1.22: A panel switch. (Courtesy of AT&T Archives)

Crossbar

A newer type of switch is the crossbar. Introduced in 1938, the crossbar switch represents an improvement over step-by-step technology. Crossbar switches are classified as a coordinate switches because connections are made by closing individual coordinate points within the switching matrix.

Horizontal and vertical bars form a grid (see Figure 1.24). The intersection of any horizontal and vertical bar can be made to touch, creating a connection. Each

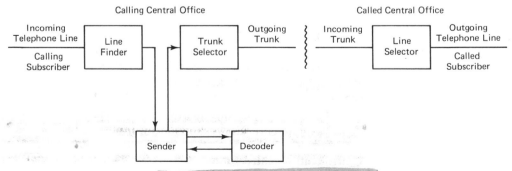

Figure 1.23: Elements of a panel switch.

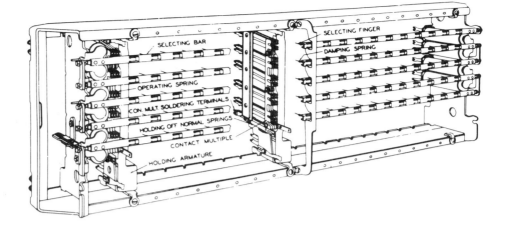

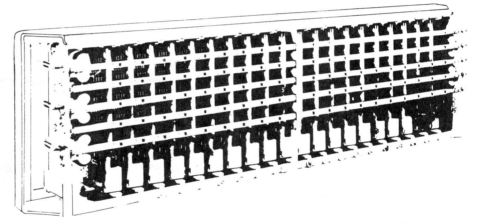

Figure 1.24: The grid of a crossbar switch. (Courtesy of AT&T Archives)

intersection corresponds to either a subscriber's telephone line or to a trunk. The hardware is designed so that several intersections may be activated along the same horizontal row of the grid. This allows many calls per grid, increasing the efficiency of operation.

The No. 5 crossbar switch, first used in 1948, is one of the most important of the crossbar switches. It incorporated significant design improvements that made it more flexible and capable than its forerunners.

Its components fall into two categories: networking hardware and common control equipment. Networking hardware includes line link frames and trunk link frames. The frames are, in turn, made up of crossbar switching grids. Common control equipment is composed of registers, markers, senders, number groups, and connectors.

The path a call takes is entirely through the networking hardware. The common control equipment is used only to set up the call. It then drops off to be available to set up other calls. Figure 1.25 shows a diagram of the No. 5's main components. When a subscriber goes off-hook, the originating register supplies dialtone and receives the dialed digits. A marker then selects an idle trunk. If the call is to another subscriber within the same exchange, it is looped from the line link frame to the trunk link frame and back to the line link frame. If the call is to another exchange, a sender is placed on the line to handle signaling over the trunk between central offices. On calls incoming from other exchanges, an incoming register is used to receive digits. A marker works in conjunction with the number group to identify the frame location of the called party. Once a call is set up, the common equipment disconnects leaving only the networking equipment on the circuit.

The design of the crossbar grid had two advantages over progressive switches. It was more compact, allowing more lines and trunks in the same amount of space. Also, it allowed for bi-directional traffic. Progressive switches could only pass calls in one direction, while crossbars could send traffic either way.

The separation of control equipment from the networking equipment also provided two major advantages. First, it allowed the components to operate somewhat independently. This was used to advantage by having only necessary common control elements brought into a call; after set up, the elements were freed to be used on other calls. Second, features could be added relatively easily by adding to or modifying the common control equipment. In other designs, such as progressive control switches, control elements were intermixed with network elements. Changes required modification to every circuit in the switch.

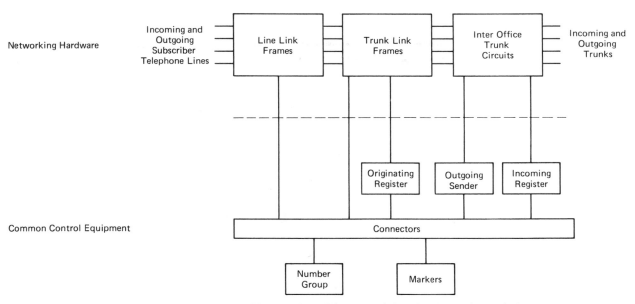

Figure 1.25:   Diagram of the No. 5 crossbar switch.

Electromechanical Switches

Electronic Switches

Electronic switches represent a major advance in switching technology. Network connections are made entirely with solid-state devices, such as transistors and microchips. Mechanical elements such as wipers, rods, and brushes are no longer needed. Since mechanical parts require a great deal of maintenance and wear out with frequent use, their elimination represents a dramatic improvement in system reliability. First used in 1965, electronic switches are still state-of-the-art. They have remained so because of their reliability and the relative ease with which they can be enhanced and updated as new ideas are introduced. Figure 1.26 shows a picture of a modern electronic switch.

The idea of separating common control from network equipment is carried over from the No. 5 crossbar switch. The concept is extended by an enhancement called **stored program control**. With stored program control, all switch functions are maintained in a program. The program consists of lines of code, like those used in computers. Call processing and routing are controlled by the program. Its code is used to make decisions about appropriate action in different situations.

A valuable aspect of stored program control is that modifications or additions can be accomplished simply by changing the code. In older switch designs, such as progressive control, control functions are closely tied to the network equipment. Changing the actions of the switch requires modifying every network connection (this usually means rewiring thousands of terminals). Changes can take weeks, or even months, to implement.

In contrast, under stored program control, changes can be put into service almost immediately. Since control and network equipment are separate, no changes need be made to network connections. Once code is written and loaded into the stored program control, the change is complete. This amounts to a tremendous saving in time and expense and a tremendous boost in system flexibility.

The network equipment associated with electronic switches is usually a frame containing thousands of terminals. Each terminal is a connection to a telephone line

Figure 1.26: An electronic switch. (Photo courtesy of Northern Telecom)

or trunk. The terminals are tied to ports in the switch. The ports are directly controlled by the common control equipment.

There are five components in the common control. The first is switch memory. The control program is stored here. Instructions are read, but data are never written to this area (unless switch functions are being modified, as described above). A second area of memory is the scratchpad. Information important to call processing is held here. For instance, data about which ports the call is using, what the numbers of the called and calling parties are, and when the call began, are stored. After the call is completed, these data are wiped clean and the area can be used for another call.

The third component is high-speed processor, used to read and execute control program instructions. The processor, in conjunction with the control program, regulates all activities of the switch. Many calls can be handled simultaneously by a process called *time-sharing*. Under time-sharing, call-processing activities are split into stages. The processor completes one stage for each call and then begins again, completing the next stage. This round-robin approach operates so quickly that call processing appears to be continuous.

Remember that only calls being set up or taken down require the attention of the processor. Once a call is up, the network equipment maintains it. A fourth component of the common control equipment is used to alert the processor when a call needs attention. This component is called a scanner. Scanners check the status of the switch's ports several times a second. When a change is detected, such as a caller hanging up after a call, a signal is sent to the processor. The processor evaluates the change and directs appropriate action. Scanners also contain registers for storing digits dialed by subscribers.

The fifth component is the distributor. Its function is to translate commands from the processor to signals that can be transmitted to network equipment or to remote switches. Table 1.3 summarizes the elements of electronic switches.

TABLE 1.3
Electronic Switch Elements

Common Control Components	Function
Semi-permanent Memory	Stores code of control program
Temporary memory (Scratchpad)	Stores data for the processing of each call
Processor	Executes program code to direct switch activity
Scanners	Monitor status of ports
Distributors	Translate processor commands and transmit signals to the network

Network Equipment	Function
Frame	Terminates lines and trunks
Terminal	Endpoints within the frame for individual lines and trunks. Tied to a port in the common control equipment.

TELEPHONE NETWORKS

Telephone networks have two parts: switches and facilities. Switches allow connections to be made to different endpoints, and facilities link locations together. As is sometimes the case, the whole is greater than the sum of its parts. A switch by itself is limited in the number of connections it can make, and a facility by itself can only link two endpoints. In combination, however, they form a network that can bring together virtually any locations in the world. This is no small feat, and the ingenuity used to create networks is worth exploring.

Networks may be either private or public, switched-circuit or packetized. **Private networks** are dedicated to use by a private group of people. The group may be as few as two people sharing a pair of lines or as many as thousands of people sharing hundreds of switches and lines. They are normally used in conjunction with Private Branch Exchanges (PBXs) and will be discussed in detail in Chapter 2. **Public networks** are available to everyone, the most common being the **Public Switched Network (PSN)**. **Switched-circuit networks** are those that set up a physical connection for calls, such as in the PSN. **Packet networks** do not set up physical connections. They send packets of information through the network using an addressing scheme to create virtual circuits. Packet networks will be discussed in Chapter 7.

The Public Switched Network

The Public Switched Network (PSN) is a combination of step-by-step, panel, crossbar, and electronic switches, using copper-wire, coaxial-cable, fiber-optic-cable, mi-

Figure 1.27: The regional bell operating companies after divestiture. (Courtesy of AT&T)

crowave, and satellite transmission. PSNs form a network that has been evolving for over 100 years. This network has been called one of the United States' greatest assets.

Ownership of the Network Ownership of the network has evolved also. Prior to 1983, the majority of it belonged to the Bell System, composed of American Telephone and Telegraph Co. (AT&T) and 23 Bell operating companies (BOCs). The remainder of the network belonged to independent telephone companies, such as General Telephone and Electric Co. (GTE), Microwave Communications, Inc. (MCI), and Sprint Communications.

In 1983, AT&T agreed to divest itself of the operating companies in exchange for the opportunity to compete in business markets, notably the computer marketplace, previously forbidden to it. The Bell operating companies were reorganized under seven regional operating companies (see Figure 1.27), and retained control of the local telephone networks in the areas they served. They kept their status as a monopolies and remained the only providers of local telephone services. They became known as local exchange companies (LECs).

AT&T retained ownership of the long-distance network. Its function was to connect the local networks to one another. Other companies, such as MCI and Sprint, compete with AT&T with their own long-distance networks. These companies came to be called *common carriers* or *inter-connect companies*. From 1984 to 1986, every telephone subscriber in the United States was given the opportunity to select a new long-distance carrier. The purpose was to give people a chance to change from AT&T if they so chose, and to more evenly distribute calls among carriers. Popular opinion was that AT&T would lose 40 to 50 percent of its marketshare. In fact, AT&T retained 70 to 80 percent in most areas, with the balance being divided fairly evenly among the others.

Network Hierarchy To better understand the operation of the network, it will be useful to look at the hierarchy within it.

When a person picks up a telephone in the PSN, she or he is connected to a **central office** owned by a local exchange company (LEC). The central office is known as an end office, because it is the usual endpoint for the network. That is, the connection from the central office to other central offices is considered a network connection, while the connection from a subscriber's telephone to the central office is called a **local loop**. Figure 1.28 shows the local loop and the hierarchy of offices within the AT&T long-distance network.

Notice that each level of the hierarchy is a different class office. For instance, end offices are class 5 while regional centers are class 1. Calls originate in the network at class 5 offices and work their way up. It is preferred that calls route laterally, for instance from a class 5 to a class 5 office. If because all connections are in use and that is not possible, then a call will route vertically one level, where it will again attempt to route laterally. This process continues until the call is completed through the network. The network is designed this way to minimize the number of offices a call must pass through before completion. The fewer the offices,

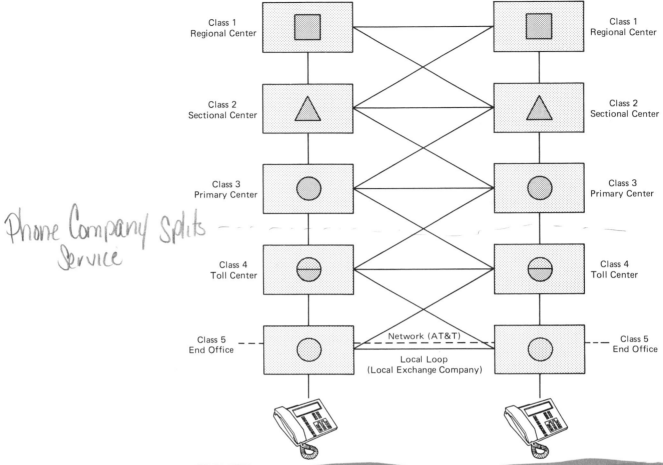

Figure 1.28: Hierarchy of offices within the AT&T long-distance network. Other carriers have similar hierarchies, but with only two or three levels.

the more likelihood of a clear connection, and the fewer the network resources that must be used.

Note also that offices are multiply connected to other offices. A toll center is connected to other toll centers. It is also connected to several end offices and several primary centers. If a call reaches a toll center, it will first try to send the call to an end office serving the called number. If it cannot, it will try to route the call to another toll center that is connected to the serving end office. If this also fails, it will route the call to a primary center as a last resort.

The pressure to keep calls as low in the hierarchy as possible results in providing the quickest call completion with the clearest line quality. It also results in needing far fewer offices as you ascend the hierarchy. For example, there are over 9,000 class 5 end offices, but there are only 12 class 1 regional centers. Not only do the number of offices change, but the trunking patterns among them change as well.

Class 5 end offices are likely to have trunks connecting them to a few other end offices. Regional centers, on the other hand, are trunked to all other regional centers.

Organization of the Network. From a telephone local calls can be made, long-distance calls can be made, even international calls can be made. Calls can be dialed directly or with the assistance of an operator. The organization of the network allows for the variety of calling patterns that can be used.

Local calling areas around the country are divided into zones known as local access and transport areas (LATAs). See Figure 1.29 for a picture of the LATAs in the southern California area. They define the boundaries of local calling service that an LEC can provide. Service between LATAs must be provided by an inter-exchange carrier.

This seems to be an odd situation. An LEC will provide service within all the LATAs in their territories, but they cannot route a call from one to the other without going through an inter-exchange carrier's network. To understand why this arrangement exists, it is important to realize that LATAs are a byproduct of the breakup of the Bell System. Their purpose is to distinguish clearly the areas that can be served by LECs and by inter-exchange carriers.

It is also worth noting that even prior to the breakup, a similar arrangement could be found. The Bell operating companies (BOCs) provided telephone service within their geographical area. From area to area, though, long-distance circuits were used that did not belong to them. The circuits were the domain of the Long Lines Department within AT&T.

The Network Numbering Plan. Every one of the millions of telephones within the PSN can be reached by dialing a number. The number is the "address" of the telephone and every one is unique. The plan for assigning them includes definitions for each of the digits and for area codes.

Two types of digits are defined. They may be

$$N = 2 \text{ thru } 9$$

$$X = 0 \text{ thru } 9$$

This means an N digit must be between 2 and 9 while an X digit can be any single digit, 0 to 9. Telephone numbers must follow the pattern:

$$\text{NXX-XXXX}$$

The first three digits are the A, B, and C digits. The A digit must be between 2 and 9, while the remaining digits may be from 0 to 9.

Under this scheme, there are at most 8 million unique numbers available. However, many more than that are needed. The problem has been solved by the use of area codes. Also known as a number plan area (NPA), the area code identifies a geographical region within which no seven-digit numbers will be repeated. Figure 1.30 shows a map of the area codes within the United States and its possessions.

1. San Francisco
2. Chico
3. Sacramento
4. Fresno
5. Los Angeles
6. San Diego
7. Bakersfield
8. Monterey
9. Stockton
10. San Luis Obispo

Lata's in the State of California

Figure 1.29a: Local Access and Transport Areas (LATAs) in Southern California. (Maps courtesy of Southern Bell)

The Los Angeles Area Lata (Courtesy of Pacific Bell)

Figure 1.29b: The Los Angeles Area LATA. (Maps courtesy of Southern Bell)

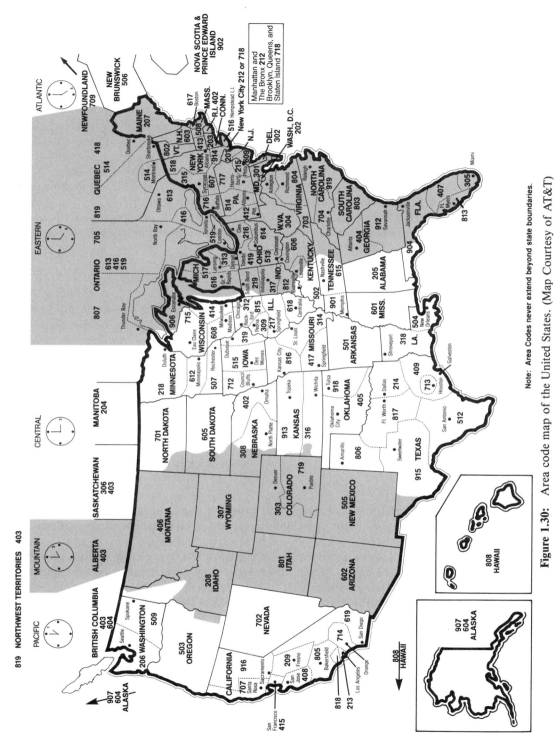

Figure 1.30: Area code map of the United States. (Map Courtesy of AT&T)

Telephone Networks 35

The digits of an area code are defined slightly differently than for a telephone number. They are

$$N = 2 \text{ thru } 9$$
$$X = 1 \text{ thru } 9$$

The area code follows the pattern

$$N \; 0/1 \; X$$

This means the first digit must be between 2 and 9, the second digit must be either 0 or 1, and the third digit must be between 1 and 9.

Dialing between locations using an area code can be done without assistance from an operator. It is known as direct distance dialing (DDD). It is accomplished by dialing

$$1 + \text{Area Code} + \text{Telephone Number}$$

The 1 preceding the area code alerts the network that a non-local call is being placed and that the next three digits will either be an area code or the exchange of the remote central office. The switching equipment distinguishes an area code by whether or not the second digit is a 0 or 1. If it is either, it is an area code; otherwise it is an exchange number.

There is an exception to the above rules. Not all locations require a 1 to be dialed before an area code. In these areas, every dialed number is examined to see if the second digit is a 0 or 1. In these areas, telephone numbers cannot have a 0 or 1 as the B digit. The pattern required for these areas is:

$$\text{NNX-XXXX}$$

International Dialing   It is now possible to directly dial telephones in other countries. While not universally available, an increasing number of countries support this system.

The international standards organization, the *Comité Consultatif Internationale Telegraphique et Telephonique* (International Consultative Committee for Telephone and Telegraph, or CCITT) has established number zones for the world. Nine regions are defined and are shown in Table 1.4. Each country has its own country code. The first digit of the country code corresponds to the zone number for its region of the world. Some examples are as follows.

Brazil	55
Israel	972
Korea	82
Luxembourg	352

South Africa	27
Switzerland	41
Tahiti	689
United States	1

Codes may be from one to three digits in length, depending upon the number of telephones in the country. Countries with the most telephones have a one-digit number, those with the fewest have a three-digit number.

International direct distance dialing (IDDD) is accomplished by dialing

$$011 + \text{Country Code} + \text{Area Code} + \text{Telephone Number}$$

The number 011 is the access code. It alerts the network that the succeeding digit(s) will be a country code. The area code is sometimes called the *routing code* or *city code*. Telephone numbers differ in length and format from country to country.

NETWORK SERVICES

Several special services are available to users of the PSN. These services range from modifications of existing voice arrangements to transmission facilities for image and data.

Wide-Area Telecommunications Service (WATS)

Wide-area telecommunications service (WATS) is a service provided by interexchange carriers intended to benefit high-volume long-distance users. It provides an alternative billing method to normal toll service.

Normal toll is based on charging customers for the duration of each call. WATS, on the other hand, is a bulk billing arrangement in which a discount is given based on the total number of hours of toll calls made during the month. AT&T, for

TABLE 1.4
Zone Numbers for Regions of the World as Defined by CCITT

Zone	Region
1	North America
2	Africa
3	Europe
4	Europe
5	South and Central America
6	South Pacific
7	Union of Soviet Socialist Republics
8	Far East
9	Middle East and Southeast Asia

instance, offers a discount after 25 hours of calling with an additional discount after 100 hours of calling. Other carriers offer variation of this. The best plan for a company will depend on its particular calling patterns. If most calls are made to Raleigh, North Carolina, and Savannah, Georgia, it is likely that checking the plans of several carriers would disclose substantial differences in pricing.

When a customer orders WATS, dedicated lines or trunks are installed from the carrier's central office to the customer's location. Bulk billing is applied only to these facilities and not to any other lines or trunks. A set monthly fee is charged that is higher than the fee for most other types of facilities. (This is why WATS is not economical for low-volume users.)

There are some limitations on WATS, also. WATS facilities are either *intra*state or *inter*state. Intrastate WATS allows calling only within the customer's state. For instance, for a customer in Chicago, Illinois, WATS rates would apply to a call to Springfield, Illinois, but not to Gary, Indiana. Interstate WATS is just the opposite. It applies only to calls to outside the customer's state. If a customer wants WATS within the state and without, two separate WATS facilities must be ordered.

Interstate WATS can provide service to part or all of the country. The continental United States is divided into five regions or bands. Band 1 includes the states nearest your home state. Band 2 covers all the states in Band 1 plus more distant ones. Band 3, Band 4, and Band 5 include successively more states, until the entire continental United States is covered. Figure 1.31 shows where the bands would be located for a company in Southern California.

Several years ago, Band 6 WATS became available. It added Alaska and Hawaii to the covered areas, making all 50 states available for the first time.

Competition has brought some changes to WATS. As with other long-distance services, AT&T has been the standard-bearer. Other carriers have chosen to distinguish themselves by loosening some restrictions. For example, the separation of intrastate and interstate WATS has been removed; a single facility can call numbers inside and outside the state. Also, a single facility can blend different WATS bands. Traditionally, a customer could order Band 3 WATS and Band 5 WATS, for instance. The facilities would be separate, and there would be a set fee for each of them and hours of toll calls would accrue separately. Blending of bands means that a customer can order a single facility, pay a single set fee and yet accrue billing for Band 3 and Band 5 as if there were two separate groups.

800 Service

The 800 service is a mirror-image of WATS. Its billing structure is the same (with discounts based on call volume), it partitions the country into bands the same way, and it separates intrastate from interstate service. It differs, however, in that it applies to incoming calls—WATS applies to outgoing calls. The 800 service is sometimes called in-WATS, while traditional WATS is called out-WATS.

Normally, the calling party is charged for the call. With 800 service, the charge is billed to the called party. This allows the calling party to place the call toll free. The 800 service is very popular with businesses that want to make it easy for customers to reach them.

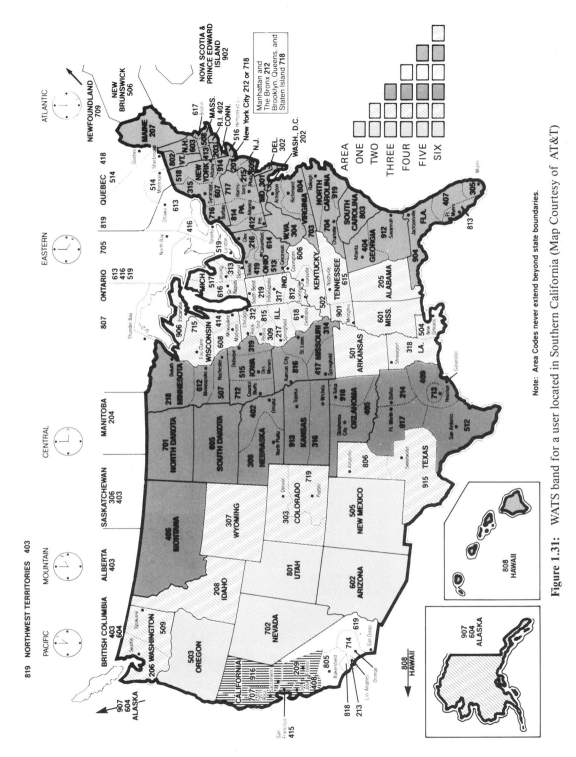

Figure 1.31: WATS band for a user located in Southern California (Map Courtesy of AT&T)

The service gets its name from the fact that all telephone numbers used have an area code of 800. It works this way: A company is given an 800 number. When someone calls the number, it is directed to a computer within the carrier's network. The computer correlates the 800 number with the company's normal telephone number and substitutes it. Billing is assigned to the company and the call is routed through the network with the substituted number.

Since its introduction by AT&T in the 1970s, the service has been well received. It has also been highly profitable and it is not surprising that several interexchange carriers now offer the service.

Foreign Exchange Service (FX)

Foreign exchange service (FX) may be thought of as a local form of WATS. It is an intra-LATA service, meaning it does not cross LATA boundaries. As such, it is offered by local exchange companies rather than by interexchange carriers.

Foreign exchange service allows a company to call an area outside its local calling area as if it were local. It is as if the company has a telephone sitting at the remote location available for its use.

In many respects, that is exactly the case. The LEC provides a line from the remote location to the company's location. A telephone number is assigned with an exchange number from the remote site. This allows the company to make outgoing calls as if it were at the remote site. It also allows people in the remote area to call a local exchange number and reach the company. Figure 1.32 shows an example of a the service.

Many companies find foreign exchange numbers useful if they have customers in a specific city or if they have a branch office. In either case, the FX line provides a convenient way for the two locations to communicate. The company pays a set monthly fee for the line plus per call charges.

DATA SERVICES

In addition to voice transmission, data services are provided through the PSN as well. There are a variety of offerings from the various LECs and interexchange carriers. Two of most common will be described here.

Analog Private Line

Analog private lines are used to provide a transmission pathway for data traffic. This capability is a special service, because the PSN was not originally designed with it in mind. Data traffic is digital in nature, while the PSN is analog. A method had to be devised to allow data traffic to share an analog network. The solution came in the form of **modems.** Modems convert digital signals to analog ones. The analog signals can then be sent through the PSN. Modems can also convert them back to digital signals.

Modems are most commonly used in pairs, one at each end of a connection.

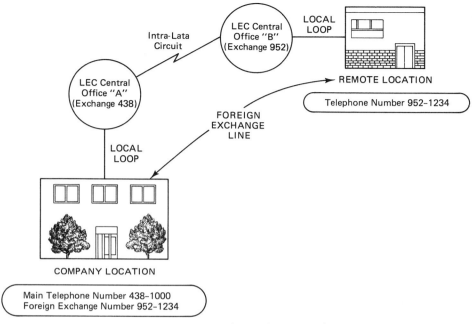

Figure 1.32: Foreign exchange service.

While one sends, the other receives. There are a great variety of modems operating at different speeds and using different transmission techniques. The modems on a circuit must be compatible for an effective transmission to take place. This often means having two of the same kind of modem or comparing specifications to ensure their compatibility.

Modems range from low speed to high speed. For data transmission of this type, low speed refers to transmission rates below 2,400 bits per second (bps) and high speed refers to transmission above that speed. At high speed, standard telephone lines are not usable because they distort the analog signals sent by modems. To correct the problem, dedicated circuits (private lines) are used. These circuits are specially designed and conditioned to handle high-speed data transmission. Once this is done, the circuits are technically no longer part of the PSN, because they can no longer be switched. However, they use the same wire, cable, fiber, microwave, and satellite facilities.

Two types of private lines are available: point-to-point and multi-point. As their names imply, point-to-point connects to two endpoints, while multi-point connects three or more endpoints. Figure 1.33 shows some sample configurations. Point-to-point is the more common of the two. It is generally used to connect one computer to another computer or to a display terminal that is off-site.

Multi-point circuits are essentially "party lines" for data equipment. Data sent on the line is received by all devices. To manage transmission, each device must be given an address. Data sent on the line must also have an address. The device with

Data Services 41

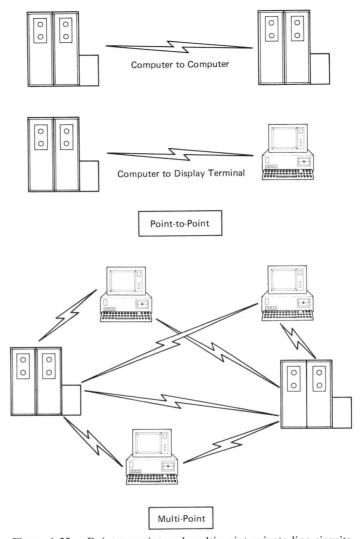

Figure 1.33: Point-to-point and multi-point private line circuits.

an address matching the data picks up the information, all other devices ignore it. In some respects, a multi-point line is like a local area network. It differs, however, in that it operates at a much lower speed and can include devices over a much larger area.

T-1 Lines

T-1 facilities are digital private lines. They differ from analog private lines in that they do not convert data signals to analog form. Instead, circuits are designed to transport the data digitally.

T-1 lines can be used for digitized voice, digitized image, or data. The bandwidth of a single facility is 1.544 megabits per second (Mbps). This is far in excess of the bandwidth needed for single voice or data channel. It is even generous for many types of image transmission.

Commonly, a T-1 line is subdivided into several smaller channels. Time-division multiplexing is used to create the subchannels, and each one handles a voice, data, or image transmission. There are 23 to 24 channels available, depending on whether a **clear channel** or **bit-robbing** transmission techniques is used.

The difference between the two techniques derives from how transmission signaling is handled. Clear channel provides 23 channels without any signaling information and a 24th channel dedicated to signaling information for the first 23. Each channel has a bandwidth of 64 kilobits per second (kbps), giving a total data throughput of 1.472 Mbps.

Bit-robbing provides 24 channels with signaling and data carried together. The least-significant bit in each byte of data is replaced by signaling information. The replacing or "robbing" of bits means that data get altered as they moves through the network. The impact of this is minimized by changing the least-significant bit. Since 1 bit (1b) out of eight is used for signaling, the data-carrying capacity of a channel is limited to 56 kbps. With 24 channels, total data throughput is 1.344 Mbps.

Bit-robbing works well for the transmission of digitized voice or image because it does not cause any noticeable change in the sound of a voice or image. Data transmission, however, is sensitive to any alteration, and therefore bit-robbing is not acceptable.

The multiplexing used with T-1 requires coordination between equipment at the customer's location and the LEC or common carrier's central office. The equipment on the customer's site, known as **customer premises equipment (CPE),** is usually a private branch exchange (PBX). See Chapter 2 for more information about them.

T-1 is widely used within the PSN to carry long-distance traffic. Commercial services are also available that use T-1. One such service is Dataphone Digital Service (DDS), and AT&T offering that transmits data synchronously at 2,400 bps, 4,800 bps, 9,600 bps, or 56 kbps without any conversion from digital to analog.

Channels can be designated as data only or as alternate voice and data. Several channels can be dedicated to a single transmission, as for example a video transmission. Today, the common carrier or LEC must be asked to change a channel's configuration. In the future, users will be able to modify T-1 channels on their own. They will be able to allocate bandwidth dynamically to meet changing needs. This will greatly enhance the flexibility and usefulness to the user. The standards under which it will be accomplished are called the Integrated Services Digital Network (ISDN). Chapter 8 discusses ISDN in detail.

Since no analog conversion is done, modems are not used on T-1 circuits. Instead, digital interface units, called digital service units (DSUs) are used to connect the data device to the network. DSUs boost the power of the data signal and pass it on to the network. The DSU is connected to a channel service unit (CSU), which is the termination point of the network. It contains circuitry to ensure that no stray electrical signals or electric currents get into the network that could damage it. Often times, the DSU and CSU are combined into one unit, a CSU/DSU.

Image Service

Image services include video and graphics transmission. The most well known is television. Companies such as the American Broadcasting Co. (ABC), the National Broadcasting Co. (NBC), and the Columbia Broadcasting System (CBS) make use of PSN facilities in private networks to distribute television programs to affiliates throughout the country. Video transmissions of this sort require a large amount of bandwidth, in excess of 3 Mbps.

Other video services include video teleconferencing. With this service, individuals or groups at two or more locations can see and hear one another. Teleconferencing equipment, which includes video cameras, microphones, monitors, and speakers, is set up at each location. The equipment is then linked together through the PSN. Actions and sounds picked up by the cameras and microphones at one location are shown on the monitors and speakers at the other locations. Video teleconferencing provides, essentially, a temporary, closed-circuit television network.

There are three varieties of video services available: full- motion video, slow-motion video and freeze-motion video. Full-motion video transmits all action and sound as it occurs. It is like watching television. Slow-motion video transmits sound as it occurs, but images are not transmitted as quickly. The result is that images appear in slow motion. Freeze-motion video transmits sound as it occurs, but images show no motion at all. They are like snapshots or still photographs. The reason for the three varieties relates to the cost of the services. Full-motion video requires the greatest amount of bandwidth and the most expensive equipment. Slow-motion

Figure 1.34: An electronic blackboard. (Courtesy of AT&T)

video requires less bandwidth, and freeze-motion less still. The latter two can be done at a lower cost, and this is their chief attraction.

Another type of equipment used to provide an image service is the electronic blackboard. Words or pictures drawn on a blackboard are displayed on a television monitor. People too far from the blackboard to see it can look at a monitor instead. In fact, the monitor can be remotely located from the blackboard with the images being transmitted over telephone lines. The blackboards work by placing a pressure-sensitive backing on the blackboard. The backing senses where the board is touched by chalk or marking pen, and these points are then displayed on the monitor. An example of how it is used would be a lecture in a large lecture hall. The words and drawings put on the blackboard by the lecturer would be displayed on a monitor, and the sounds of the lecturer's voice would be heard on a speaker (the voice transmission is carried over a separate channel).

Probably the most common video service, aside from television, is facsimile. With facsimile, images of text or graphics are transmitted over telephone lines to a remote location. Typically, a paper document is placed on the facsimile machine. A remote machine is called, and the two machines transfer the image of the document. Facsimile has become increasingly popular in recent years. It is being incorporated in new ways. For instance, facsimile boards for personal computers are available. These boards transmit data directly from the computer to a facsimile without a paper document required at the originating end. Facsimiles are also beginning to be incorporated in other types of office equipment. For instance, a facsimile and photocopier combined in one machine or a combination facsimile and laser printer.

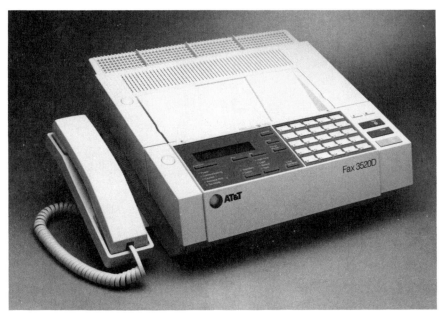

Figure 1.35: A facsimile machine. (Courtesy of AT&T)

CELLULAR TELEPHONE SERVICE

Cellular telephone service allows mobile use of the PSN. While normal telephones connect to the network via wires, cellular telephones connect via radio waves to antennas located throughout an area. This allows users to move freely, for instance to ride in a car, while talking on the telephone.

The use of radio waves with telephones is not new. It is used with marine telephone service, mobile telephone service, and even with cordless telephones. All of these services have the disadvantage of a limited number of calls that can be handled at one time. This stems from the limited number of radio frequencies available for use. Cellular telephone service gets around the problem by using low-energy transmitters that broadcast relatively weak signals. They carry only about a mile and then become too weak to be useful. This allows a single frequency to be reused, so long as the transmitters are at least a mile apart. Figure 1.36 shows how frequencies can be reused without interfering with one another. The area covered by a single transmitter is called a cell, and from this comes the name of the service. Cells are arranged in a honeycomb fashion so that entire cities or areas have access to the service.

When a person makes a call from a cellular telephone, a connection is made to the nearest transmitter. As the person moves out of range of the transmitter, the call is automatically transferred to the transmitter in an adjoining cell. The parties on the call are not aware of the transfer. To them, it seems like a normal, continuous connection. The call will be transferred again and again as the caller moves from cell to cell. In this way, the caller can move freely while the call follows from place to place.

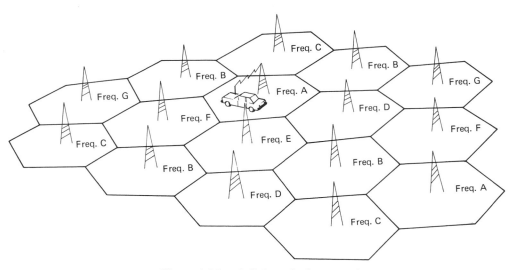

Figure 1.36: Cellular telephone service.

CHAPTER SUMMARY

Voice communication is based on the telephone and the Public Switched Network. These two components combine to create a powerful and flexible system allowing communication over vast distances. Originally, calls through the network used analog transmission techniques and rotary dialing. In recent decades, digital transmission techniques and tone dialing have been used as alternatives. Switching within the network began as a manual operation performed by humans. Manual switching has, for the most part, been replaced by automatic switching performed by machines. Calls are switched through the PSN, which is composed of local loops and the long-distance network. Local exchange companies (LECs) control the local loop, while common carriers control the long-distance network. Local exchange companies have a monopoly on telephone service within a local access and transport area (LATA). Common carriers, on the other hand, compete to provide connection service between LATAs. Several types of transmission services are available with the PSN. Wide-area telecommunications service (WATS), inbound WATS (800 service), and foreign exchange service (FX) provide pricing advantages. Analog private line and T-1 services provide data and image transmission through the PSN.

KEY TERMS

Amplitude	Line
Analog	Local access and transport area (LATA)
Bandwidth	
Bel	Local loop
Bit-robbing	Modem
Central office	Modulate
Clear channel	Oscillate
Crest	Packet network
Customer premises equipment	Pitch
Decibel	Private network
Decibel loss (dB loss)	Progressive control
Digital	Public network
Dual-tone multi-frequency (DTMF)	Public switched network
Exchange	Sidetone
Facility	Sine wave
Frequency	Stored program control
Frequency-division multiplexing	Switched-circuit network
Hertz	Switching

Switch
T-1
Time-division multiplexing (TDM)
Time-sharing

Timeslot
Trough
Virtual circuit
Volt

REVIEW QUESTIONS

1. What is a Strowger switch?
2. Give an example of a progressive control switch.
3. Give an example of a coordinate switch.
4. Give an example of a stored program control switch.
5. What does CPE mean? What is it?
6. What is the PSN?
7. If a T-1 line has a bandwidth of 1.544 Mbps and clear channel transmission provides throughput of 1.472 Mbps, what is the difference in capacity used for?

TOPICS FOR DISCUSSION

1. What are some benefits we take for granted that would not exist without telephones and the public switched network?
2. Has the breakup of the Bell System helped or hurt business in the United States? What has the impact been on individuals and residential users?
3. How can video teleconferencing be used in business? In education? What advantages can teleconferencing provide over other ways of "meeting"?
4. What new uses can cellular telephones provide that are not practical with immobile, corded telephones?

The PBX

CHAPTER 2

OBJECTIVES

Upon completion of this chapter, you should:

- understand what a PBX does
- understand the evolution of the PBX
- understand basic PBX architecture
- be able to list and describe PBX features
- understand what a private network is

INTRODUCTION

PBX is an abbreviation for *private branch exchange.* A PBX "switches" calls in the same way as central office equipment. The difference is that a PBX switches calls for private groups, while central office equipment switches calls for the public. Another difference is that PBX equipment is located at the user's premises, while central office equipment is located the local exchange company's (LEC's) central office.

PBX is just one of the several names given to this type of equipment. It is also known as *private automatic branch exchange* (PABX), *computerized branch exchange* (CBX), and *integrated branch exchange* (IBX). In most cases, a manufac-

Figure 2.1: A PBX.

turer of switching equipment has chosen a variation to distinguish its product from those of other manufacturers.

DEVELOPMENT OF THE PBX

The Public Switched Network (PSN) is owned and operated by the American Telephone and Telegraph Company (AT&T), the Bell operating companies (BOCs), General Telephone and Electric Co. (GTE) and scores of small, local telephone companies. Operation of the network is regulated by the Federal Communications Commission (FCC), an agency of the federal government, and by state utilities commissions, agencies run by each state.

Key Equipment

For many years, only equipment owned by these companies could be used in the Public Switched Network. Service was provided by connecting telephones at user locations to switching equipment at a central office. Each telephone had a separate

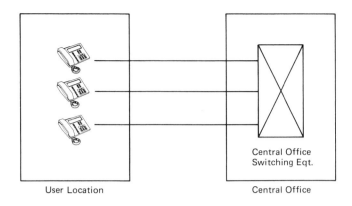

One telephone is connected to each line.

Figure 2.2: Traditional telephone service arrangement.

line to the central office. Costs could be saved if more than one telephone could be connected to a line. If five users could share two lines, then the cost of three lines could be saved. The first attempt to implement this idea was with key systems. Key systems allowed a line to appear on several telephones at once. Each telephone had buttons, or "keys." Each key was connected to a line. Depressing a key connected the user to that line.

Additional features became available, too. In addition to one line appearing on several telephones, several lines could appear on one telephone. If two or more users connected to a line at the same time, they would be conferenced together. Calls could be placed on hold. Often, a button would be dedicated for use as an intercom. Pressing it would connect the user to another user's telephone rather than to a line.

Figure 2.3: A key telephone set. (Photo courtesy of Northern Telecom)

Development of the PBX 51

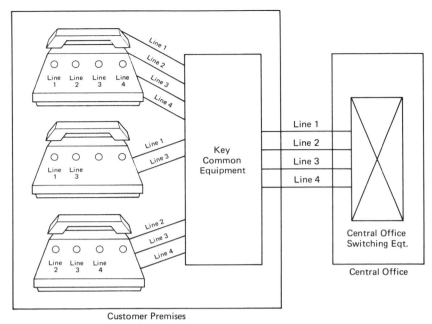

Figure 2.4: Key equipment using common equipment on a user premises.

With key equipment, all telephones were connected to a cabinet known as **common equipment.** The common equipment contained the hardware to connect the appropriate lines to each telephone. It also housed the equipment to create an intercom and to put calls on hold.

Common equipment was located on a customer's premises (*customer* referring to a customer of the telephone company). It came to be known as **customer premises equipment (CPE).**

Private Branch Exchange (PBX)

As technology evolved, the economic benefits and features provided by CPE led to the development of the PBX. The PBX allowed many more users than a key system, allowed more sophisticated features, and, most importantly, allowed flexible routing of calls. The PBX is functionally more like central office equipment than key equipment. In fact, it performs many of the same functions as central office equipment, albeit for a private group of users rather than for members of the public.

A PBX may support thousands of telephones. Calls may be switched among any of them. Users have access to hundreds of features. PBXs may be linked together into networks. It is easy to understand why this machine has been popular since it was first introduced and why it continues to be a powerful tool for the telecommunications industry.

BASIC PBX ARCHITECTURE

The *raison d'être* and primary function of a PBX is to switch calls. Switching is the ability to set up, take down, and change connections as needed. Switchboard operators performed this function in the early days of telephony. Operators would listen to callers' requests and then connect them to the correct line to complete a call. The PBX has automated this task. A PBX reads the digits dialed by a user and interprets them to determine the correct connection.

Other functions besides switching have been incorporated into PBXs. Modern PBXs keep track of the calls made so that costs and traffic usage patterns can be compiled. PBXs can support data communications as well as voice communications. PBXs can also be linked together to form private networks.

There are four functional parts to a PBX: the common control, the line interface (line side), the switching matrix, and the trunk interface (trunk side).

The Common Control

The common control oversees the operation of the PBX. It controls the path that calls take through the switching matrix. It also controls user access to features and monitors the "health" of system components.

The common control is usually a microprocessor designed for the specialized function of controlling and monitoring hundreds of calls simultaneously. In this regard, a PBX can be viewed as a specialized computer.

The Line Side

The line side connects all telephones and other user equipment to the PBX. Normally this includes a telephone at a user's desk, cable connecting the telephone to

Figure 2.5: A manually-operated cordboard switchboard.
(Photo courtesy of Northern Telecom)

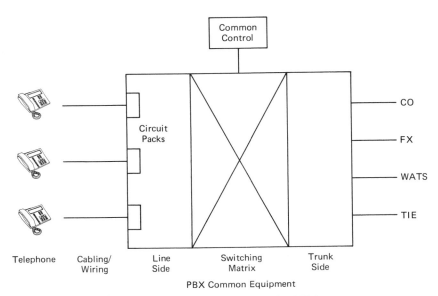

Figure 2.6: Basic elements of a PBX.

the common equipment, and a line-circuit pack in the common equipment. In addition to telephones, other items such as data terminals, computers, modems, answering machines, paging equipment, dictation equipment, and loud bells are often connected to the line side of a PBX. These allow special requirements to be satisfied.

The cable connecting the peripheral equipment to the common equipment is **twisted pair.** It is copper wire that usually runs from a walljack, through the wall, to a circuit pack in the common equipment.

The line circuit is an integral part of the common equipment (see Figure 2.7). It connects the cable from peripheral equipment to the switching matrix. Four to eight line-circuit interfaces are on a line-circuit card. The cards are located in a carrier that can hold from 10 to 20 line circuit cards. The carriers are housed in a cabinet that can hold four to six carriers. Each cabinet, then, has the capacity to handle from 160 to 960 users. If greater capacity is needed, additional cabinets can be brought in. This modular "building block" approach allows PBXs to cover a wide range of sizing requirements.

The Switching Matrix

The switching matrix is the heart of the PBX. In fact, the word *switch* is often used interchangeably with the term *PBX*, reflecting the importance of the switching function. The switching matrix has the capability of connecting a line to any other line or to any trunk. It is this capability that makes the PBX a powerful, yet flexible, tool.

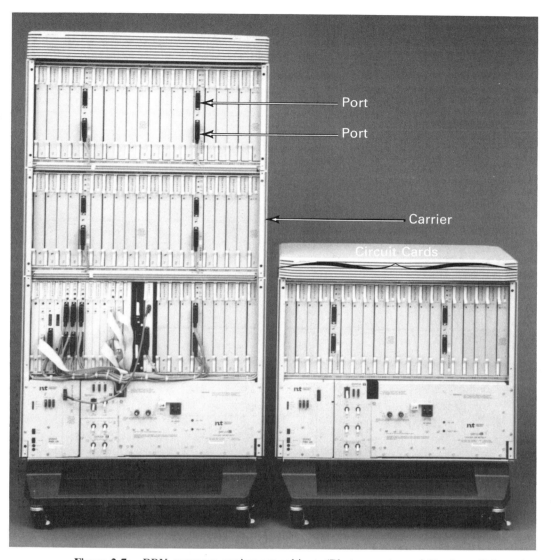

Figure 2.7: PBX common equipment cabinet. (Photo courtesy of Northern Telecom)

The Trunk Side

Trunks connect a PBX to points outside the switch, normally a LEC central office. Most PBXs support a wide variety of trunks, including central office trunks, tie trunks, WATS trunks, and T-1 trunks. The trunk side of the PBX connects these facilities to the switching matrix.

Basic PBX Architecture 55

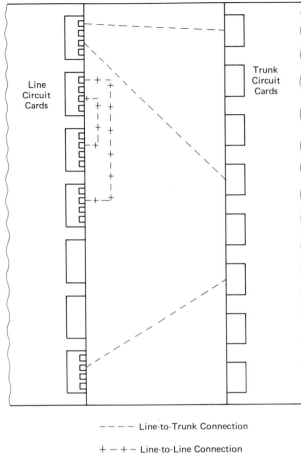

Figure 2.8: Switching matrix.

EVOLUTION OF THE PBX

PBXs have gone through several generations of change. They have increased in sophistication and capability, though their basic function has remained the same: to switch calls.

First-Generation PBXs

The first-generation PBXs were switchboards. A switchboard can be thought of as a manually operated switching matrix. Cords were plugged into jacks to create a talkpath between two endpoints. Once the **talkpath** was set up, a conversation could take place. An average switchboard could set up about 40 talkpaths simultaneously. All connections were analog, and only one conversation could be carried on a talkpath.

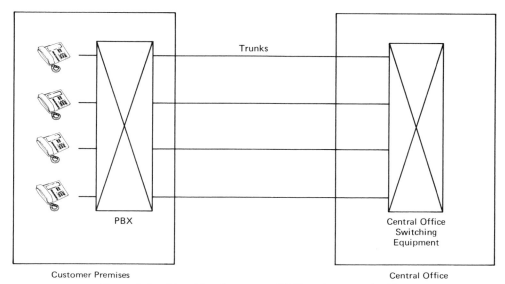

Figure 2.9: Trunking between a PBX and a central office.

Early Second-Generation PBXs

The key distinction between first- and second-generation PBXs is their transition from a manually operated switching matrix to a mechanically operated switching matrix. Early second-generation PBXs used mechanical relays to set up talkpaths. The relays established a physical path between two endpoints without the assistance of an operator. This was accomplished by the user dialing digits on the telephone. The dialing generated electrical pulses corresponding to the dialed digits. The PBX common equipment responded to the pulses by closing appropriate relays, thus completing the talkpath. As technology evolved, mechanical relays were replaced by electromechanical relays; they performed the same task, but were more reliable.

Later Second-Generation PBXs

First-generation and early second-generation PBXs required a separate physical path for each talkpath. This is referred to as **space-division switching.** A new technique was developed allowing multiple talkpaths over a single, high-speed physical path. The technique was called **time-division switching** or **time-division multiplexing (TDM).**

Time-division multiplexing slices the physical path into timeslots. Figure 2.10 shows a diagram of the technique. Four timeslots are depicted, though in reality PBXs usually have hundreds of timeslots. Every fourth slice is reserved for timeslot A. One call occupies this timeslot. Similarly, timeslots B, C, and D can each carry a call. A total of four calls can be carried making a total of four talkpaths over one

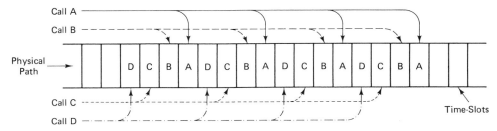

Figure 2.10: Time-division multiplexing.

physical path. The physical path is now called a bus, due to its increased capacity to carry calls and due to the high speed at which it operates.

Not only is it necessary to divide the physical path into slices, it is also necessary to slice the call itself. With space-division switching, the analog call can flow intact through the physical path. Time-division multiplexing takes the path and creates timeslots, which may be thought of as the boxcars of a train. To be carried, an analog call must be sliced into pieces small enough to fit into the boxcars.

Slicing is accomplished by a technique called **sampling.** Sampling takes successive pieces, or samples, of an analog call. Each one is placed on a timeslot through the switching matrix and is reassembled at the other end. If the samples are frequent enough and the speed through the switching matrix is fast enough, the reassembled call sounds normal to the human ear. The process is conceptually similar to the making of a motion picture. Action is recorded in still frames on film. When the frames are played back at proper speed through a projector, the action seems normal and uninterrupted.

Experiments have shown that an acceptable sampling rate is 8,000 times per second. At this rate, the human ear perceives no difference between a sampled call and an uninterrupted analog call.

A sampling technique used by late second-generation PBXs is **pulse amplitude modulation (PAM).** Amplitude, in this context, is a measure of pitch. When people talk during a telephone conversation, the pitch of their voice varies up and down creating a varied analog signal. Under PAM, the signal is sampled 8,000 times per second and each one is assigned an amplitude approximating the true amplitude. Figure 2.11 shows a comparison of an analog signal and a PAM-sampled signal.

Time-division multiplexing requires a PBX to keep track of the timeslots occupied by each call. A great deal more coordination and synchronization among the PBXs elements is needed than that required by space-division switching. To meet the need, **stored program control** was developed. Under stored program control, the PBX is microprocessor-controlled rather than mechanically-controlled. Functions are stored in a program allowing a much greater level of complexity than could previously be achieved. Furthermore, the program could be easily changed when necessary. With stored program control, the operation of a PBX becomes very similar to the operation of a computer. In fact, one view of PBXs is that they are nothing less than specialized computers.

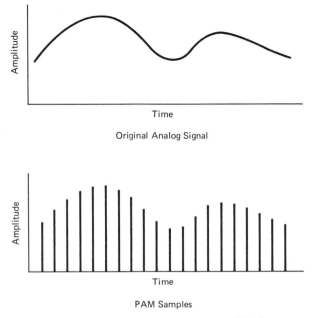

Figure 2.11: Pulse amplitude modulation.

Third-Generation PBXs

Third-generation PBXs were the first to have data-transmission capabilities designed into them. Earlier switches provided some data communications, but as an add-on feature, with limitations on speed and protocol. Digital data signals would be converted to analog form with a modem. They appeared to the switch as normal voice traffic.

Two design characteristics distinguish third-generation PBXs from their predecessors: the use of a digital switching matrix and the use of a non-blocking architecture.

A digital switching matrix allows data signals to pass through without first being converted to analog form. Voice signals, on the other hand, are initially analog and must be converted to digital form before passing through the matrix. This is done by the use of a **codec.** A codec is a microchip that performs analog-to-digital conversion. The word derives from "coder-decoder."

Most codecs in use today convert signals to a digital encoding scheme called **pulse code modulation (PCM).** PCM is similar to PAM in that the analog signals are sliced into samples. Instead of assigning an amplitude to the sample, however, one of 256 digital codes is assigned. The codes are 8-b binary representations of the amplitude of the sample. They are electronically identical to a byte of data generated by a computer. In fact, the codes pass through the switch as data. At the receiving end, a codec converts the digital signal back to analog.

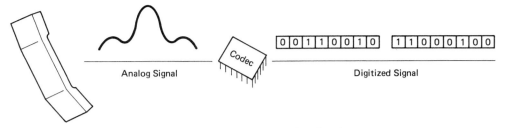

Figure 2.12: Digitizing voice signals.

What are the reasons for converting a voice signal to digital form and then converting it back? There are three major advantages. First, it allows voice and data traffic to be handled in a uniform way. This simplifies the design of a switch and reduces hardware costs by allowing a single transmission scheme instead of requiring separate schemes for voice and data. Second, voice and data traffic can be multiplexed over the same physical path, allowing for simultaneous voice and data transmission. Third, the quality of voice transmission is improved because digital signals are not susceptible to noise and distortion, as are analog signals (see Figure 2.12).

Voice signals can be digitized at one of two locations, at the user's telephone or at the common equipment. Figure 2.13 shows the two methods.

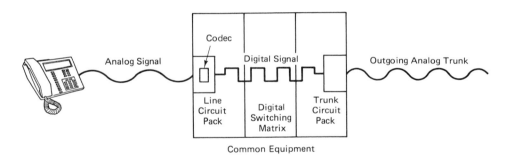

Voice Signal Digitized at the Common Equipment

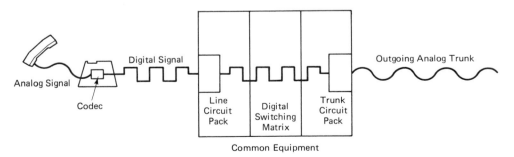

Voice Signal Digitized at the Telephone

Figure 2.13: Digitizing methods.

If digitized at the telephone, a codec installed in the instrument provides the conversion. The digital voice signal can be multiplexed with computer signals through the same cable from the telephone to the common equipment. Figure 2.14 shows how this can be done. An interface (e.g. an RS-232C interface) is provided for a computer terminal. The data signals are interleaved with the digitized voice signals and passed to the common equipment. The voice and data signals are separated back into voice and data streams that pass through the switching matrix independently. This is an efficient way to handle combined voice and data traffic.

Some users have no need for data transmission. In these cases, it is more economical to digitize the voice signals at the common equipment than at the telephone. Less-expensive analog telephones can be used, and the codecs located in the common equipment can provide conversion for several telephones instead of just one. The need to handle data results in the third-generation PBX's second distinguishing characteristic: the use of a **non-blocking** architecture. **Blocking** versus non-blocking architecture refers to the availability of talkpaths through the switching matrix. With a blocking architecture, the switching matrix can be a bottleneck for traffic trying to get through the PBX. Figure 2.15 shows how this happens. A PBX may have 1,000 lines and 100 trunks connected to it. The switching matrix may be limited to 100 talkpaths. That means only 100 of the 1,000 lines may be in use at one time. The 101st user that tries to make a call will find no talkpath available through the switching matrix. The call will be blocked.

When a user is blocked, only silence is heard; there is no dial tone, intercept tone, or recording. Most users, after a few moments, will hang up the telephone and try again. This time the call should go through.

Blockage is classified according to **grade of service:** P.01 grade of service means that one call in 100 will be blocked, P.02 means that two calls in 100 will be blocked, P.03 means three calls in 100 will be blocked, and so on. The designation in P1 means 100 calls in 100, or all calls, will be blocked. The normal range of blockage for voice traffic is P.01 to P.05. In this range, the average user will virtually never be aware of blocking. This is because blocking is spread among the entire group of users. For instance, in a system with 1,000 lines and a P.01 grade of service, a user will be blocked once in 100,000 tries. The Public Switched Network is designed for P.01 grade of service.

$$\frac{1 \text{ call blocked}}{100 \text{ tries}} \times \frac{1 \text{ user}}{1{,}000 \text{ users}} = \frac{1 \text{ call blocked}}{100{,}000 \text{ tries}}$$

When data traffic is introduced, the situation changes. While voice calls tend to be short in duration (the average voice call lasts less than five minutes), data calls tend to be long in duration. When a user makes a typical data call, a session is established between the user and a host computer. It may take two to three minutes to set up the session. Then, the user may work a few minutes to several hours, depending on the task being done. It is not uncommon for workers to set up a

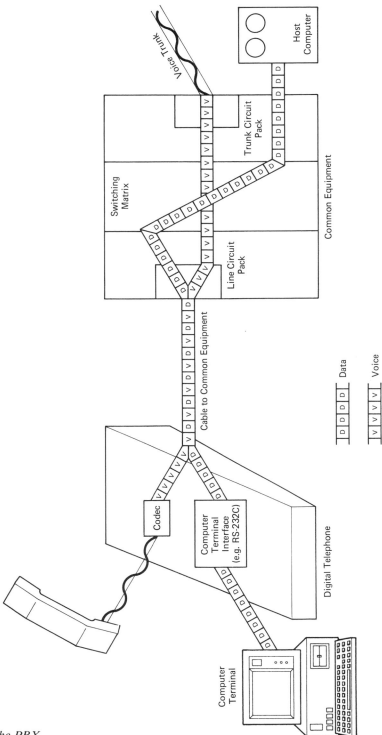

Figure 2.14: Voice and data multiplexed at the telephone.

62 Chapter 2 The PBX

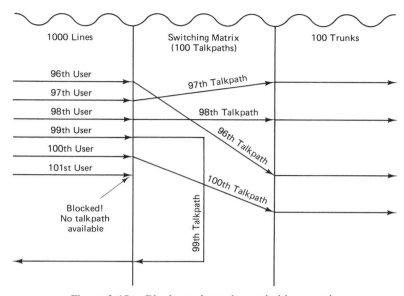

Figure 2.15: Blockage through a switching matrix.

session when they come to work in the morning and leave the session up until they go home in the evening. Though they may not be interacting with the host the entire time, it is more convenient to make one data call and leave it up than to make several calls as needed during the day. For these reasons, data calls tend to last substantially longer than voice calls.

The long duration of data calls increases the capacity required in a PBX. For example, to provide a P.01 grade of service for voice traffic, a PBX with 450 lines would need 100 talkpaths. To provide P.01 grade of service for data traffic, the PBX would need 250 talkpaths.

Fourth-Generation PBXs

Fourth-generation PBXs take the integration of voice and data one step further by incorporating local area networks. The technology used, whether CSMA/CD, token ring or packet switching, would be inherent in the PBX. To date, there are no PBXs with these capabilities built in. There is much interest in the blending of local area networks and PBXs, and technologically it is a logical progression, so true fourth-generation PBXs can be expected in the future.

Table 2.1 shows a comparison of the different generations of PBXs.

PBX FEATURES

From a few simple capabilities on early machines, the number of features available in PBXs has mushroomed to over 100 standard features plus scores of custom fea-

TABLE 2.1
Generations of PBXs

1st Generation	2nd Generation	3rd Generation	4th Generation
Analog talkpaths	Analog talkpaths	Digital talkpaths	Ditigal talkpaths
Manually controlled switchboard	Mechanical or Electro-mechanical relays	Non-blocking design Time Division Multiplexing (TDM)	Local Area Network (LAN) technology incorporated
	Space Division Multiplexing (SDM) or Time Division Multiplexing (TDM)	Pulse Code Modulation (PCM) or	
	Stored program control	Adaptive Differential Pulse Code Modulation (ADPCM)	
	Pulse Amplitude Modulation (PAM)		

tures. There are far too many for most users to learn and use, and most will find a handful that are particularly helpful and use them repeatedly.

Standard Features

Features available on most PBXs are:

Authorization Codes. Authorization codes are numbers, such as 123456 or 864753, which must be dialed before a telephone call can be made. Usually, each group within an organization is given an authorization code. The chief use of authorization codes is to group calls for billing purposes. For example, when someone makes a call, it can be easily put with other calls having the same authorization code. All calls from the group can be put together. Authorization codes also act as a form of security. If someone tries to make a call and does not know an authorization code, the call will not go through. The negative side of authorization codes is their inconvenience; every time a call is made, extra digits must be dialed.

Automatic Call Distribution. Incoming calls are routed to a group of attendants. The calls are evenly divided among all the attendants and statistics can be kept of the number of calls received and the length of each call.

Automatic Callback If a user is on the telephone and another user is trying to call, the calling user can be alerted when the other user's telephone is free. In fact, the PBX will connect the two users automatically when both extensions are free.

Automatic Route Selection or Least-Cost Routing. This feature involves the ability of the PBX to look at the number to be dialed and to select the most economical way to place the call.

Bridged Call. The bridged call feature allows a telephone extension to appear on two or more telephones. In this way, users can share the extension in the same way that lines are shared in a key system.

Call Forwarding. Calls that would come to a user are forwarded to another location. This feature is often used when someone is away from the office for several hours. Calls are forwarded to a secretary or receptionist. The calls ring at the secretary/receptionist's telephone and a message is taken. *aka: hunting*

Call Park. Allows a call to be placed on hold by one telephone and taken off hold by a different telephone. This feature is useful when a call must be moved to a different location and transferring is inconvenient. For example, a person needs to refer to information that is in the company library down the hall. The call may be "parked" while the user walks down the hall to the library. In the library the call can be retrieved and the conversation completed.

Call Pickup. When someone else's telephone is ringing, the call pickup feature brings the call to your telephone. This is very useful when a telephone starts ringing and the person is not there to answer.

Call Waiting. Normally, when a person is on the telephone and a second person tries to call, the second party gets a busy signal. With call waiting, the second party's call "camps on" the first person's line instead of getting a busy signal. The first person hears a tone indicating that there is a call waiting. When the first person completes the first telephone call and hangs up, the camped on call is put through automatically.

Centralized Attendant Service. When more than one PBX is linked together in a private network, it is possible to place all attendants at a main location rather than having attendants at each location. The attendants can assist all users on the network by performing as operators providing directory assistance, completing calls, and helping users having trouble.

Conferencing. Conferencing involves the ability to connect from two to six users to the same call so that all may participate in a conversation.

Dictation. This feature allows the connection of the PBX to dictation equipment. This enables users to dictate letters and correspondence from their telephone without special equipment.

Direct Inward Dialing. Direct inward dialing lets calls from the Public Switched Network terminate directly at a user's telephone rather than at an attendant's console.

Direct Outward Dialing. Allows users to place calls directly to the Public Switched Network without the assistance of an attendant.

Hold. The hold feature allows a person to suspend a telephone call temporarily. A user can disconnect from a call to activate a PBX feature or place another call, and then return to the call. The held person hears nothing while on hold, although some systems have music-on-hold, which plays music until the conversation is resumed.

Host Computer Access. The ability to connect a host computer to the PBX is provided the user. Data users can dial up a connection to the computer. Several methods of connection are available, among them, the digital multiplexed interface (DMI), the computer-to-PBX interface (CPI), and the EIA RS-232C interface.

Hunting. Hunting is the ability to look, or "hunt," for an idle extension when the dialed extension is busy. A list of extensions is assigned to a hunt group. When a call is made to an extension that is busy, the call "hunts" to the next extension in the group. The call will hunt to every extension in the group until an idle extension is found.

Intercom. This feature provides a dedicated connection between two users. Intercoms are most often used between people who work closely together, such as secretaries and bosses.

Modem Pooling. A group of modems is made available to data users on the PBX. Users need not have their own modems.

Paging. The connection of a loudspeaker paging system to the PBX allows access to the paging system through a user's telephone.

Power Failure Transfer. In the event of a power failure, two-way central office trunks are connected to designated telephones. These telephones can place and receive calls while other telephones will be without service until power is restored.

Protocol Conversion. This feature gives the user the ability to interface data equipment that use different protocols. For example, an asynchronous computer terminal using ASCII protocol could interface to a host computer using a bisynchronous protocol. The PBX provides the conversions necessary to allow the data devices to communicate.

Queueing. If all the trunks in a group are in use, a call can be placed in queue to wait for one of the trunks to become free. A queue is an area of the PBX where information about a call can be stored until the call is completed.

Redial. This feature "remembers" the last telephone number that was dialed and redials it.

Remote Access. The ability to use features from a telephone that is not within the PBX is afforded by this feature. For example, a user who is out of the office and calls in from a pay phone would be able to listen to voice mail messages or dictate a memo.

Speed Dialing. Pressing two or three numbers causes an entire telephone number to be dialed. This is convenient for frequently called numbers or for emergency numbers. For example, a user's telephone might be programmed to dial the telephone number 619-555-6789 when the digits 1—2—3 are pressed, or the police could be dialed when the digits 9—1—1 are pressed. Some speed-dial numbers are reserved for individual users; others, such as 911, can be shared by everyone on the system.

Station Message Detail Recording. The PBX records information about calls. A report is generated listing all the calls made by each user listing the date, time, duration, and place called. The information is used to manage telephone costs.

Transfer. This feature provides the ability to send a call to another user. If a user answers a call that is for someone else, transferring the call sends it to the other person's telephone.

Voice Mail. Voice messages can be left when a user is away from the telephone. It is like providing an answering machine for each user.

Custom Features.

Custom features are developed when a customer has a unique requirement that cannot be fulfilled by standard PBX features. Since the features will not be used by anyone else, they involve negotiations with the vendor and can be costly.

Most PBXs have the flexibility to be customized. The federal government, for instance, has a specialized private network called AUTOVON (for *auto*mated *v*oice *n*etwork). Most vendors provide the custom hardware and software used by this network. More typical may be a customer who wants telephones with a special ringing pattern or an attendant's console that displays a special message.

PRIVATE NETWORKS

A private network connects different locations to allow communications among them. For example, the Yummy Cheez-Whipped Mashed Potato Co. has sixteen locations across the United States. The company has a private network that allows its users to call one another easily and without incurring long-distance charges. Users can send **facsimile** transmissions and data files over the network as well.

The key features of a private network are:

1. it is used by a private group
2. access to it is limited and controlled
3. two or more locations are linked
4. Public Switched Network **facilities** are not used.

A private network, by definition, is available only to a private group. The group usually consists of the employees of a company. Sometimes, it is not cost effective to include all company locations in the network; small sites may be left out. Other times, two or more companies share a network, allowing the cost of the network to be shared.

Access to the private network is limited and controlled. Unlike the public network, which virtually all telephones can use, a private network can only be accessed through specific nodes. Telephones not connected to one of the nodes do not have access to the network. Companies control the nodes and, in this way, control access to the network. One exception is networks that allow remote access. Remote access allows a user to dial into a node from a telephone in the Public Switched Network. Once dialed in, the private network capabilities are available to be used.

A private network has at least two nodes. Normally, the nodes will be at different locations. Some large installations, however, may have too many stations to be contained in a single node; in these situations, a single location may have multiple nodes. Theoretically, there is no maximum number of nodes that may be part of a private network. Practically, though, the equipment used in the nodes can communicate with only a limited number of other nodes. Some large private networks may have 30 or more nodes. An average private network will have from 5 to 15 nodes.

No Public Switched Network facilities are used to connect nodes. The Public Switched Network, by definition, is available to the public at large, while a private network is not. A private network uses dedicated facilities.

When Does a Private Network Make Sense?

Usually, the decision to install a private network is based on economics. If the cost of private facilities is less than the cost of the Public Switched Network, then a private network is justifiable.

There are two ways of procuring private facilities. The first option is leasing from a common carrier. Costs will vary from carrier to carrier, so it is worthwhile to compare several. Be sure a carrier can serve all needed locations. Most carriers can connect St. Louis to New York City, but not all will have facilities to connect to Sioux Falls, South Dakota.

It may be desirable to lease from two or more carriers to get connections to all locations and to keep costs down. This approach can work, but there is a catch. A private network may have two or three routes from one location to another. Figure 2.16 shows an example. There may be a direct link between nodes A and B. There may be an alternative route from A to C to B. If the leg from A to B is leased from one carrier, the leg from A to C leased from a second carrier, and the leg from C to B leased from a third carrier, it becomes difficult to determine which carrier to call in case of trouble. Most communications managers will attest to this being often a serious problem. When a circuit is in trouble, it needs to be fixed quickly. Wasting precious time identifying who to call can be an intolerable situation. The best approach when multiple carriers are used is to choose a single one for all links between two specific nodes.

The second way of procuring private facilities is to install your own. This usually means microwave. Setting up a microwave private network involves purchasing or leasing microwave equipment and obtaining rights-of-way. Microwave equipment is discussed below. Rights-of-way are needed for the paths the microwave signals take. Just as permission is needed to walk across someone's land, a right-of-way is needed for a microwave path to cross someone's land. Getting permission is a legal issue that is usually handled by a company's legal or contracting department.

An important consideration when installing private facilities is maintenance. When facilities are leased, maintenance is done by the common carrier. When purchased, it becomes the responsibility of the company.

Maintenance may be handled in one of three ways: a company's employees can maintain the equipment; the company can enter into a separate maintenance agreement with a third party maintenance company; or the company can lease microwave equipment while owning rights-of-way.

When employees maintain the equipment, the company usually has them

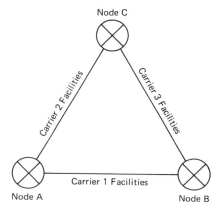

Figure 2.16: Leased carrier facilities.

Private Networks

trained by the equipment vendor, and the vendor supplies spare parts and backup technical support. This option can be the most economical one for large networks, but is not usually the best choice for smaller networks.

The second method is to contract for maintenance with a third party. There are companies that do nothing but maintain equipment made by others. The advantage of using them is that they generally provide the service at less cost than the equipment vendor. This must be balanced with the company's reputation and with its ability to maintain all locations.

A third choice is available when rights-of-way permissions are owned but the microwave equipment is leased. Because the equipment is leased, the equipment vendor provides maintenance.

Elements of a Private Network

There are two parts to a private network: nodes and facilities. *Nodes* are the source of the network's intelligence and switching capabilities, while *facilities* are the connections that tie locations together.

Nodes are the hub of a private network. They are switching points for calls and contain the intelligence needed to route calls. Normally, a PBX or a Centrex provides the capabilities. Calls coming into the node from another location are examined. If the call is for a user in the node, the call is put through to the correct extension. If it is for a user at another node, possible routes are examined and the call is placed over the best available one. Receiving a call and passing it on to another node is called **tandeming**.

Nodes are also the points that provide user access to the network. Normal access is through a telephone or data terminal. It could also be via a remote access port located at the node.

Regulating which employees get telephones, data terminals, or remote access numbers allows control over access to the network. Most users will have telephones allowing them to talk to other users on the network. Some will have data terminals allowing them to communicate with computers and other data users on the network. Remote access provides an entry point for users not directly connected to a telephone or data terminal at the node. Anyone who has a remote access number can enter the network. For this reason, care must be taken that remote access numbers are given only to those who need them.

Facilities, the second main element of a private network, link the nodes together. They are the media that carry the voice and data signals from node to node. They may be cable, microwave, or satellite.

Cable facilities include copper wire, coaxial cable, and fiber-optic cable (see Figures 2.17, 2.18, and 2.19). Copper wire is the least expensive and the most prevalent. Originally used for telegraph, it was later used for the Public Switched Telephone Network and is found wherever telephones are found. It consists of a strand of copper surrounded by an insulator. The insulator, usually made of plastic or polyvinyl chloride, keeps electrical signals that pass through the wire from mixing with signals in other wires. The wires are used in pairs and often are referred to as

Figure 2.17: Twisted pair copper wire within a cable. (Photo courtesy of Northern Telecom)

twisted pairs. Twisted pairs are bundled together into larger cables containing 25, 100, 400, or more pairs.

Of the three kinds of cable facilities, copper wire has the least capacity for carrying communications signals. The capacity for carrying signals is called bandwidth. Think of it as a pipe. The larger the pipe, the more that can pass through it. Copper wire, with a relatively small bandwidth, can be thought of as a relatively small pipe. A copper-wire T-1 facility uses four wires and can carry 23 voice conversations.

Coaxial cable has greater bandwidth than copper wire. A single coaxial cable can carry over 13,000 voice conversations. It is made up of a copper core surrounded by an insulator and a metal sheath. Information travels along the copper core. The metal sheath keeps electromagnetic interference and other electrical pulses from reaching the core. This gives a better transmission quality than can be achieved with twisted pair.

Fiber-optic cable has the greatest bandwidth of the types discussed. A single fiber-optic strand can carry over 100,000 voice conversations. Fiber-optic cable does not use copper as its transmission medium. Instead, a strand of glass or plastic is used. Light passes along the strand carrying information with it. The strand is made of the purest glass or plastic available. Impurities distort the light passing through and diminish the fiber's bandwidth. The light is usually generated by a laser. Because no metal is involved, fiber-optic cable is impervious to electromagnetic interference and most other types of electrical disturbance.

The second kind of facility is the microwave facility. Microwave facilities do not send signals through a wire or fiber. Microwaves are a form of electromagnetic wave. In relation to the rest of the electromagnetic spectrum, they are the high-

Figure 2.18: Multiple coaxial cables within a larger cable. (Photo courtesy of Northern Telecom)

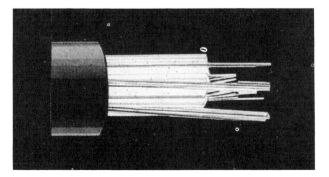

Figure 2.19: Fiber-optic cable. (Photo courtesy of Northern Telecom)

frequency end of the radio band. The signals travel through the air but are not affected by rain or fog. As such, they are well suited to transmit information.

Microwaves must travel in a straight line, however. That means transmission must be line-of-sight. There can be no obstructions, such as mountains or buildings, that would block the signals.

The components of a microwave system are microwave radios, repeaters, and towers. The radios are specially designed to transmit and receive in the microwave range of the electromagnetic spectrum. They work in pairs, one transmitting and one receiving. The radios are "aimed" at one another to provide the strongest signal between them. Transmissions can go in either direction, allowing for two-way communications.

Repeaters are used where the distance between radios is too great for signals to pass. A microwave signal can travel about 10 miles before it **attenuates** to the point that a receiving radio would not recognize it. Repeaters take a signal from a radio and regenerate it. Repeaters are spaced about every 10 miles to keep signals moving through the network.

Towers are often used to place radios high above the ground so that there will be a clear line-of-sight. A tower may house several radios. Figure 2.20 shows an example of the use of a tower.

The third type of facility involves satellites. Satellite transmission is another form of microwave transmission. Microwaves are sent to a satellite orbiting the earth. The satellite regenerates the signals and sends them to their destination. Satellites are very useful when signals must travel long distances or must pass over large bodies of water (see Figure 2.21).

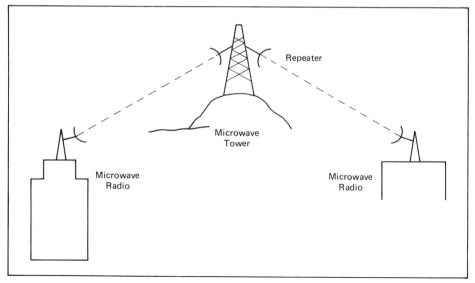

Figure 2.20: Microwave facilities.

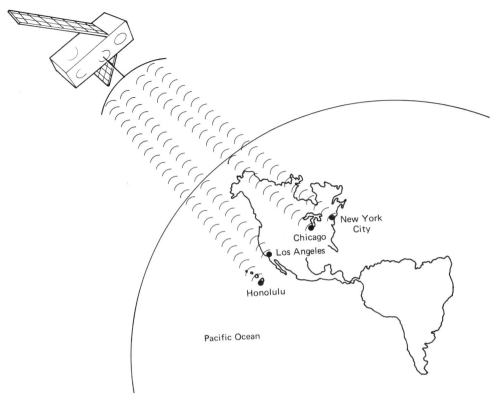

Figure 2.21: Satellite facilities.

Satellite facilities consist of earth stations and satellites. The facilities are extremely expensive to build and maintain, and it is rare to find a company that owns its own satellite network. Most often, satellite facilities are leased from carriers such as Comsat or Satellite Business Systems (SBS). A typical network arrangement is shown in Figure 2.22.

An earth station connects to a network node. It sends and receives signals between the node and a satellite. The great distance between earth station and satellite necessitate expenditure of fairly large amounts of energy to boost outgoing signals and amplify incoming ones. Large, concave antennas help by trapping relatively weak incoming signals (see Figure 2.23).

The interplay of earth stations and satellites makes it possible to send a signal from earth to space more than once before reaching its final destination. The first earth station sends the signal to a satellite. The satellite sends the signal to a second earth station. The second earth station sends the signal to another satellite, which sends the signal to a third earth station. The third earth station is connected to a network node and delivers the signal to its intended endpoint. Each trip the signal

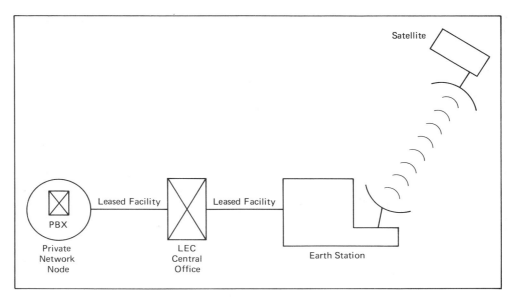

Figure 2.22: Typical network arrangement using satellite facilities.

Figure 2.23: An earth station.

Private Networks 75

makes into space is called a *hop,* and the second earth station is referred to as an intermediate station (see Figure 2.24). Theoretically, a signal could make an unlimited number of hops.

The great distance from earth to satellite can be both a benefit and a problem. It makes signal transmission over long distances easy. However, it also takes much longer for signals to traverse the distance. The delay is not usually noticeable on a voice call, but it can be a serious problem on a data call. A single **satellite hop** may have no apparent effect, but two or three hops may disrupt the call. The reason is this: when data are sent, the receiving end is expected to acknowledge receipt of the data. The transmitting end waits a predefined number of milliseconds for the acknowledgment. If none is received within the required time, the transmitting end usually resends the data. If no acknowledgment is received after several transmission attempts, the transmitting end disconnects from the call.

The delay introduced by satellite hop causes the equipment on each end of the connection to become unsynchronized. The receiving end sends back an acknowl-

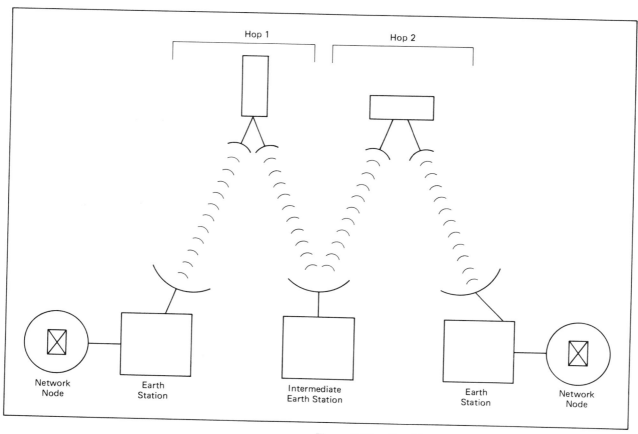

Figure 2.24: Satellite hop.

edgment, but the transmitting end does not receive it in time. In fact, the transmitting end may be resending the data. If so, it will miss the acknowledgment, and the receiving end will get a second copy of the data. Eventually, the call will be disconnected due to the lack of synchronization.

This is a common problem with satellite transmission. One solution would be to increase the time a transmitting station will wait for an acknowledgment. The problem with doing this is that it may not be possible to extend the interval long enough. A better solution is to limit the number of hops that a call can take to one or two. Most satellite carriers can program their circuits to limit the number of hops.

CHAPTER SUMMARY

PBXs are switching machines located on a user's premises. A user has more direct control over telephone features and configurations than over a telephone system based at an LECs central office. As a result, users are generally more involved with the repair and maintenance of PBXs than with central-office-based systems.

PBXs range in size from a few dozen lines to several thousand lines. Some smaller systems are designed to grow incrementally into their larger cousins. There are four basic elements of a PBX: the common control, the line side, the switching matrix, and the trunk side. Each component coordinates with the others to allow for the routing and completion of calls.

Features are also an important part of PBX systems. There are usually hundreds available, among them call transfer, conferencing, call forwarding, intercom, and speed dialing. These features give the PBX a rich array of capabilities from which users can select the most beneficial.

PBXs can also be used to set up private networks. Private networks link locations that are geographically separated. Sophisticated PBXs can choose between routing a call over a private network or the PSN, depending on the most economical path.

KEY TERMS

Attenuate	**Pulse amplitude modulation**
Bandwidth	**Pulse code modulation**
Blocking	**Sampling**
Bus	**Satellite hop**
Codec	**Space-division switching**
Common equipment	**Stored program control**
Customer premises equipment (CPE)	**Talkpath**
Facsimile	**Time-division multiplexing**
Grade of service	**Time-division switching**
Non-blocking	**Twisted pair**

REVIEW QUESTIONS

1. Describe the different generations of PBXs.
2. What are the basic elements of a PBX?
3. How do PAM and PCM differ? What are their similarities?
4. When is a PBX non-blocking?
5. Describe time-division multiplexing.
6. What is the function of a codec?
7. What percentage of calls are blocked if a facility operates at P.05?

TOPICS FOR DISCUSSION

1. End-user telephone equipment has evolved from individual telephones to mechanical key equipment to automated PBX equipment. Speculate on how telephone equipment will change in the future. Then, survey opinions in current telecommunications publications and compare your views with them (see Appendix A).
2. PBX vendors advocate the use of their equipment for data transmission in addition to the more traditional voice transmission. Discuss the strengths and weaknesses of transmitting data through a PBX as opposed to transmitting data through a local area network.
3. Prior to the Carterphone decision in 1968, only telephone companies could manufacture and sell equipment connecting to the PSN. How does this differ from the situation today? How would users be affected if things had not changed?
4. Concerns have been raised that PBXs can be used to bypass LEC central offices and connect directly to common carriers. This eliminates a major function provided by LECs. Discuss the impact of allowing businesses to bypass the LEC.

Data Communications and Computers

CHAPTER 3

OBJECTIVES

Upon completion of this chapter, you should:

- be able to explain simplex, half-duplex, and full-duplex transmission
- understand some of the major data transmission codes, including ASCII, EBCDIC, and Baudot
- be able to discuss the major forms of error checking, including use of a parity-bit, vertical-redundancy check, longitudinal-redundancy check, checksum, and cyclical-redundancy check
- understand the differences between terminals and microcomputers
- be aware of the differences between mainframe computers, minicomputers, and microcomputers
- be able to discuss the basic components of a computer system

INTRODUCTION

Since the major emphasis in this book is on integrated voice and data networks, it is critical that you have a basic understanding of data communications. In this chapter we'll review some basic concepts, including how data is stored and transmitted, how errors are detected and then corrected, and how terminals and computers process data. This chapter will provide you with the information you need to understand the material on networks and micro-mainframe communications that follows.

THE NATURE OF DATA

Most of the information or **data** that we will be examining throughout this book will be in digital form. Since computers are **binary** in nature and utilize a base 2 numbering system, data or **bits** (*bi*nary digi*ts*—abbreviated "b") will be in form of a 1 or a 0 (zero). Many computers consider 8 bits to be a **byte,** or a discrete item of information.

Since humans are used to thinking in decimal, it takes a little training to translate numbers from decimal into binary. It helps, though, to remember the concept of place value. The decimal value 111 really means the following:

$$1 * 100 = 100$$
$$1 * 10 = 10$$
$$1 * 1 = 1$$
$$\overline{111}$$

Since a computer thinks in binary, each 1 or 0 represents a power of two. Let's look at a typical binary number and convert it into its decimal equivalent.

$$11001$$
$$1 * 2^0 = 1 \times 1 = 1$$
$$0 * 2^1 = 0 \times 2 = 0$$
$$0 * 2^2 = 0 \times 4 = 0$$
$$1 * 2^3 = 1 \times 8 = 8$$
$$1 * 2^4 = 1 \times 16 = \overline{16}$$
$$25$$

$$11001_2 = 25_{10}$$

Converting from Decimal to Binary

A decimal number can be converted into its binary equivalent by repeatedly dividing the number by 2 and using the 1 or 0 remainder as the binary place holder. Let's convert the decimal number 17 into its binary equivalent:

$$2 \overline{)17} 1$$
$$2 \overline{)8} 0$$
$$2 \overline{)4} 0$$

$$2 \overline{) \, 2 } \quad 0$$

$$2 \overline{) \, 1 } \quad 1$$

$17_{10} = 1$
100001_2

Does this work? Let's convert the binary number back into decimal:

$$
\begin{aligned}
1 * 2^0 &= 1 * 1 = 1 \\
0 * 2^1 &= 0 * 2 = 0 \\
0 * 2^2 &= 0 * 4 = 0 \\
0 * 2^3 &= 0 * 8 = 0 \\
1 * 2^4 &= 1 * 16 = \underline{16} \\
& 17
\end{aligned}
$$

There are other number systems used by computer programmers, especially octal (base 8) and hexadecimal (base 16). Octal can be thought of simply as groups of three binary digits. Using the binary/decimal conversion chart below, let's take an octal number and see if we can read it and then convert it into its decimal equivalent.

```
001    1
010    2
011    3
100    4
101    5
110    6
111    7
```

11011011_2
011 011 011
 333_8

$$
\begin{aligned}
333_8 = 3 * 8^0 &= 3 * 1 = 3 \\
3 * 8^1 &= 3 * 8 = 24 \\
3 * 8^2 &= 3 * 64 = \underline{192} \\
& 219
\end{aligned}
$$

$333_8 = 21910$

The hexadecimal number system is built around a base 16, which consists of the numbers 1 through 9 and the letters *A* through *F*. A decimal number can be converted to its hexadecimal equivalent by dividing it repeatedly by 16 and using the remainders as hexadecimal place holders, much the same way we converted from decimal to binary. To convert binary numbers to hexadecimal, we simply group four binary numbers together instead of the three numbers we grouped together for octal or base 8 number system.

The Nature of Data

$$4F7_{16} = ?_{10}$$

$$7 * 16^0 = 7 * 1 = 4$$

$$F\,(15_{10}) * 16^1 = 15 * 16 = 240$$

$$4 * 16 = 4 * 1000 = \underline{4000}$$

$$4244$$

$$4F7_{16} = 4244_{10}$$

Binary	Decimal
100	4
$1 \times 2^2 + 0 \times 2^1 + 0 \times 2^0$	
$4 + 0 + 0$	
111	7
$1 \times 2^2 + 1 \times 2^1 + 1 \times 2^0$	
$4 + 2 + 1$	

Modes of Transmission

In **simplex** transmission, data flows in only one direction. One example of this approach is a person listening to a radio station. The music flows from the radio station's transmitter to the listener's radio receiver. The people at the radio station are not aware of anything the listener says in response to the music received. A second example of simplex transmission would be some of the messages we have broadcast into space. Will intelligent life "hear" the message? We really don't know because the communication mode is strictly one-way, or simplex.

A second mode is **half-duplex** transmission. With this method, information is sent in only one direction at a time. Half-duplex transmission works very much along the same principles as a CB (citizen's band) radio. One trucker may send a message to his friend and then conclude by saying "over." The friend then broadcasts his message. Half-duplex transmission can utilize more than one set of wires to avoid the time it takes to clear a line before transmitting in the other direction. Despite the extra set of wires, however, transmission is only in one direction at a time.

The third possible transmission mode is **full-duplex,** which consists of simultaneous transmission in both directions. There are a number of ways to accomplish this task including using four-wire lines or dividing up a channel so that a portion is used for transmission in either direction.

Types of Transmission

Data can be "packaged" in different forms for transmission. With **asynchronous transmission,** each alphabetical character or number is sent along with a special "start" bit that notifies the receiving device that a byte of information is now arriving. The byte of information is followed by a special "stop" bit that announces

Figure 3.1: Asynchronous transmission.

that this represents the end of transmission for that particular byte of information. Asynchronous transmission by its very nature is "bursty," rather than steady, and sometimes it is referred to as **start/stop transmission.** Figure 3.1 illustrates how asynchronous information travels.

A much more efficient method of transmission would be if a device could indicate a specific time that it would be sending information and then indicate how much information could be expected. When two devices synchronize their clocks and agree on exactly how much data will be received, we call this method of transmission **synchronous.** Synchronous transmission is far more expensive than asynchronous transmission, but when speeds far greater than 2,400 bps are required, may be the only alternative.

As Figure 3.2 illustrates, synchronous transmission consists of **frames.** Each frame packages its different types of information into various fields. Each field has a certain length (in bytes) so that the receiving device can interpret the meaning of each of these fields. In Chapter 5 we will examine several examples of frames used in synchronous transmission in the context of IBM mainframe computer operations.

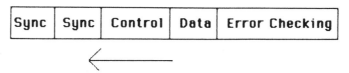

Figure 3.2: An example of synchronous transmission.

DATA TRANSMISSION CODES

What Is A Data Code?

Computers keep track of information by providing a **data code,** or designated bit pattern, that represents each character. The **alphabet** used by a specific coding scheme refers to all the characters that have corresponding bit patterns, not merely the 26 letters of the alphabet. In one data alphabet, for example, a space is represented by the decimal number 32. Obviously spaces must be transcribed along with the words between them so that a normal sentence can be reassembled from the computer's memory and printed out in a form that a human being can understand.

The number of bits per character is critical, and it varies from data code to data code. If 7 bits are used for each character, then there are 2^8, or 256, different characters that can be represented. In a similar vein, a data code that uses 8 for each character can represent 2^7, or 128, different characters.

Morse Code

Everyone who has ever seen a western movie is familiar with the telegraph office and the dots and dashes that constitute our first real data code. Samuel Morse developed his code in conjunction with his telegraph. A dot is a single short bit while a dash is a single long bit.

Morse decided in the name of efficiency that the most commonly used letters in the alphabet should have the shortest code. The most common consonant in the English alphabet, *T* is represented by a single dash—, while the rare letter *Q* is represented by a "dash-dash-dot-dash" (— — · —) combination.

Baudot Code

During the last century, Emile Baudot invented the Baudot distributor, a device capable of transmitting five pieces of information at a time. Baudot's scheme has become the basis of the modern telex, and this coding method is recognized as International Telegraph Alphabet #2. Since only 5 bits of information can be transmitted at a time, this would appear to mean that there are only 2^5, or 32, different letters or numbers that could be represented. The code uses special control codes, a *figures-shift* and a *letters-shift,* to indicate that what follows will either be figures or letters. These control codes extend the range of possible characters to 58. Figure 3.3 illustrates the Baudot code.

The Baudot code is still used in telex transmission as well as by the United Press International and Associated Press new services. There are inherent limitations to this Baudot code, however. There is no way of ensuring that the transmission is correct, since there is no **parity check,** a concept we will discuss later in this chapter. Because the control codes indicate whether letters or numbers are to follow immediately, a mistake in correctly identifying one of these can result in a massive misreading of entire portions of a transmission. Since a human must identify this error either at the transmitting end or at the receiving end, the process cannot be automated entirely.

American Standard Code for Information Interchange (ASCII)

The **American Standard Code for Information Interchange (ASCII)** represents a method of coding characters and numbers in a 7-bit format with an eighth bit used for parity, or error, checking. Despite the fact that virtually every computer publication and government publication describes the "ASCII standard," there are actually several variations of ASCII, depending upon users' needs. It is very important that the two devices that wish to communicate with each other use the same ASCII dialect.

Figure 3.4 illustrates the "standard" ASCII code. ASCII provides the coding

Baudot Code

Character shift		Binary code
Letter	Figure	Bit: 4 3 2 1 0
A	—	11000
B	?	10011
C	:	01110
D	$	10010
E	3	10000
F	!	10110
G	&	01011
H	#	00101
I	8	01100
J	'	11010
K	(	11110
L	)	01001
M	.	00111
N	,	00110
O	9	00011
P	0	01101
Q	1	11101
R	4	01010
S	bel	10100
T	5	00001
U	7	11100
V	;	01111
W	2	11001
X	/	10111
Y	6	10101
Z	"	10001
Figure shift		11111
Letter shift		11011
space		00100
Line feed (LF)		01000
Blank (null)		00000

Figure 3.3: The baudot code.

necessary for data-link control characters such as STX (start of text) and EOT (end of transmission), graphic control characters such as CR (carriage return), BS (back space), and FF (form feed), and alphanumeric characters. Since ASCII utilizes a 7-bit format, there are 2^7 possibilities, or 128 different control characters, graphics control characters, and alphanumerical characters, that can be represented. Using much the same technique that is used with Baudot code, it is possible to use special *shift out* and *shift in* characters to extend the range of characters that can be represented. An alternative to this approach is to eliminate the parity bit and use an 8-bit code, which then extends ASCII's range to 256 characters.

Extended Binary-Coded Decimal Interchange Code (EBCDIC)

Extended Binary-Coded Decimal Interchange Code (EBCDIC) is an 8-bit code developed by IBM that can represent 2^8, or 256, different characters and numbers.

American National Standard Code for Information Interchange

Bits				7	0	0	0	0	1	1	1	1
				6	0	0	1	1	0	0	1	1
4	3	2	1	5	0	1	0	1	0	1	0	1
0	0	0	0		NUL	DLE	SP	0	@	P	`	p
0	0	0	1		SOH	DC1	!	1	A	Q	a	q
0	0	1	0		STX	DC2	"	2	B	R	b	r
0	0	1	1		ETX	DC3	#	3	C	S	c	s
0	1	0	0		EOT	DC4	$	4	D	T	d	t
0	1	0	1		ENQ	NAK	%	5	E	U	e	u
0	1	1	0		ACK	SYN	&	6	F	V	f	v
0	1	1	1		BEL	ETB	'	7	G	W	g	w
1	0	0	0		BS	CAN	(	8	H	X	h	x
1	0	0	1		HT	EM	)	9	I	Y	i	y
1	0	1	0		LF	SUB	*	:	J	Z	j	z
1	0	1	1		VT	ESC	+	;	K	[	k	{
1	1	0	0		FF	FS	,	<	L	\	l	|
1	1	0	1		CR	GS	-	=	M	]	m	}
1	1	1	0		SO	RS	.	>	N	^	n	~
1	1	1	1		SI	US	/	?	O	—	o	DEL

Example:
Bits: P* 7 6 5 4 3 2 1

1 1 0 0 0 0 0 1 = letter "A" (Odd Parity)
0 0 1 1 1 0 0 0 = number "8" (Odd Parity)

P* = Parity Bit

Figure 3.4: ASCII character code.

Pronounced "ebb- see-dik," this coding approach is found in IBM's minicomputers and mainframe computers; consequently, non-IBM equipment frequently requires translating devices such as protocol converters to convert back and forth from ASCII to EBCDIC. Figures 3.5 and 3.6 illustrates how standard EBCDIC code provides for more special control codes.

Incompatibilities Between ASCII and EBCDIC

It should be clear from the figures illustrating both ASCII and EBCDIC that these two coding schemes are incompatible. EBCDIC does not use parity checking, because mainframe computers and minicomputers use a much more sophisticated error checking scheme, which we will examine later in this chapter.

This basic incompatibility is a major issue for telecommunications managers,

Figure 3.5: EBCDIC code.

Bit Positions 4, 5, 6, 7	Second Hexadecimal Digit	00				01				10				11				
		00	01	10	11	00	01	10	11	00	01	10	11	00	01	10	11	
		0	1	2	3	4	5	6	7	8	9	A	B	C	D	E	F	
0000	0	NUL	DLE	DS		SP	&	-						{	}	\	0	
0001	1	SOH	DC1	SOS		RSP		/		a	i	~		A	J	NSP	1	
0010	2	STX	DC2	FS	SYN					b	k	s		B	K	S	2	
0011	3	ETX	DC3	WUS	IR					c	l	t		C	L	T	3	
0100	4	SEL	RES/ENP	BYP/INP	PP					d	m	u		D	M	U	4	
0101	5	HT	NL	LF	TRN					e	n	v		E	N	V	5	
0110	6	RNL	BS	ETB	NBS					f	o	w		F	O	W	6	
0111	7	DEL	POC	ESC	EOT					g	p	x		G	P	X	7	
1000	8	GE	CAN	SA	SBS					h	q	y		H	Q	Y	8	
1001	9	SPS	EM	SFE	IT				\	i	r	z		I	R	Z	9	
1010	A	RPT	UBS	SM/SW	RFF	¢	!	¦	:					SHY				
1011	B	VT	CU1	CSP	CU3	.	$	,	#									
1100	C	FF	IFS	MFA	DC4	<	*	%	@									
1101	D	CR	IGS	ENQ	NAK	(	)	_	'									
1110	E	SO	IRS	ACK		+	;	>	=									
1111	F	SI	IUS/ITB	BEL	SUB	|	¬	?	"								EO	

Bit Positions 0,1 — 00, 01, 10, 11
Bit Positions 2, 3 — 00, 01, 10, 11
First Hexadecimal Digit — 0–F

since their data communications often include both types of data coding, and since they often have a need to transfer information on a microcomputer (ASCII) to their mainframe computer (EBCDIC). As we shall see later in this chapter, it is possible by means of a device such as a protocol converter or through micro-mainframe communications hardware and software, to accomplish this conversion from one data coding scheme to another.

ERROR CHECKING

Parity Checking

When we discussed ASCII coding, we pointed out that ASCII uses 7 bits of information to describe a control character, character, or number. The eighth bit is used as **a parity bit.** We refer to this parity method of error checking as **vertical redundancy**

ACK	Acknowledge	IT	Indent Tab
BEL	Bell	IUS/ITB	Interchange Unit Sep./Intermediate Text Block
BS	Backspace		
BYP/INP	Bypass/Inhibit Presentation	LF	Line Feed
CAN	Cancel	MFA	Modify Field Attribute
CR	Carriage Return	NAK	Negative Acknowledge
CSP	Control Sequence Prefix	NBS	Numeric Backspace
CU1	Customer Use 1	NL	New Line
CU3	Customer Use 3	NUL	Null
DC1	Device Control 1	POC	Program-Operator Comm.
DC2	Device Control 2	PP	Presentation Position
DC3	Device Control 3	RES/NEP	Restore/Enable Presentation
DC4	Device Control 4	RFF	Required Form Feed
DEL	Delete	RNL	Required New Line
DLE	Data Link Escape	RPT	Repeat
DS	Digit Select	SA	Set Attribute
EM	End of Medium	SBS	Subscript
ENQ	Enquiry	SEL	Select
EO	Eight Ones	SFE	Start Field Extended
EOT	End of Transmission	SI	Shift In
ESC	Escape	SM/SW	Set Mode/Switch
ETB	End of Transmission Blk	SO	Shift Out
		SOH	Start of Heading
ETX	End of Text	SOS	Start of Significance
FF	Form Feed	SPS	Superscript
FS	Field Separator	STX	Start of Text
GE	Graphic Escape	SUB	Substitute
HT	Horizontal Tab	SYN	Synchronous Idle
IFS	Interchange File Sep.	TRN	Transparent
IGS	Interchange Group Sep.	UBS	Unit Backspace
IR	Index Return	VT	Vertical Tab
IRS	Interchange Record Sep.	WUS	Word Underscore

Figure 3.6: EBCDIC code chart.

Figure 3.7: Even parity in an ASCII transmission.

checking (VRC). When using the parity bit method of error checking, there can be an **even parity** or an **odd parity**. With even parity, all 8-bit units have an even number of 1-bits. This means that if the 1-bits in the seven bits that comprise actual character being sent sum to an odd number, a 1-bit is added in the parity bit position to ensure that an even number of 1-bits will be sent. Figure 3.7 illustrates an even parity ASCII character.

Similarly, if the seven bits representing an ASCII character sum to an odd number of 1-bits and we are using **odd parity** for error checking, we would then place a 0-bit in the parity bit position to ensure that an odd number of 1-bits will be sent. Figure 3.8 illustrates an odd parity ASCII character. When ASCII information flows from one computer to another, either from a direct cable connection or over a telephone line with modems converting the analog signal to a digital signal, it is imperative that both computers use a communications program to establish certain standards, such as the speed of transmission and the type of parity to be used.

Figure 3.8: Odd parity in an ASCII transmission.

As we pointed out earlier in this chapter, there is one very serious flaw in the parity bit approach to error checking. If two or more bits are transposed or damaged in such a way that the information is altered yet the total number of 1-bits remains the same, parity bit checking will not discover that there is an error in transmission. The faster the transmission rate, the more likely that this condition may exist. In an era in which transmission speeds continue to increase, this "blind spot" represents a major limitation to parity bit checking.

Longitudinal Redundancy Check (LRC)

The even and odd parity schemes are sometimes referred to as vertical redundancy checks, because they check one bit at the end of each byte of information that is sent. It is also possible to use a **horizontal redundancy checking** scheme that will check for errors after an entire block of information has been sent. A **longitudinal redundancy check (LRC)** adds a **block check character (BCC)** at the end of a block of data.

Cyclic Redundancy Checking

One of the most common examples of a longitudinal parity checking scheme is cyclic redundancy checking (CRC), which is usually associated with EBCDIC code. Also known as CRC-16, this scheme uses a complex algorithm that consists of dividing an integer by a prime number and noting the remainder. This remainder is calculated at the receive end of a data transmission and compared with the remainder calculated at the transmit end. If the two numbers do not match, then an error in transmission has occurred.

XMODEM

A very simple error-checking scheme commonly used with microcomputers is known as XMODEM. Devised by Ward Christensen, this method consists of having the receiver send an NAK signal (negative acknowledge character). The transmitter then begins transmitting information in blocks of 128 bytes followed by a checksum that consists of the total of all the 128 bytes in a message. If the receiver finds that the checksum displays the correct number and that the sequential block number is one more than the previous block received, it sends an ACK (acknowledgment) signal back to the transmitter, which then sends the next block of information. If an error is detected, the receiver sends an NAK signal, which results in the block being retransmitted.

While XMODEM is a relatively simple method of error checking that is effective for microcomputer data communications, it is not sophisticated enough for communications among large computers. Another drawback is that it requires manual operation at both ends. In other words, people must monitor the error-checking process; it cannot be completely automated.

Another major limitation of XMODEM protocol is that it requires that both computers be able to receive and send an 8-bit byte. Some communications channels are limited to 7 bits per character, since an eighth bit is used for special control purposes. These channels are incapable of using the XMODEM protocol.

There are some extentions to the XMODEM protocol that overcome some of built-in limitations. The XMODEM CRC protocol adds a cyclic redundancy check (CRC) by replacing the 1-byte checksum with a 2-byte code. This improves error detection to approximately 99.9%.

Kermit

Frank Da Cruz and Bill Catchings of Columbia University's Center for Computing Activities developed a protocol named after the famous muppet found on the television show *Sesame Street*. **Kermit** sends blocks of data just like the XMODEM protocol, but it does not require an 8-b byte block or "packet." The default Kermit packet is 80 characters long and ends with a "end of line" character such as a carriage return.

Kermit is a particularly valuable communications protocol because it is able to transfer information among a wide range of computers, including those that use EBCDIC rather than ASCII code. It is available for over 100 machines, ranging from microcomputers to the Cray supercomputer.

DATA TERMINALS AND COMPUTERS

Terminals

For tens of thousands of people, the tool they use most at work consists of a computer terminal. Also known as a **cathode ray tube (CRT)** or a **video display terminal (VDT),** the computer terminal serves as an interface between the human user and the computer. While terminals can differ substantially (as we shall see shortly), they do share certain characteristics. They have a keyboard that can generate a complete alphanumeric character code set. They also must have a screen that can display the full alphanumeric character code set. Finally, they must have the ability to send and receive information from a computer. This information may be transmitted over phone lines and then received by the terminal's built-in modem, or it may be received via a direct cable link to the computer.

First introduced in 1965, terminals have become more and more sophisticated. The industry standard for the microprocessor controlling the terminal's operation has progressed from the Intel 8088 to the Intel 80386, with the Intel 80486 waiting in the wings. Frequently microprocessor-based programs reside in ROM (read-only memory) or in PROM (programmable read-only memory). These programs enable the terminals to provide special features as well as display certain character sets.

A **dumb terminal** is a terminal with very limited capabilities. Generally, these terminals can only display and send data. While they are lowest-cost terminals, they often lack essential features, such as the ability to program certain keys for special functions and the ability to save information.

A **smart terminal** enables the operator to format the screen in various ways. It also enables the user to program keys to perform certain functions. Smart terminals can save several screens of information and then send the entire message via synchronous transmission. Windowing is another feature found in a smart terminal. This means that the screen can be divided into several different portions, each of which can contain different types of information, such as text, graphics, and menus and/or commands.

Since a smart terminal is addressable, this means that a host computer can send and receive information specifically designated as going to or coming from that particular terminal. The host computer can send a special screen format, such as an entry order form to a terminal that then can save this form and use it whenever necessary to input information. The host computer can specify that only certain fields of information can be received by a specific terminal.

A smart-terminal user, therefore, has a lot more independence vis-a-vis the host computer than has a dumb terminal. The user can work at his or her own pace and move the terminal's cursor to whatever point on the screen is desired, without instructions from the host computer.

Asynchronous terminals are inexpensive. They transmit the characters typed on their keyboards immediately to the host computer, since there usually are no buffers present. A user typing the letter *h* will cause the ASCII code for *h* to be sent immediately to the host computer. Similarly, the host computer's asynchronous transmission appears on the terminal's screen immediately after being sent.

Figure 3.9 illustrates a typical asynchronous ASCII terminal. The WY-150 for example, contains a 14-in. (diagonal) flat-screen green phosphor display. The terminal has 1188 x 416 pixel resolution.

Even this simple asynchronous ASCII terminal has many features that at one time would have been reserved for much more expensive terminals. It is able to display underline, reverse video, dim certain blocks of text, and provide a blinking cursor. It is also able to **emulate,** or imitate, certain other terminals, such as the ADDS Viewpoint A2 and the TeleVideo 925. This terminal emulation feature is important, since certain host computers may only be set up to transmit to certain types of terminals.

Asynchronous and Synchronous Terminals

Terminals are capable of either asynchronous or synchronous transmission. As we pointed out earlier in this chapter, asynchronous transmission tends to be in "bursts" rather than a continuous flow of information. An ASCII terminal will send 8 bits of information (1 byte) to represent a single character. In effect, each time a typist

Figure 3.9: An ASCII asynchronous terminal. (Photo courtesy of WYSE)

types a character, 8 bits of information flows from the terminal to the attached computer. Notice, though, that the information is not a smooth, continuous flow. A typist usually cannot sustain a steady 100 words per minute for a long period of time.

Earlier in this chapter we examined synchronous transmission, which can be thought of as a continuous flow of data. This transmission mode requires a different type of terminal, because the transmission is built around a unit know as a *frame,* which contains a number of fields of information. Some information is required to synchronize the timing between the terminal sending the data and the computer receiving it. Other fields are needed for error checking, a key feature in data transmission.

Non-ASCII Terminals

The world of IBM mainframe computers is a world in which these computers talk only to non-ASCII terminals. IBM's mainframe computers transmit in synchronous mode to IBM 3270 terminals, which are **synchronous terminals.** The 3270 terminal contains memory that serves as temporary storage area, or buffer. Information is stored in these buffers until the user presses a key to display the information. It is also possible to use non-IBM asynchronous terminals that are connected to special protocol converters that make the IBM computer believe it is sending information to 3270 terminals.

Terminal Configurations

Terminals can be connected to computers in a number of different ways. One method is a **point-to-point connection.** As we mentioned briefly in Chapter 1, there is a direct connection between each terminal and its computer, as illustrated in Figure 3.10. Just like a very narrow street on which only one car can travel at a time, this point-to-point connection presents a data flow problem. Who determines when the terminal or the computer gets to use the line for data transmission?

One method used of handling this dilemma is **contention.** Both the terminal and the computer *bid* for the right to use the communications line that they share. This "bid" constitutes a request to the other unit for permission to use the line. Assuming the line is free and the other unit doesn't have anything to send, permission is granted. We call this method *contention* because in a sense both host computer and terminal contend for the right to use the same line.

A second method commonly used to determine whether terminal or host computer gets to use the communications line is *first come first served*. This is conten-

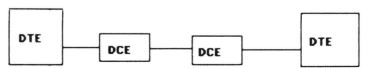

Figure 3.10: Point-to-point terminal configuration.

tion in its purest sense, since the first device to start using the communications line continues to use it until finished. When it finishes sending, it receives an acknowledgment signal from the other unit indicating that the message was received. If there is no response, the sending device assumes an error in transmission and resends the message.

A second type of terminal configuration is a **multi-point connection.** As illustrated in Figure 3.11, several terminals can share a common communications link with their host computer. Usually the host computer is considered the primary station, while the terminals are considered the secondary stations. The primary station has total control of this communications line. It uses it when it wishes to send information to the terminals. When it does not have information to send, it polls the terminals one at a time by checking to see if they have information to send. Each terminal responds to this poll message either by sending a positive acknowledgment indicating it has information to send or a negative acknowledgment indicating that it does not have information to send.

A multi-point connection has some major advantages and some significant disadvantages. It is an efficient method of using resources, since terminals are not capable of sending information continuously at the maximum rate that a host computer can accept it. Another major advantage of this configuration is that devices such as modems can be shared. Only one modem need be used at each end of multi-point connection, as seen in Figure 3.11.

A major disadvantage of a multi-point configuration is that terminals do not have immediate access to a host computer. They have to wait their turns along with whatever other terminals wish to communicate with the host computer. A second major disadvantage of this approach is that it is not possible to use inexpensive dumb terminals, since the terminal must have large enough buffers (i.e., must be smart enough) to retain the information until it gets its turn to send.

Desirable Terminal Features

There are certain features in a computer terminal that are desirable. These features are discussed below.

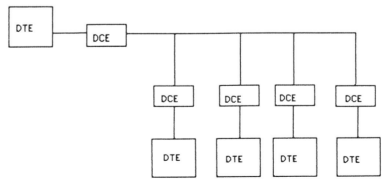

Figure 3.11: A multi-point connection. Multi-point connections enable terminals to share a modem.

1. Clearly it is desirable for a terminal to be able to emulate (imitate) as many other types of terminals as possible, so that a company can mix and match them with existing terminals attached to various computers.
2. A programmable keyboard enables users to change the keyboard for special tasks. One user might prefer the Dvorak keyboard layout, which is designed to place vowels on one side of the keyboard and consonants on the other for faster typing.
3. Different display formats can be essential. Some terminals can be set to display 80 columns, 132 columns, or some other programmable format.
4. Color flexibility can be a major issue. The Wyse 350 terminal, for example, offers 16 different predefined palettes that can be invoked from the terminal's keyboard by the user or by the host computer. It offers 64 different colors displayed in eight-color palettes.
5. **Ergometric** features can be critical. Ergonomics is the science concerned with the physical and psychological effect of equipment on human users. Terminals with CRTs that can be adjusted up or down to a comfortable eye level and keyboards that contain cables long enough to position the keyboard comfortably are ergonomic considerations.

COMPUTERS IN THE OFFICE

A telecommunications manager must be as knowledgeable about as computers as about telephone systems. Today, the distinctions between mainframe computers, minicomputers, and microcomputers are blurring, but there are still generalizations we can make about each type of computer system. Since it is very common in a large company to have mainframe computers, minicomputers, and microcomputers linked together and linked to the company's PBX system, it is important that a telecommunications manager become familiar with some basic computer terminology and concepts. In this section of the chapter we will examine the basic building blocks of a computer system and will differentiate among the various types of computers.

Generally we differentiate between **mainframe computers, minicomputers,** and **microcomputers** on the basis of speed and size. We refer to the number of instructions a computer can execute in a second as a MIP (millions of instructions per second). While a microcomputer may execute 100,000 instructions per second or perhaps with the very fastest models 1 to 2 MIP, a mainframe can execute more than 10 MIP—that is, 10 million instructions per second.

Another major distinction between a mainframe computer and a smaller computer is that mainframes are usually housed in a special computer room designed specifically for the machine. This room will have specialized wiring, air conditioning, humidity control, and security features for that particular computer. A computer operator's console (or screen) allows the operator to observe the many different programs that this mainframe computer is capable of executing simultaneously.

Mainframe computers are able to process large blocks of information at one time. The number of bits a processor can handle at one time is known as its **word size.** A mainframe has a large word size that can consist of 32 bits or even 64 bits.

Permanent storage is virtually unlimited on a mainframe computer. Many secondary storage units have shelves for storage of libraries of tapes containing various programs and data. In some cases a computer operator will manually mount the appropriate tape, while in other cases the process is automated in much the same way that a jukebox automatically selects the correct record and loads it for playing.

Dozens and sometimes even hundreds of users' terminals can be connected to a mainframe computer, such as the one pictured in Figure 3.12 . At Mervyns, the decision was made to perform most data processing functions on its mainframe computer. Employees access this computer by use of dumb terminals. Mainframe computers perform the "number crunching" required by many programs that would simply take too long on slower machines.

The ultimate mainframe computer is known as the **supercomputer.** Supercomputers are cooled in liquid nitrogen to increase conductivity and decrease heat. They are distinguished from regular mainframe computers by their enhanced speed, often in the range of billions of instructions per second. Speed such as this is quite expensive, and supercomputers have cost as much as $17 million dollars. Companies such as Cray, Control Data, Hitachi, and Fujitsu make supercomputers that use a technique called **parallel processing,** in which several processing units divide up the work to increase speed. Figure 3.13 pictures a supercomputer.

Mainframe Computers

A mainframe computer is composed of several parts: one or more input device(s), a central processing unit (CPU), secondary storage, and one or more output devices. We'll examine each of these key elements and then view the entire system in operation.

Figure 3.12: An IBM S/370 model 145 mainframe computer. (Photo courtesy of IBM Corp.)

Figure 3.13: A Supercomputer. (Photo courtesy of Scientific Computer Systems Corp.)

Input Devices

There are a number of ways to input information into a computer. The most common way is through a terminal keyboard. Each time a key is pressed, the corresponding numeric code for that particular letter or number is transmitted to the computer for processing. An optical scanner is another possible input device. It converts the letters on a printed page to their ASCII equivalent and transmits that information into the computer. Figure 3.14 illustrates a typical optical scanner.

The Central Processing Unit (CPU)

The **central processing unit** is the very heart of a computer. It is composed of a main memory, a control unit, and an arithmetic-logic unit (ALU). Figure 3.15 illustrates a complete computer system.

The main memory in a computer consists of temporary storage. This temporary storage is reserved for programs that are being executed, for data that is to be processed, and for data that has been processed and is awaiting routing to a second-

Figure 3.14: An optical scanner.

ary storage device for permanent storage. When a computer is turned off, all material in its main memory is lost. In this respect, a computer's main memory is like a blackboard that begins the day completely clean. During the day instructors write material on the board, and this material is stored permanently in students' notebooks. After the information has been presented by the instructor and processed by the students (who then store it), the bell rings and the board is cleaned to make more room for the next instructor's information. While some material may remain in a corner of the blackboard during the day, the board is cleaned completely at night when the classroom is closed ("turned off").

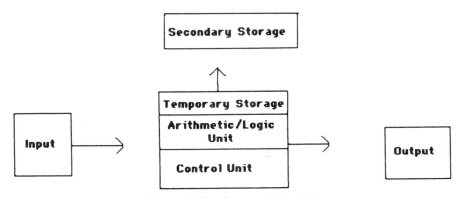

Figure 3.15: A computer system.

We learned earlier in this chapter that data resides in a computer in binary form as bits. In main memory this binary information is stored in sequential memory locations that each contain their own address. You might think of these memory locations in main memory as hotel rooms. A telecommunications convention is scheduled for a hotel on Tuesday. Each company is assigned a block of rooms beginning with a room number. AT&T's personnel, for example, might be housed in a block of rooms beginning with room 3500.

These people are housed (or *stored*) only temporarily while the actual convention is running. The very next day, when the plumber's convention is running, room 3500 will contain a different person. To carry this analogy just a bit further, at any given time the front desk at the hotel can identify who is in which room and can summon that person to the front desk by making a call to the room. Similarly, we will see very shortly that the control unit of the computer also keeps track of which programs are residing at which addresses in memory and also can summon or "fetch" a copy of the information or even the program instructions residing within that particular block of addresses. Each program that is being executed in the computer is stored in a certain portion of memory with its own unique address. The payroll program currently running, for example, might be housed in main memory beginning at memory location 3000. When the payroll program is no longer running, its memory locations are overwritten with data and other program instructions that the computer currently needs.

The second main constituent part of a CPU, the control unit, is responsible for coordinating the computer's activities. When a program requests that a certain task be performed, the control unit routes the data requested to the appropriate location in main memory. The control unit also ensures that a program that is to be executed is loaded into the primary memory area.

Since a program consists of a series of instructions that must be executed in sequential order, the control unit makes sure that the instruction that is to be executed is stored in a special holding area, or **instruction register,** while the address in main memory for the *next* sequential instruction that will be executed is stored in a **memory address register.** This process ensures that the correct instruction will be fetched when its turn to be executed arises. When data needs to be processed, including adding, multiplying, dividing, or subtracting, the control unit routes this data to the arithmetic-logic unit, where these tasks are performed. Finally, when the processed information needs to be stored permanently in a secondary storage device such as a disk drive or tape drive, the control unit routes this data to the appropriate device.

The third main constituent part of a CPU, the **arithmetic-logic unit (ALU),** contains the circuitry responsible for all arithmetic and logical operations. In addition to adding, subtracting, multiplying, and dividing data, the ALU performs the logical operations of determining whether numbers are equal to, greater than, or less than other numbers. The ALU also uses registers for temporary storage of information. A special register known as an **accumulator** is used for temporary storage of the results of computations. The ALU also uses several general-purpose registers for the temporary storage of calculated values that will be used in other calculations.

How the CPU Executes Program Instructions

There are four basic steps that take place whenever the CPU executes a program instruction. We refer to these steps as comprising a **machine cycle.** The first step is for the control unit to fetch the instruction, which is stored in main memory. The control unit decodes this instruction and then issues instructions for the necessary data to be moved from main memory into the ALU for processing. These two steps comprise the instruction time or I-time for a computer.

The control unit directs the ALU to process the information, and this task is accomplished. The final step is for the control unit to order that the results of this calculation be stored in memory or in a register. We refer to these two final steps as the execution time or E-time for a computer.

Secondary Storage

For permanent storage of information, mainframe computers use a variety of devices. A **direct access storage device (DASD)** such as a magnetic disk drive provides each record with a unique address. This means that when the computer wishes to load a particular file that it has saved, it can go directly to that particular address without having to search through all its files until it finds the specific file it needs to load.

Magnetic disks represent the most common form of secondary storage for large computers at this time. Data is recorded on an oxide-coated metal platter as a series of electronic spots. The platters turn at several thousand revolutions per minute while a read/write "head" much like a record player's arm moves to the exact location of a particular file. The head can then copy the location's contents (bit patterns) and bring this information into main memory. Some of the larger disk drives can hold as much as 2.5 billion characters. It is possible to attach several of these drives to a mainframe computer.

Mainframe computers also use magnetic tape for storage. This storage medium provides sequential storage, which means that a computer would have to search through the entire tape to find a particular file. This problem is kept from getting worse by keeping the various storage tapes (a library or databank may consist of thousands of them) specifically labeled. IBM has a tape unit that holds rolls of magnetic tape around 2 in. wide. Each of these rolls can hold around 50 million characters.

One of the newest forms of secondary storage is the optical disk, also known as the laser disk, or CD-ROM. An optical disk drive such as the one illustrated in Figure 3.16 uses a laser to pit the surface in binary fashion. A laser mark indicates a 1-bit, while the lack of such a mark indicates a 0-bit. What makes these optical disks so appealing is the vast amount of information they can hold. A single platter can hold the equivalent of 550 megabytes of information. A system such as one supplied by FileNet is capable of supplying up to 204 optical disks in a complete library accessed by four separate optical disk drives. These optical disks are particularly effective if they are used in conjunction with optical scanners or digitizers so

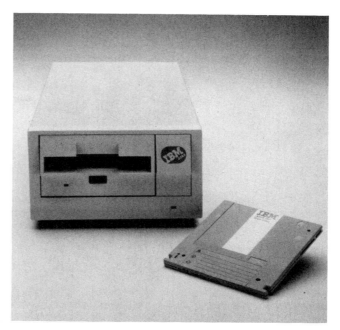

Figure 3.16: An optical disk drive and optical disk cartridge.

that information stored in other forms can be inputted directly into these devices, thus eliminating the need for retyping.

Output Devices

Another main component of a computer system is an output device. Once information is processed, a user may need to see this data. One very common form of output device is the cathode ray tube, or CRT. We also refer to this device as a terminal screen or VDT (video display terminal). In any case, the terminal's screen provides a unit with a visual display of the computer's output.

A second major output device is the high-speed printer. A printer such as the Genicom 4490 pictured in Figure 3.17 is capable of printing 1,400 lines per minute. These printers, quite appropriately, are known as "line printers."

Other Mainframe Peripherals

We have discussed a number of external devices (peripherals) that can be connected to a CPU, including printers and optical scanners. In a sophisticated data-processing environment, mainframe computers usually have several other types of peripherals connected to them. Let's briefly examine some of the major devices you are likely to encounter in the computer room of, say, a Fortune 500 company.

Computers in the Office 101

Figure 3.17: A high-speed line printer.

Protocol converters are devices that allow one computer to talk with another computer even if they use a different method of storing and transmitting information. A protocol converter functions much the way that a professional translator functions during government trade negotiations. It ensures that both computers receive information in a form that they can understand. We will examine this whole issue of protocols in Chapter 6.

Front-end communication processors are a special type of computer that has been programmed to perform many specialized control and/or processing functions. It processes a good deal of information that otherwise would have to be processed by the mainframe, or "host," computer. This concept of "off-loading" work onto a front-end communication processor rather than slowing down a mainframe computer's productivity is very similar to the role of an administrative assistant to a senior executive. The administrative assistant is responsible for performing many routine tasks so that the senior executive can concentrate his or her specialized skills on more challenging and important assignments.

Minicomputers

Minicomputers developed during the late 1960s because of the expense and inconvenience of mainframe computers. Many departments longed for their own smaller computer, one that they had access to at any time without having to wait for time on the mainframe computer. Also, many departments felt that their computing needs were relatively simple and did not require the power of the mainframe computer.

A minicomputer such as the HP 3000 or IBM AS400 contains the same basic elements of a mainframe computer, but it is generally smaller in size and slower in operation. It is not capable of supporting as many users as a mainframe computer. Minicomputers are frequently found at the departmental level in educational and scientific institutions as well as in small businesses. Larger companies link these minicomputers to their mainframe computers to realize the maximum benefits of both kinds of machines.

Generally minicomputers offer a wide range of business software as well as scientific programs. They usually provide some level of software compatibility among their range of machines so that a DEC user, for example, can move up when she or he outgrows a smaller machine without having to start from scratch with all new programs. Figure 3.18 illustrates a typical minicomputer in operation.

The Microcomputer

While our example company, Mervyns, has chosen to use dumb terminals and its mainframe computer for data-processing applications, many Fortune 500 compan-

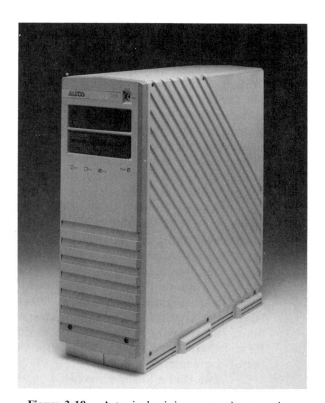

Figure 3.18: A typical minicomputer in operation.

ies have chosen to replace many of their terminals with **microcomputers.** In order to understand why they have made this decision, we'll take a closer look at the microcomputer's development, its basic components, and its capabilities.

During the early 1970s, the earliest microcomputers were sold in kit form. They had very little processing power and no programs available for them. Hobbyists bought these units to learn about electronics, while people already in the mainframe computer industry bought these units because many were fascinated by the prospect of owning their own computers.

The company likely to dominate the microcomputer industry at this time was Radio Shack (now Tandy Corp.). Radio Shack had stores in virtually every major community and had a reputation for standing behind their products if they required servicing. Since the ability to distribute a product and a reputation for reliability and dependability are two prime criteria for success in the computer industry, why didn't this company succeed at dominating the entire microcomputer industry?

The first Radio Shack TRS-80 Model I computer was offered for sale in the late 1970s. The company had chosen to seal the unit. This meant that anyone who wanted to add something to this machine or to modify the existing hardware violated the machine's warranty. This approach is known as "closed architecture." A conventional printer, for example, would not work properly with this computer. The company had also chosen to sell only its own programs in its stores. If money were to be made in this new fledgling industry, Radio Shack was determined to get as large a share as possible by requiring its customers to buy virtually all computer hardware and software from it.

At the same time that Radio Shack was offering its TRS-80 Model I computer, two young men started a company called Apple in a northern California garage. While the machine featured graphics and color capabilities that were better than the TRS-80 Model I, its major distinction was that this computer featured "open architecture"—the computer could be opened by removing its cover. Removing the cover revealed a number of empty slots that could be filled with circuit boards designed to enable the computer to do everything from send information to a printer to play music on an electronic keyboard. Apple freely provided schematic diagrams of its computer and encouraged companies to design circuit boards for the machine as well as write programs to work on it.

Because of its open architecture, Apple quickly began to move ahead of Radio Shack in terms of available hardware options and the number of programs available to run on it. Still, up until around 1977, there were a number of educational programs but no major business programs available that would encourage companies to use the computer.

This situation changed dramatically with the introduction of an electronic spreadsheet program called VisiCalc. This program permitted users to create very large ledgers that could permit the instant recalculation of various possibilities. Soon, the ability of this program in conjunction with the Apple computer to perform "what-if" analyses resulted in grudging acceptance of the Apple within a number of companies.

While Radio Shack remained a determined competitor, Apple clearly dominated the microcomputer industry until IBM's entry in 1981 with the IBM PC (personal computer). Built around a much more powerful microprocessing chip, this computer was capable of addressing substantially more memory, performing calculations much more quickly, and utilizing disk drives with much more storage capacity. It had one even more important advantage, IBM's name and reputation.

The IBM PC featured open architecture much like the Apple's. The original model contained five empty "expansion slots." IBM made the concept of a personal computer acceptable to Fortune 500 companies, which soon began to replace terminals with IBM PCs. People liked the idea of having their own computer that was available for processing at any time. Soon a number of programs, including the very powerful electronic spreadsheet program Lotus 1-2-3, made the microcomputer easier to use and in some ways more versatile than the mainframe or even the minicomputer. Since these programs were new, they offered many more features than comparable programs available on mainframes or minicomputers.

In 1984 Apple attempted to recapture the microcomputer market by offering the Macintosh computer. This very small unit featured "user friendly" screen menus as well as hand-pointing device known as a "mouse." Since all programs used the same menu structure, Apple proclaimed that its users would be spared the harrying experience of IBM PC users who often found that each new program required memorizing an entirely new set of commands. Figure 3.19 illustrates the unusual appearance of the Macintosh.

The year 1987 saw both Apple and IBM release new computer models. Interesting enough, both companies have come closer together in their conception of what a personal microcomputer should look like and what it should be capable of doing. The IBM PS/2 series of computers features the same small-sized disks (3.5 in.) as the Macintosh. They also feature user-friendly menus linked to a mouse.

Figure 3.19: The Apple Macintosh SE. (Photo courtesy Apple Computer Corp.)

While the IBM models became more Macintosh-like, Apple's newest Macintosh (the Macintosh II) featured an IBM-like appearance and the same expansion slot approach that characterized the IBM PC. With the addition of a special circuit card placed within one of the expansion slots, the Macintosh II became capable of running IBM PC programs.

And so in a little over 10 years microcomputers have achieved a certain uniformity based on standards imposed upon the industry by the popularity of the models by IBM and Apple. Part of the response to this popularity has been the appearance of IBM-compatible computers and "clones" made by companies other than IBM. Microcomputers in general have increased in processing power until a machine such as the Compaq 386 has processing power that exceeds that of most minicomputers.

While microcomputer hardware has become more and more sophisticated, the software has lagged far behind. Today, there are very few programs that are able

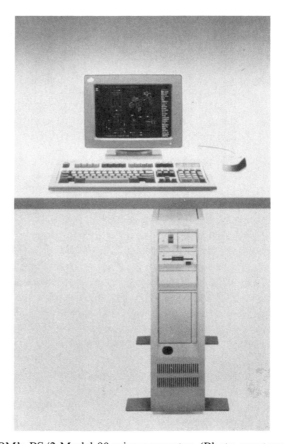

Figure 3.20: IBM's PS/2 Model 80 microcomputer. (Photo courtesy IBM Corp.)

to utilize the power of a 32-bit microcomputer. Microsoft and IBM have offered a new **operating system** that may change this situation. An operating system is a set of programs that control a computer's operation. The operating system programs must be written specifically for a computer's unique hardware, since it is responsible for such critical activities as sending correct information to a printer, storing information correctly on disk, and loading information from disk into main memory.

IBM and Microsoft offer two versions of an operating system called OS/2 and OS/2 Extended Edition. When enough software appears running under these two operating systems, IBM's PS/2 series of computers will be able to perform many tasks at the same time (multitasking) as well as use virtually all available secondary storage as if it represented main memory. In other words, these two new operating systems enable microcomputers to utilize their hardware prowess to offer many features at one time found only on a mainframe computer. Figure 3.20 illustrates IBM's PS/2 Model 80 microcomputer.

CHAPTER SUMMARY

While computers "think" in binary, programmers find it easier to work with decimal, octal, or hexadecimal number systems. Information can flow in one direction (simplex), alternately in both directions (half-duplex), or in both directions simultaneously (full-duplex). Data may be coded a number of different ways, including ASCII (most microcomputers) and EBCDIC (most of IBM's larger computers). Information may be transmitted one character at a time (asynchronous) or in a continuous stream once both computers have synchronized their clocks (synchronous).

A dumb terminal generally is limited to displaying information and transmitting information to a computer, while a smart terminal permits screen formatting, saving of information, windowing, and various other functions. Terminals may be asynchronous or synchronous.

Computers are classified according to size, speed, and cost into the categories of mainframe, minicomputer, and microcomputer. The fastest mainframes are called supercomputers. Minicomputers are powerful enough to provide multiuser operations at the departmental level, while microcomputers are rapidly becoming almost as powerful. The heart of a computer is its central processing unit, which contains the control unit, the arithmetic-logic unit, and main memory. Secondary storage can be in several forms, the latest development being optical disks.

KEY WORDS

Accumulator
Alphabet
American Standard Code for Information Interchange (ASCII)
Arithmetic-logic unit (ALU)
Asynchronous terminal
Asynchronous transmission
Binary
Bit

Block check character (BCC)
Byte
Cathode ray tube
Central processing unit (CPU)
Contention
Cyclic redundancy checking (CRC)
Data
Data code
Direct-access storage device (DASD)
Dumb terminal
Emulate
Ergometric
Even parity
Execution time (E-time)
Extended Binary-Coded Decimal Interchange Code (EBCDIC)
Frames
Front-end communications processor
Full-duplex
Half-duplex
Horizontal redundancy checking
Instruction register
Instruction time (I-time)
Kermit
Longitudinal redundancy check
Machine cycle
Mainframe computer
Memory address register
Microcomputer
Minicomputer
Multi-point connection
Odd parity
Operating system
Parallel processing
Parity bit
Parity check
Point-to-point connection
Protocol converter
Simplex
Smart terminal
Supercomputer
Synchronous terminal
Synchronous transmission
Vertical redundancy checking (VRC)
Video display terminal
Word size
XMODEM

REVIEW QUESTIONS

1. Convert the following decimal numbers to their octal equivalents: 15, 456, 800
2. Convert the following hexadecimal numbers to decimal: 7E, 3C4
3. Convert the following decimal numbers to their binary equivalents: 145, 32, 174
4. Determine whether each of the following numbers is in even parity. If not correct it:

   0   1110101
   1   1101010
   0   0101011

5. Use the charts in this chapter to write the ASCII and EBCDIC equivalents for the following:
   A, a, 9, T, L, h, H

6. What is the letter represented in EBCDIC as 10100010?
7. What is parity check? What is the basic weakness of this approach?
8. Define *simplex, half-duplex,* and *full-duplex* transmission.
9. Compare and contrast the abilities of a dumb terminal and a smart terminal.
10. What are some major advantages of synchronous transmission over asynchronous transmission?
11. Describe the differences between point-to-point and multipoint connections.
12. How do we differentiate among mainframe computers, minicomputers, and microcomputers?
13. Describe the major components of a central-processing unit.

TOPICS FOR DISCUSSION

1. What problems does a telecommunications manager face if a company decides to provide employees with microcomputers rather than terminals?
2. In terms of company efficiency, what sorts of programs do you think should remain the domain of the mainframe computer? What types of programs should employees run on their microcomputers?
3. Talk to a telecommunications manager or data processing manager and try to discover the major security problems associated with terminals versus those security problems found when employees have their own microcomputers.
4. Talk to a retail computer salesperson about the relative costs of a dumb terminal, a smart terminal, and a microcomputer. Price an IBM PC, an Apple Macintosh, and an IBM PS/2. Each computer should have approximately 640K of RAM (random access memory), a monochrome monitor, and two disk drives.
5. Visit a local computer dealer and identify the hardware and software required to permit ASCII-to-EBCDIC conversion.

Local Area Networks

CHAPTER 4

OBJECTIVES

Upon completion of this chapter you should:

- be able to discuss the major components of a local area network
- know the differences between disk servers and file servers
- understand the advantages and disadvantages of twisted-pair wire, co-axial cabling, and optic-fiber cable
- know the major network architectures
- be able to explain the characteristics of the major IEEE 802 network standards
- know the major features of the IBM Token-Ring Network
- understand the concept of system fault tolerance
- understand how bridges connect networks

INTRODUCTION

Once microcomputers became a common sight in most offices, the next logical step was to link these machines together so that companies could share information and resources. In this chapter we will examine a number of different types of local area networks, that is, networks limited geographically to a single location. We will learn what components are required for a successful network, how the networks transmit and receive information, and how different networks are able to communicate with each other.

THE MAJOR COMPONENTS OF A LOCAL AREA NETWORK

A local area network (LAN) links computers together to share hardware and software resources over a limited geographical area. Generally a LAN will contain the following components:

- a file server or a disk server to store network programs and data files
- a network interface card in each network workstation
- cabling that connects the network interface cards of all the network workstations
- special network software

Disk Servers and File Servers

During the mid 1980s, many companies began installing LANs in order to share hardware and software resources—only to discover that in some cases the financial savings were not as substantial as they had hoped. Software simply had not kept up with hardware advances, so that most of the programs these companies wanted to network were in reality single-user MS-DOS programs designed to run on a single IBM PC or compatible; these programs, designed to run on versions of MS-DOS prior to version 3.1, lacked the protection necessary to permit more than one user to access information from the same file at the same time—that is, they lacked **record locking.** However, some solutions to this problem arose. One solution came in the form of the **disk server,** from such companies as Corvus and 3Com.

A disk server, as illustrated in Figure 4.1, organizes network user's files into **volumes.** In addition to user volumes that limit the ability to read/write and delete the files contained there to the users who created these files, it also contains some **public volumes** that are accessible for read-only purposes by everyone on the network. A disk server is adequate for networks where productivity software such as word processing and electronic spreadsheets constitutes a majority of the network workload, since users usually do not share their data files for these types of programs. Disk servers have also been used by Corvus and 3Com for their networks connecting radically different computers having different operating systems, such as the Apple Macintosh and the IBM PC. We will defer looking at this Apple-IBM connectivity issue until the next chapter, which will deal with the entire issue of interconnecting unlike networks.

When a workstation logs on to a disk-server-based LAN, the disk server transfers a copy of its file **allocation table (FAT)** to the RAM of the workstation so that it will know which files reside on its particular volume. Every time that the particular

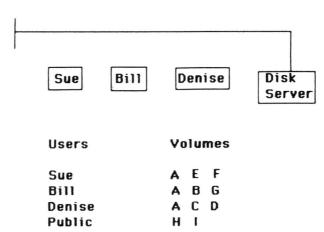

Figure 4.1: A disk server–based network.

user makes changes in his or her files, the disk server will update its FAT to maintain accuracy. One problem with a disk server based network is that even if a user has been given access to a colleague's files residing on another volume, he must know the precise volume letter as well as the file name to access this information.

Disk servers are not an efficient use of hard-disk space, since single-user programs must be copied into the volumes of each user who wants to access the program. There are a number of legal issues that have not been resolved in such a situation. Are these single-user programs licensed for one user at a time on one computer? If the answer is yes, than how does a company prove that only one user at a time was actually using, say, Lotus 1-2-3?

A **file-server**-based LAN, on the other hand, completely eliminates volumes. The file server needs only one copy of each program, and it decides where to store the program. The file server maintains its own file allocation table and simply requires a network user to specify a file name. It checks a table of user access to specific files, and if the user is entitled to use the requested file, it delivers the file quickly and efficiently. Sophisticated file-server software such as Novell's NetWare permits several file servers to be "mapped," so that a user need only request a file by name and the file server he or she has logged on to connects with the appropriate file server and transfers the desired file.

File servers often are **distributed** rather than **centralized**. Distributed file servers located at various points on a network can speed up processing considerably, since a particular department can access its data from a file server located nearby, rather than wait in line while a centralized file server would handle requests in a queue or waiting-list table. Another benefit to distributed file servers is that an area such as accounting can have its file server turned off and locked up at the end of the day so that unauthorized network users cannot try to access confidential information.

Network Interface Cards

Each network workstation on a local area network contains a **network interface card**, which slides into a expansion slot on the computer's motherboard in the case of IBM PCs and compatibles) or is already built in (in the case of an Apple Macintosh network). This circuit card usually contains its own microprocessor as well as RAM to process and temporarily store information reaching it from other workstations on the network. As we shall see shortly, information comes in the form of packets of bits that must be interpreted using the appropriate procedure. Figure 4.2 illustrates an Excelan network interface card's architecture. Because the protocol processing is handled by the 80186 microprocessor, the local area network's speed or throughput is maximized, since the workstation's microprocessor can continue to run application software.

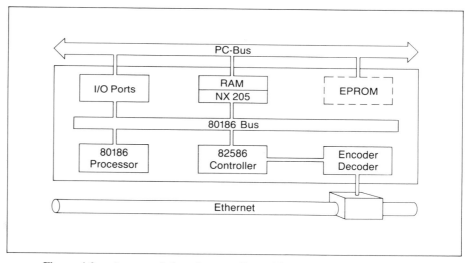

Figure 4.2: A network interface card's architecture. (Courtesy Excelan)

NETWORK CABLING

Twisted-Pair Wire

The least-expensive medium for local area networks is **twisted-pair wire**, known to most people as common telephone wire. As Figure 4.3 illustrates, two insulated wires are twisted together so that each wire faces the same amount of disruptive "noise" from the environment. Twisted-pair wires are bundled in groups ranging from 4 to 3000 pairs, with local area networks generally opting for the 25-pair groups. Twisted-pair wire is also distinguished by size, having an American Wire

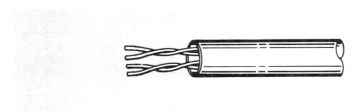

Figure 4.3: Twisted-pair wire. (Courtesy Howard W. Sams Co.: Schatt, *Understanding Local Area Networks*)

Gauge (AWG) number based on the wire's diameter. Local area networks tend to use 22-and 24-gauge sizes.

Most existing buildings already have common unshielded twisted-pair cabling installed for telephone use. This unshielded variety can be used for local area networks if only a small number of devices are going to be linked together and if the building environment does not contain electrical interference. One twisted-pair wire can carry voice transmission, while another can carry data messages. Both IBM and AT&T offer unshielded twisted-pair cabling for their networks.

Shielded twisted-pair wire contains two insulated copper wires surrounded by a copper sleeve. This shielding significantly reduces electrical interference while permitting faster transmission for greater distances. While many local area networks used to promise only 1-Mbps transmission speed, some newer twisted-pair networks promise up to 10-Mbps speed. Later in this chapter we'll examine a way of achieving this speed with twisted-pair wire running Ethernet. IBM recommends its own shielded twisted-pair cabling scheme for companies cabling a new building.

There is no question that twisted-pair offers a number of advantages to a manager contemplating a local area network installation. It is inexpensive, likely to be present in an existing building, and adequate for a relatively small network not subject to electrical interference. A major disadvantage, though, is that as more and more twisted-pair wires are used to add additional capabilities to a network, maintaining the cabling can become difficult. Each pair, after all, must be labeled and traced through a morass of cabling so that problems can be identified and corrected quickly. Figure 4.4 illustrates how a company with an IBM Token Ring Network keeps a detailed cabling chart describing how the individual network workstations are connected together within a centralized wiring closet.

Baseband Coaxial Cable

Baseband Coaxial cable represents an inexpensive, proven medium for local area networks. Baseband coaxial cable is a single channel analogous to a garden hose. As Figure 4.5 indicates, baseband coaxial cable consists of a copper conductor surrounded by inner and outer layers of insulation. It is capable of a high rate of data transmission ranging from the industry-standard Ethernet 10 Mbps to some proprietary local area networks that transmit data at 80 Mbps.

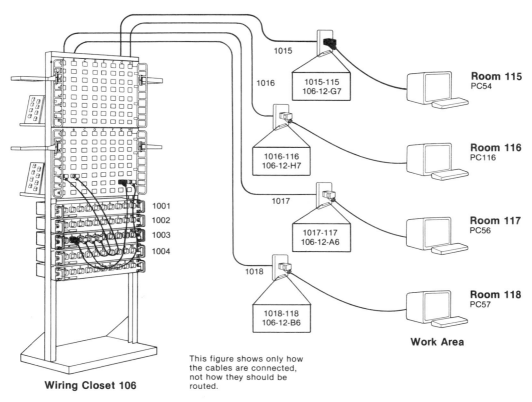
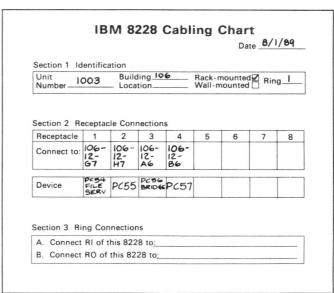

Figure 4.4: A wiring closet and detailed IBM cabling chart. (Courtesy IBM Corp.)

Network Cabling

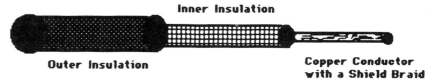

Figure 4.5: Baseband coaxial cable. (Courtesy Howard W. Sams Co.)

One of the leading manufacturers of local area networks utilizing baseband coaxial cabling is 3Com Corp. This company offers both a thick and a thin type. The thick coaxial cabling has an outer diameter of 0.405 in., the thin model has an outer diameter of 0.193 in. The thick variety is much more difficult to work with and almost impossible to bend, but it offers the major advantage of greater protection from electrical interference. Network managers are able to combine the best features of both types of baseband coaxial cabling and "mix and match" the two varieties by using special adapters to link them together.

Broadband Coaxial Cable

A local area network utilizing **broadband coaxial cable** operates much the same way as a commercial cable television system. The broadband cable carries multiple communication services because it is capable of transmitting simultaneously several types of data, including voice and video signals, along different frequencies. Using frequency-division multiplexing a cable's 300-to-400-MHz bandwidth is divided into several separate channels.

A network with broadband coaxial cable transmits over a radio frequency in analog form. Digital information is converted by a modem located on a network workstation's network interface card. It is transmitted at one frequency to a central site on the network, known as the "headend," from which the signal is transmitted at a different frequency to the station that is supposed to receive the message. If a local area network consists of one set of coaxial cable, then the bandwidth of this cabling is subdivided into "forward" and "reverse path" frequencies.

An alternative method of achieving two-way communications on a LAN using coaxial cabling is to use a dual-cable system. One set of cabling carries only inbound signals, while the other set carries only outbound signals. Because this method is quite expensive, it is not used as much as forward/reverse-path frequency subdividing.

There are a number of advantages to broadband coaxial cabling. It lends itself to a treelike structure in which it is easy to branch in several different directions. Such flexibility is necessary in many multi-building complexes. Broadband coaxial cabling also is less sensitive to electrical interference than twisted-pair or baseband cabling because the electrical noise usually exists at a lower bandwidth level than the frequencies used for signal transmission.

A network manager must plan carefully if he or she plans to install broadband coaxial cabling. He or she may want to assign several "guardbands," frequencies that will not be used for signal transmission but will be used to keep signals transmitted in adjacent bandwidths from overlapping. Some of the newer Ethernet and token-bus products filter out any overlap that might result from adjacent bandwidths.

Since there is little standardization in the network industry concerning the assignment of specific broadband band widths for specific network functions, a network manager must anticipate that future network services will require the setting aside of specific channels, often in 6-MHz increments, for those purposes.

Optic-Fiber Cabling

Optic-fiber cabling consists of pure glass drawn into a core surrounded by cladding, a layer of glass with a lower refractive index than the glass in the core. A laser or a light-emitting diode (LED) sends a signal over the optic-fiber cabling. A photodiode at the other end receives the message and translates it back into a digital or analog signal. Figure 4.6 illustrates how this process works.

Optic-fiber cabling is much too expensive to become the cabling of choice for most installations, but it does offer a number of advantages over twisted-pair and coaxial media. Since optic fiber transmits data in the form of light, it is immune to electrical interference, difficult for unauthorized users to intercept, and capable of being transmitted at speeds in excess of 100 Mbps.

Local Area Network Architecture

We have looked at the key components of a local area network, including the file server/disk server, the network interface card, and the media. We will examine some management issues relating to the choice of network software later in this chapter. Local area networks can be configured in a number of different patterns, which we will refer to as network topology or **network architecture**.

The **star** network topology utilizes a centralized file server with cable radiating from this source to each individual network workstation. Figure 4.7 illustrates this

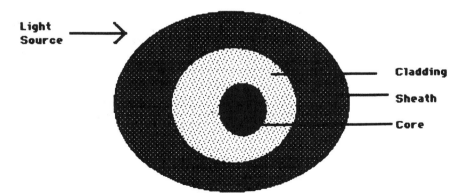

Figure 4.6: Fiber-optic cabling. (Courtesy Howard W. Sams Co.: Schatt, *Understanding Local Area Networks*)

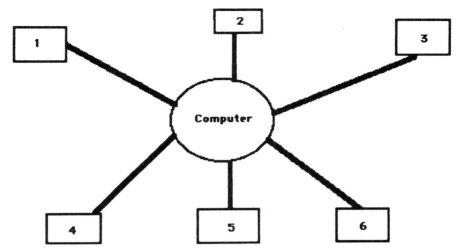
Figure 4.7: A star network.

design. This centralized approach has many advantages for a network administrator. It is very easy to add and delete workstations while continuing normal network operations, since the path data takes for other network workstations is not disrupted. Another major advantage to this approach is that it provides a network administrator with a great deal of information and control over network operations. Since all network requests for data as well as all communications from one workstation to another workstation go through the file server, it is easy to compile detailed reports on network usage as well as records on specific files requested by specific network users. It is easy to determine peak traffic patterns as well as determine users who are trying to view unauthorized files.

A major weakness of star topology is the very same limitation of any centralized computer or phone system. If the centralized file server fails, then the entire network fails. An alternative that remains fairly unpopular as yet because of its complexity is the **mesh topology**. Figure 4.8 illustrates how this network architecture results in each workstation being connected to every other workstation. One problem with this arrangement is that it does present some difficulty whenever a new workstation needs to be added or deleted. This is particularly true if workstations are distributed in different departments that may be housed on different floors.

For relatively small networks a bus topology can prove ideal. The **bus** is a straight-line data highway that carries the network information from one network workstation to another. Multiple buses can be linked together, so that while a LAN initially may consist of only the accounting department, additional departments can be cabled as buses and then linked together. A little later in this chapter we'll examine some industry-wide standards for a bus network and look at how precautions are taken to avoid data collisions along this network path. At this point, though, it is important to realize that there are distance limitations on most bus networks. No matter how fast the network may transmit information, significant distances be-

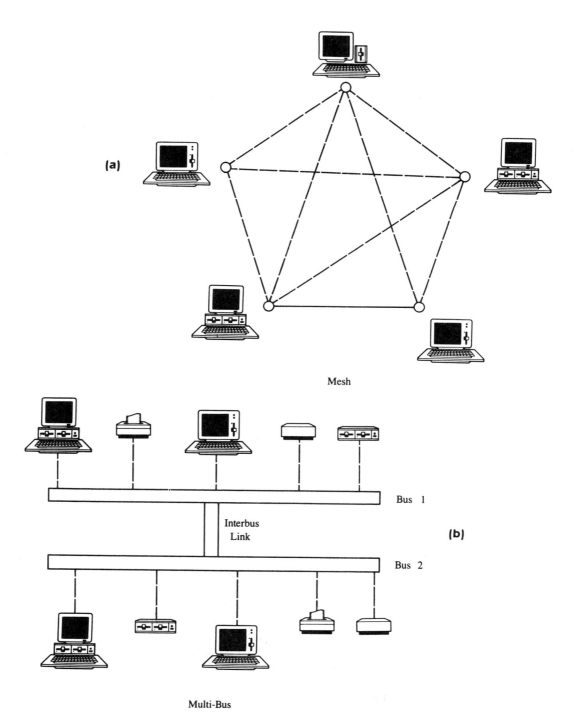

Figure 4.8: A mesh network and a multi-bus network.

Network Cabling 119

tween workstations that need to communicate are bound to affect overall network performance. Figure 4.8 illustrates a multi-bus network.

The Ring Topology

The **ring topology** has gained popularity because IBM chose it as its major LAN architecture. A closed ring generally uses a **token,** a specific bit pattern that indicates that a workstation has permission to send information. Later in this chapter we will take a close look at IBM's Token-Ring Network and see how traffic control is maintained. Rings can be bi-directional as well as uni- directional; the newest fiber-optic networks take advantage of bi-directional cabling to achieve data transmission speeds of over 100 Mbps. Figure 4.9 illustrates a typical ring network with workstations numbered in sequential order.

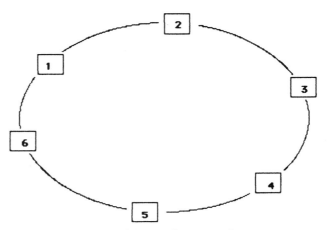

Figure 4.9: A ring network.

IEEE 802.3: THE "ETHERNET" STANDARD

Background

In February 1980 the Institute of Electrical and Electronics Engineers (IEEE) Computer Society established the IEEE Project 802 committee and charged it with developing a set of standards for local area networks. It defined a **local area network** as a data network confined to a moderate-sized geographic area such as a building or a large warehouse. By specifying the means (the protocols and interfaces) by which devices would communicate with each other over such a network, the IEEE could ensure that such networks would be vendor-independent. It would be possible for end users to mix and match network products as long as all vendors adhered to the same standards. A set of standards would also ensure a large enough potential market for chip manufacturers to manufacture the microprocessing chips necessary to power these networks. The committee was to base its standards on the Open Systems Interconnection (OSI) model first three layers: Physical, Medium access control, and Logical link control as described in Chapter 6.

There was at least one complication, however. In the early 1970s scientists at Xerox's Palo Alto Research Center turned from the traditional broadband coaxial cable method of data transmission and developed a baseband coaxial cable local area network. By 1980, Xerox, joined by Intel Corp. and Digital Equipment Corp., issued the first Ethernet technical specifications. The influence of these three companies provided a de facto standard that presented a problem for the IEEE 802 committee. By selecting Ethernet as the local area network standard the committee would make it difficult for such formidable computer powers as IBM to compete, and it would stifle any alternative types of local area networks. The committee decided to issue several alternative sets of standards that would cover different types of local area networks, one of which would closely resemble Ethernet.

The 802.3 set of standards describes a very fast (10-Mbps) bus coaxial cable packet-switched local area network spanning about 1 kilometer. Such a network lacks centralized control, because each workstation uses its network adapter card's built-in intelligence to recognize packet addresses as information flows along the bus, or data highway.

Data Collision Detection and Avoidance

A data-bus network such as the IEEE 802.3 standard is a **contention network**, since workstations will compete or contend with each other for the right to send messages. Imagine if you will a large neighborhood that contains a one-lane highway. If Ms. Smith, living at 800 Elm Street, wishes to drive to 1600 Elm Street, she would step to the curb and look down the street to see if there is any oncoming traffic. She might even shout "Watch out, here I come!" before getting into her car and driving out of the driveway. There is a critical time period during which another driver further down the block—Mr. Jones, perhaps—might initiate exactly the same process without being aware of Ms. Smith's intention to drive to 1600 Elm. The result, of course, could very well be a collision. Similarly, if Ms. Smith and Mr. Jones were simultaneously sending data over a network, they might experience a data collision. The information contained in both packets would be destroyed.

Standard bus networks feature a possible solution for sharing a network among several workstations. The solution is a protocol often referred to as CSMA/CD, or "carrier sense multiple access with collision detection." CSMA means that a workstation is aware (that is, has a sense of) signals being sent from other workstations in a situation where more than one device has access to the same line. With CSMA, each workstation "listens" to see if there is any other traffic on the network before sending its packet. CD, or "collision detection," means that workstations actively "listen" to find out if their message arrives safely whenever they send. If two workstations under CSMA/CD protocol send simultaneously, they both listen and hear the collision. They then wait a random period of time (different for each workstation) before trying to retransmit.

A second protocol sometimes used is CSMA/CA, which stands for "carrier sense multiple access with collision avoidance." This means that workstations listen *before* sending, to avoid altogether the possibility of a data collision. In either CSMA/CD or CSMA/CA there is still the possibility of data collision, since we are still dealing with a contention network.

With the Ethernet networking scheme, something else happens. If a collision does occur, the controller found on the Ethernet interface card transmits a **jam signal,** which notifies all network users that there has been a network collision and that workstations wishing to use the network should delay a random period of time before trying to transmit. A binary exponential backoff algorithm is used under Ethernet. Each workstation has retransmission slot times in steps that are intervals of time slightly longer than the time necessary to make a round trip of the network channel. There are fewer collisions on an Ethernet network, and almost peak efficiency if data packets are relatively long compared to the collision interval. Very heavy traffic of small packets will virtually eliminate the advantages of CSMA/CD and result in a *de facto* CSMA/CA network.

The Ethernet Packet

Figure 4.10 depicts an Ethernet packet. The 8-byte Preamble field provides the information necessary for synchronization, an alternate series of 1s and 0s ending with two consecutive 1s. The Destination Address field is a 48 bit field that indicates which workstation should accept the packet. A 1 broadcast as the first bit indicates a logical group of recipients, that is, a group of workstations that are specified to receive the packet. If all 1s are broadcast in this field we are looking at an all-station broadcast ("Everyone leave the office because we're having an earthquake").

The 48-bits Source Address field contains the unique network address of the workstation that has sent the packet, while the 16-bits Type field indicates the higher level protocol type associated with the packet which will determine how the network adapter card interprets the information found within the packet.

In Chapter 6 we will take a close look at some of the major standards, or *protocols* that govern the computer industry. Among the major higher-level protocols associated with Ethernet today are Xerox Network System (XNS) and Transmission Control Protocol/Internet Protocol (TCP/IP). XNS was developed by Xerox for handling network messaging and logical handling session management by devices utilizing Ethernet. TCP/IP was developed by the U.S. Department of Defense for handling network interfacing over the Advanced Research Projects Agency Network (ARPANET). Some vendors' Ethernet network adapter cards are chips

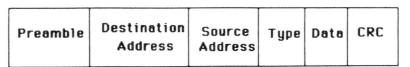

Figure 4.10: An ethernet packet.

designed for both XNS and TCP/IP protocols. These chips can complicate an Ethernet network, since the two cards are generally incompatible. The TCP/IP protocol is becoming more and more popular with Ethernet customers because of the ease of connecting 802.3 networks to host DEC computers running their VAX operating system and to computers running the Unix operating system.

The Ethernet packet contains two critical fields beside the Type field. The Data field can vary from 46 to 1500 bits; the minimum Data field length ensures that a small data packet will be distinguishable from simple debris caused by a data collision on the network. Finally, the 32 bit CRC field provides a redundancy check to ensure that data has not been lost during transmission.

Differences Between Ethernet and the IEEE 802.3 Standard

Some corporations have been "mixing and matching" conventional Ethernet nodes and newer IEEE 802.3 workstations. While the two packet forms are very close, they are not identical. The difference can result in serious network problems. The Ethernet packet, for example, has a 2-byte field called Type, which indicates the protocol type, while the IEEE 802.3 frame has a 2-byte field called *length*. Since the numbers used in the Ethernet type field are generally higher than those found in the 802.3 length field, 802.3 specifications generally include a note that packets with a length greater than a certain value can be discarded. The two different types of frames can also be monitored by using protocol analyzers.

The important thing to remember between the two very close technologies is that mixing and matching is dangerous. Using an Ethernet cable with a 802.3 transceiver can introduce noise to a connection and can corrupt data. Sometimes nodes outfitted with 802.3 transceivers can become overloaded trying to handle Ethernet messages broadcast from Ethernet transceivers.

Running Ethernet on Baseband Coaxial Cable

The IEEE 802.3 standard specifies every aspect of this coaxial cable bus network, including a 50-ohm coaxial cable and a maximum of 100 transceivers or taps per cable segment with no two transceivers within 2.5 meters of each other. While single cable segments may be up to 1500 feet in length, it is possible to exceed this distance if repeaters are used to amplify the signal.

Running Ethernet with Broadband Coaxial Cable

In September 1985 the IEEE standards board met and approved a 10-Mbps broadband coaxial cable standard. What makes this decision so appealing to many companies is that it is now possible to support other local area networks, such as token-ring, on different frequencies while sending video and voice signals on their own frequencies. Each workstation's network adapter card contains a radio-frequency (RF) transceiver that uses its modem to modulate the Ethernet signals onto their own network frequency or channel. This network can cover a distance of approximately 4 miles.

Perhaps the most appealing aspect to a broadband implementation of Ethernet is that it provides a more reliable network with fewer data collisions. It also provides more efficient notification of the data collisions that do take place, while still using the CSMA/CD approach. On a baseband Ethernet network, data collisions are detected by transceivers by sensing an unusually high level of DC voltage on the coaxial cable. On a broadband Ethernet network, however, RF modems reserve a separate collision enforcement frequency that provides a much more efficient method of notifying workstations of data collisions.

Running IEEE 802.3 Networks with Twisted-Pair Wire

Recently a number of leading Ethernet vendors, including 3Com Corp. and Digital Equipment Corp. have been offering what they hope the IEEE 802.3 committee will adopt as a new standard: 10-Mbps Ethernet running over unshielded twisted-pair telephone-type wiring. The 3Com products use a special adapter, called a Pair-Tamer, which permits Ethernet signals to use an unused pair in a bundle of twisted-pair wire. When the wire reaches a wiring closet within an office a second PairTamer connects the telephone wire with 3Com's Multiconnect repeater, which in turn can connect with other devices or wiring closets within a network. The products 3Com offers even make it possible to connect twisted-pair wire running Ethernet cabling with other media, including coaxial and fiber-optic cabling. Figure 4.11 illustrates this phenomenon.

IEEE 802.4: The Token Bus

IEEE 802.4 specifies a bus network similar to IEEE 802.3's bus topology, but one in which only one workstation tries to use the network at a time. We call this a

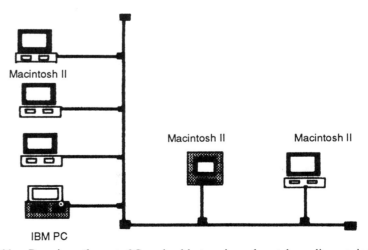

Figure 4.11: Running ethernet. 3Com is able to mix and match media running ethernet.

non-contention network, since workstations are not actually contending or fighting among themselves for network access. Even though the network is cabled in a bus topology, the bus forms a logical ring in which each workstation is assigned a logical position. The workstation assigned the very last position is followed by the workstation assigned the first position. Each workstation is aware of the logical workstation following it and proceeding it.

Imagine, if you will, a tribal council meeting at which several elders of an Indian tribe sit in a circle. A chief stands and raises a carved piece of wood that symbolizes his authority, his right to speak. All the other speakers fall silent as he begins. When he finishes he simply passes his symbol of authority to the elder next to him, who rises to speak. The control packet known as a token operates in much the same way. Only the workstation controlling the token has permission to speak, and all other workstations must wait their turn to control this packet.

A workstation with the token has it for a specified period of time, during which it may transmit one or more frames of information and poll other stations for their responses. A workstation that has very important network functions can be assigned additional time frames in a table that allocates this. There are actually four different priorities of access, with the highest level (level 6) receiving the maximum amount of network bandwidth for its highest-priority transmissions. Data can be classified into priority levels of 6, 4, 2, and 0. After a workstation has broadcast its level 6 data and no other workstation has level 6 data to broadcast, the workstation may broadcast its level 4 data. The same principle holds for broadcasting level 2 data and level 0 data. Each workstation controlling the token must also permit new workstations to join the logical ring. During its time period with the token, a workstation issues a solicit-successor frame, which is an invitation for a workstation to join the network at a location between itself and the next node in the logical ring. If there is no response to this invitation, the token-controlling workstation passes the packet to the next workstation in its logical ring. If a workstation does request entrance, then the token is passed to it so that it may set all its parameters and linkages before the next workstation in the logical ring gets its chance.

Figure 4.12 illustrates the frame structure for an IEEE 802.4 token bus. A Preamble field is used to establish sychronization between sender and receiver. The Start delimiter field indicates the start of the frame, while the Frame control field indicates whether a data frame or a token frame is being dealt with. The Destination and Source address fields are identical to the Ethernet frame format we have already discussed. The Data field is followed by a Frame Check Sequence identical to the IEEE 802.3 standard and an End delimiter indicating the end of the frame.

IEEE 802.4 also specifies that its packets travel along a broadband medium consisting of 75-ohm community-antenna-television (CATV) coaxial cable. This means that there can be several different channels transmitting different types of information on different frequencies simultaneously. Broadband cabling also offers a couple of other major advantages over the baseband cabling with regard to Ethernet and the IEEE 802.3 standard. Taps can be placed anywhere on a broadband network, while Ethernet's baseband standard usually requires a minimum distance of 2.5 meters between nodes. In addition to broadband's lower sensitivity to noise, it can stretch much further than baseband. With repeaters a broadband network

| Preamble | SD | FC | DC | SA | Data | FCS | ED |

- SD — Start Delimiter
- FC — Frame Control
- DA — Destination Address
- SA — Source Address
- FCS — Frame Check Sequence
- ED — End Delimiter

Figure 4.12: A token bus frame.

can extend up to 25 miles. Depending upon the number of channels used, data rates between 1 and 10 Mbps are possible.

Manufacturing Automation Protocol (MAP)

Closely based on the IEEE 802.4 token bus standard, **Manufacturing Automation Protocol (MAP)** is a set of specifications designed to permit devices from various manufacturing and data-processing suppliers to be linked on a communications network. Originating with General Motors in the early 1980s, MAP represents an attempt by GM to encourage vendors to develop and manufacture compatible equipment. To General Motors the major advantages of MAP include reduced software expenses (since software does not have to be written for proprietary hardware), greater flexibility in selecting vendors, reduced training costs, and the assurance that there will always be plenty of suppliers available.

IEEE 802.5 The Token Ring

The IEEE 802 committee developed a standard that became known as IEEE 802.5 for a non-contention local area network, a network that used the ring topology to transmit a token, or frame, containing information. As we observed with the IEEE 802.4 token bus standard, a token consisting of a predetermined bit pattern can be used by only one workstation or network node at a time. The transmitting workstation physically alters this token's bit pattern, which announces to all other workstations that this token is being used. The token, now transformed into a frame of information containing the message to be sent, is transmitted around the ring until it reaches its destination.

Messages sent along a token-ring network are received by each network workstation, which in turn checks to see if it is the frame's correct destination. If it is not, the workstation acting as a network repeater retransmits the frame to the next network workstation. Finally, the destination workstation receives the message and copies it to its internal memory before retransmitting the frame back to the sending workstation. The frame makes its way around the ring back to the sending workstation, which observes that the message was successfully copied and then resets the token so that it is available for another station to use.

Built into the IEEE 802.5 standard is the concept of a monitoring station, a station that is designated to check the network for error conditions such as no free token. If something should happen to this monitoring network node, another station is capable of replacing it and assuming its function.

Since token-ring topology dictates that the frame travel around the ring by being retransmitted from workstation to workstation, there has to be a mechanism to prevent the entire network from failing if one workstation is out of service. Also, there has to be a way of connecting and disconnecting network workstations without having to physically break the ring and halt all messages.

Wire centers serve this function of linking active workstations to a token-ring network. Figure 4.13 illustrates how a wire center functions very much like a multiple outlet. Workstations can be added or deleted from the network by plugging or unplugging them from a wire center.

Even with wire centers it is possible for messages to become garbled while being transmitted from workstation to workstation around a network. To ensure accuracy as we just indicated, the workstation that originally sent the frame then re-transmitted the frame verifies that the message was received correctly before removing the frame and transmitting a new token. The key to understanding how this process works is to examine not just the real token ring network's hardware but the actual frame it transmits and the fields of information contained therein. Since the most commercially popular example of the IEEE 802.5 standard is IBM's Token-Ring Network, we'll use it as an example of how such a system performs.

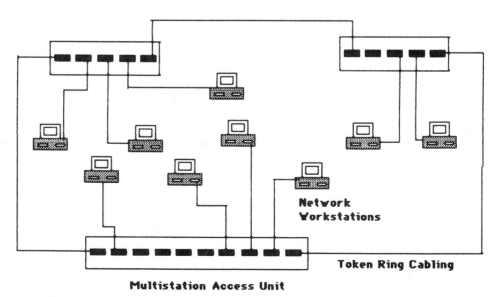

Figure 4.13: Wire centers. Wire centers are vital parts of a token-ring network.

IBM'S TOKEN-RING NETWORK

In October 1985, after several years of public anticipation, IBM announced its 4-Mbps Token-Ring Network. This network looked more like a star topology than a ring, because workstations radiated from the network's wire centers, which were known as **multistation access units.** Figure 4.14 illustrates four different multistation access units, each of which can connect up to eight PCs to the ring, but in this example each wire center has a single PC workstation connected by a cable known as a **lobe.**

On IBM's Token-Ring Network each workstation contains a Network PC Adapter Card (or Network PC Adapter Card II, for the IBM AT) and appropriate cabling from the card to a wire center designated by IBM as a multistation access unit. The IBM Token-Ring **NETBIOS** program permits programs written to utilize the NETBIOS (Network Basic Input/Output System) interface to operate on the Token-Ring Network by translating NETBIOS requests into appropriate network

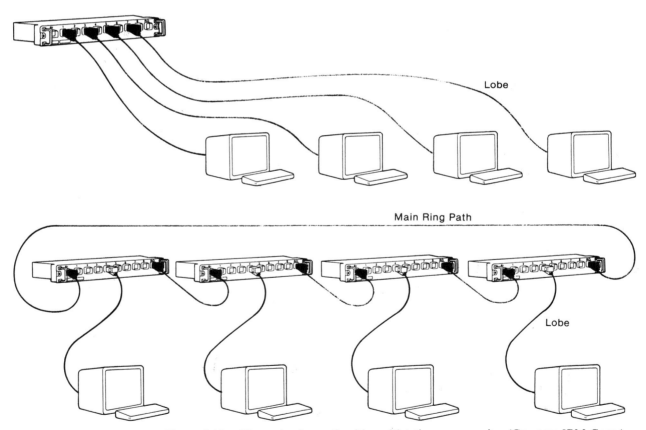

Figure 4.14: The main ring path with multistation access units. (Courtesy IBM Corp.)

protocol requests. The cabling connecting the network can range from type 3 unshielded telephone wire to include optic fiber and IBM data-grade cabling. The promised 4-Mbps transmission speed may not be achieved using the unshielded telephone wire if there is any interference. Another limitation of using inexpensive telephone wire is that it will support a maximum of only 72 workstations, compared to the 260 workstations that can be supported on one ring with data-grade coaxial cabling. This type of network offers a major advantage for larger companies because it is possible to send large amounts of data without loss of transmission speed or degradation of data over substantial distances—the latter being a concern because of the token-ring's method of retransmitting information from workstation to workstation.

Under IBM's PC LAN program each workstation can serve simultaneously as a file server and as a workstation; in other words, this software results in a distributed rather than a centralized file server. Unfortunately, the network software permits a workstation's dual function by using a "time slicing" technique in which it divides up the workstation's microprocessor time into very small increments and apportions this time to both functions. This can result in an unsatisfactory performance for both the file server and the workstation. A network manager probably would want to minimize this loss of speed by avoiding distributed file sharing during peak network traffic periods.

IBM's 16Mbps Token-Ring Network

IBM also offers a 16-Mbps version of its token-ring network. This version requires its own special adapter cards, which come with 64K of on-board RAM. This extra memory allows larger frame sizes and more concurrent sessions when the adapter is used as a server. The larger frame sizes permit high-volume RAM-to-RAM transmissions such as images.

This high-speed version of a token-ring network comes at price, though, since it requires IBM type 1, 2, or 9 shielded cabling and cannot use the inexpensive unshielded twisted-pair wire. Who needs 16-Mbps speed? Most experts feel that the primary function of such speed is to serve as a **backbone network**, a kind of super router that connects several different networks together. We will discuss this concept in more detail in Chapter 5.

The IBM Token-ring Network Token and Frame Formats

As illustrated in Figure 4.15, IBM's Token-Ring Network follows IEEE 802.5 specifications in both its token and its frame formats. The token itself is a fixed 24-bit pattern. Note that the Access Control portion of the token contains bit patterns providing key information on the status of a workstation's priority to use the network, the status of the token itself, information for the Monitor workstation, and reservation priority. Since a token-ring network transmits either a token or a frame, the token bit in the 3-byte access control field is set to 0 for a token and set to 1 for a frame.

A network workstation releases its token after each transmission because it is

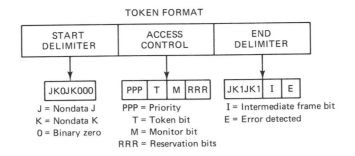

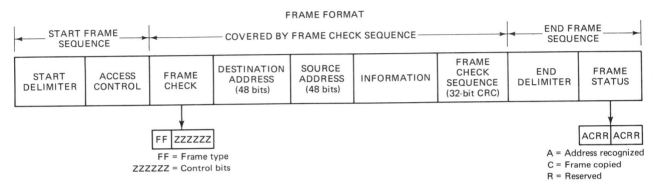

Figure 4.15: The token and frame formats on IBM's token-ring network.

not permitted to broadcast continuously, regardless of its high priority. The monitor bit is broadcast as a 0 in all frames and tokens except the monitor itself, to ensure that no frame monopolizes the ring and no token has a priority greater than 0. Thus every workstation has equal access to the network. But what could happen if a workstation sets a token bit to "busy" and then fails to begin to broadcast? The malfunctioning workstation would not be able to reset the "busy" bit, and presumably the network would continue to send this same message forever. To prevent this condition from ever occurring, a token-ring workstation serving as the Monitor sets the monitor count bit to 1 in any busy token that passes it. If a busy token arrives back at the monitor with the monitor count already set to 1, this indicates that the token has not been removed appropriately by the sending workstation. The Monitor destroys the frame and generates a free token.

When a workstation is ready to continue a message's path through the network by rebroadcasting its bit pattern, it examines the reservation bits (RRR). If it has a message of its own that it wants to send and its priority is higher than the present sender, it raises the value of this 3-bit field to its own level assuring that its message is the highest one waiting to be transmitted. Further down the network a second workstation with even a higher priority might change this bit pattern to reflect its own priority level and thus "bump" the previous workstation from its reservation for the next available token. The priority established in the "PPP" Priority field reflects the workstation's priority to use the network. If a workstation observes that its priority is higher than the one already reflected in the "RRR" Reservation field, it raises this value to its own level and thus reserves the next token for its own use. The address of the bumped workstation goes into a memory location that serves as queueing area for displaced workstations seeking access to the network but forced to wait their turns.

Adjacent to the access control field we have been discussing is the Frame Check field. The frame type can be designed for media access control (MAC) and thus contains a 00 or can be specifically designed for logical link control (LLC) designated by a 01. This information is important because it determines how the workstation receiving this information will decode it. If the information is designated for the MAC, it might contain additional higher-level addressing information regarding its destination service access point (DSAP), where the information will go within the receiving workstation, and its source service access point (SSAP), where precisely it originated from within the originating workstation. There also will be an indication in the first bit of the SSAP byte whether the information is a command or a response.

If the Frame check field indicates that the information is designed for the LLC, then the information may be a standard message or a **datagram**, that is, a piece of information not requiring any acknowledgment.

Let's assume that a workstation on the Token-Ring network receives the token and transmits a frame of information. Figure 4.15 also displays an IBM Token-Ring Network frame format. The source and destination address 48-bit fields are designed to convey a number of different kinds of information. If the first bit is set to 1, then the message is a group broadcast for everyone on the network. An initial zero bit, on the other hand, indicates that this is a message addressed to a specific workstation. The second bit in the address field indicates whether the address is global (0) or local (1). *Local* refers to another node on the same network. Each workstation has a unique 48-bit address obtained by the manufacturer from the IEEE and burned into the PC Adapter ROM chip. Locally administered addresses are assigned by the local network administrator; these addresses override the universally administered addresses found on the ROM chip. The address fields have been designed to accommodate the addresses of workstations that exist on other rings; in fact, the first 2 bytes of these two fields are designated for a workstation's ring number.

The Frame Status field is used by the workstation originating the information frame to determine if the workstation designated to receive the message actually recognized its own address. The originating workstation sets the Addressed Recognized bit to 0, while any other station on the ring sets it to 1 if it recognizes the address as its own. The originating workstation also sets the Frame Copied bit to 0. When the receiving workstation copies the frame into a read buffer it sets this bit to a 1. When the frame returns to the originating workstation, it checks this key Frame Status field to see if the frame was recognized and copied correctly.

Error Checking on the Token-Ring Network

The Frame check Sequence field is responsible for error checking for the Frame Check, Destination Address, Source Address, and Information fields. Bit seven of this field is the error-detected bit. The workstation originating a frame sets this bit to 0, while the first station to detect a transmission error sets it to 1. The first station to detect this error flag counts the error and prevents other stations from also logging this error. This method helps to locate where an error has taken place. The workstation serving as the Monitor scans the network for transient and permanent errors. Transient errors are logged by various workstations on the network as "soft error conditions." These errors generally can be corrected by retransmitting the frame. Permanent errors. on the other hand, can disrupt network operations. If a frame comes back with the indication that a destination workstation has not recognized its own address and copied the frame, the Monitor workstation can use the bypass circuitry built into the multistation access units to bypass the defective station and to maintain the network's integrity. A workstation "downstream" from a defective workstation can send out a special signal called a "beacon," which contains its own address as well as the address of the workstation immediately "upstream" from it on the network, presumably the defective workstation, which becomes known as the NAUN ("Nearest Active Upstream Neighbor"). When the NAUN has copied eight of the beacon frames it removes itself from the network and runs a set of self-diagnostics. If these tests do not detect a malfunction, the NAUN then reattaches itself. If the network is still not functioning after the NAUN and beacon workstation have tested themselves and found themselves to be operative, then the network will require manual repair by the system administrator.

Sometimes there may be a complete break in the network's cabling. Since each workstation has its own self-diagnostic tests, it is relatively easy for it to identify a break in its own cabling and inform the Monitor workstation.

The functions of the Monitor workstation are provided by the IBM Token-Ring Network Manager Program. It provides continuous monitoring of the network for errors as well as the corrective actions we have been discussing. It also logs network error and status information for report generation. It is this program that provides the mechanism for the creation of addresses for each workstation as well as the establishment of passwords.

The NETBIOS Interface to Token-Ring Network

Business users (as opposed to programmers) of IBM's Token-Ring Network do not want to concern themselves with how various programs communicate over their network; they simply want the communications to be effortless and invisible. As IBM moves to connect its microcomputer workstations on its Token-Ring Network with its minicomputers and mainframe computers, there is an awkward phase during which users do need to be aware of the mechanics of application-to-application communications. In this section we'll take a look at one of two major Token-Ring Network interfaces, NETBIOS (Network Basic Input/Output System). In the next chapter, on micro-mainframe communications, we'll examine the second interface, APPC/PC (Advanced Program-to-Program Communications interface).

As we indicated earlier, many programs are written to address specifically the IBM PC's NETBIOS. To utilize these programs on Token-Ring Network it is necessary to use both the Token-Ring Adapter Handler Program and the NETBIOS Interface Program. Figure 4.16 shows that an application program written to utilize NETBIOS will have its requests translated through the NETBIOS Program and routed to the Token-Ring Adapter Card, where IEEE 802.2 data link control protocols will be followed as well as IEEE 802.5 token-ring protocols. The data eventually will be sent through the Token-Ring Network at 4-Mbps. Once again, remember that if the user has remembered to load all the appropriate interface programs before using the application program that requires a NETBIOS interface, this entire process takes place without a need for user intervention.

PC-DOS, OS/2, NETBIOS, and IBM Local Area Networks

In order to understand the importance of NETBIOS interfaces in a local area network, it is critical that you consider the role of the operating system. An operating system functions very much like a traffic officer during rush hour. It is responsible for coordinating all the computer's input/output requests, including requests for programs to be loaded into RAM, for files to be saved to disk, for files to be renamed or deleted, and for files to be printed. In addition, the operating system handles all command processing. When a user types an incorrect command ("DIT" instead of "DIR", for example), the command processor portion of the operating system needs to be able to recognize that this is not a legitimate command and to be able to provide an error message. When Microsoft developed PC-DOS for IBM's

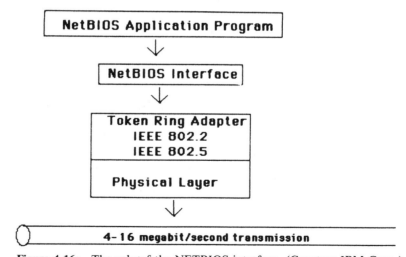

Figure 4.16: The role of the NETBIOS interface. (Courtesy IBM Corp.)

Personal Computer, it created a powerful operating system that was designed for a single user to perform a single task at a time.

As primitive local area networks using IBM PC's developed during the early 1980s, the single-user limitations of PC-DOS became a major problem for companies. Early versions of this operating system supported primitive file locking, in which one user on a network could use a particular file and lock out a second user, who would not be able to access information within this file at the same time as the first user and thereby risk destroying the file's information. Imagine, if you will, two students rushing to a blackboard to record their answers simultaneously on the same area of blackboard space: someone's answer would surely be destroyed. With PC-DOS 3.1, record locking became possible. This advance meant that two users could enter information in the same inventory file at the same time, as long as they did not try to access a particular inventory item record at the same time. The second user would be told by the program that someone else is using that particular record and that it would be available shortly. While record locking was a major advance, it had to be implemented by the particular application program that made the proper PC-DOS calls, and it also had to be supported by whatever network software was used.

IBM microcomputer networks running under PC-DOS require each PC workstation to have a network adapter card containing an 8K read-only memory (ROM) chip known as the NETBIOS(Network Basic Input/Output System). The NETBIOS functions as a high-level communications interface so that application programs could simply address specific network services required and not worry about network protocols.

The release of PC-DOS 3.1, with its record-locking capability, was a boon to programmers. When writing network applications in which several users would need to access a particular file at the same time, programmers were given the option of writing commands directly to PC-DOS as well as writing commands to the IBM PC's NETBIOS. It is much easier from a programming perspective to write input/output commands to PC-DOS (which offers an effective interface between NETBIOS and application programs) than to directly address NETBIOS. The Redirector program works closely with PC-DOS to handle input/output commands from an application program.

A SHARE program contained in PC-DOS 3.1 also contributes to programmers' ability to create multiuser network application programs. This program permits users to specify a "sharing mode" for a particular file. Users can be given different access levels, including "read/write" and "deny write," which enables programmers to write-record locking applications.

Unfortunately, even with the multiuser programming "hooks" available under PC-DOS 3.1, companies found that they were performing multiuser operations with what was really still a single-tasking operating system. On IBM's Token-Ring Network its Local Area Network Program runs as a task under PC-DOS on a file server. Each workstation's request for service is queued up and handled sequentially by the file server in what is, in effect, a series of single tasks that slows up network performance. Other network vendors have taken different approaches around this single tasking limitation. The 3+ networks by 3Com utilize a DOS emulator in the file server that uses a time slicing method to provide multitasking. Novell's NetWare imposes a software shell around DOS that intercepts DOS input/output commands and sends them directly to the NETBIOS or to hardware. This means that special Novell hardware drivers must be used and that the file server must be formatted completely under NetWare. The result of this approach, according to virtually every objective assessment, has been greatly enhanced network performance.

In 1987 IBM announced the PS/2 series of microcomputers, which utilized what it called "microchannel architecture," powerful new hardware that promised true multitasking in the future when an appropriate operating system would be available for a new generation of Intel 80286- and 80386-based computers. IBM and Microsoft also developed and released OS/2 (Operating System 2) and OS/2 Extended Edition. Designed to overcome the inherent single-tasking limitation of DOS, OS/2 promised a **protected mode**, a mode built into the operating system to permit several different application programs to run at the same time, in a true multitasking approach, while protecting the integrity of each program. Thus, if one program "crashed," it would not affect other programs' performance.

Microsoft and 3Com then indicated that under OS/2 their LAN Manager would provide a number of network services, including security, the maintenance of network statistics, and the ability to communicate with new PS/2 workstations under OS/2 and with older PC workstations under DOS. LAN Manager is an extension to OS/2 that enhances the operating system's kernel, the layer that links the operating system to network hardware. LAN Manager uses "named pipes," a standard interface across a local area network, so that any workstation is able to access the entire network's resources. For example, under this scheme it is possible for an application operating on one workstation to display results on the monitor of another workstation. It is also possible using named pipes for an application program to run under three different modes: on a single workstation under OS/2, as a multitasking application on a network file server, and as a distributed application that shares processing under more than one network server and workstation.

There is a problem with a LAN Manager that is based on an operating system such as OS/2 that was not developed with security in mind. An OS/2 LAN file server, for example, does not require a log-in identification for entrance, permits data to be manipulated with simple OS/2 commands, does not provide security for network users at the same level, permits non-dedicated servers to access to the network through application programs, and permits log-in upgrades with almost unlimited access to network files and accounts. In other words, while a new operating system such as OS/2 may have a number of advantages, it may not be ideal for network management purposes.

Unfortunately, to complicate matters even more, IBM announced that under OS/2 Extended Edition it will offer a Communications Manager as well as a Database Manager and Presentation Manager. There is no indication just how compatible LAN Manager will be with Communications Manager.

For the next few years there will continue to be a good deal of confusion involving standards for local area networks utilizing IBM microcomputers. It is clear that the announcement of OS/2 and OS/2 Extended Edition will usher in an era of true multitasking local area networks, but it is not yet clear which standard or standards will emerge. In any event, it will take a considerable period of time for software publishers to issue OS/2 versions of their application programs. Thus, companies will have some time to consider which direction they wish to take: follow IBM's lead and select a 100% IBM solution involving all IBM hardware and software; follow the lead of Microsoft and 3Com and assume that LAN Manager will become a *de facto* standard for LANs; or select an alternative vendor such as Novell that apparently has chosen to follow its own direction with the proviso that it will continue to offer the corporate world the highest level of network performance.

A Bridge Between Two IBM PC Networks

As mentioned earlier in this chapter, IBM's PC Network was designed for small-office needs or limited departmental functions and is therefore unlike the Token-Ring Network. Many companies installed the broadband version of this network only to discover later that they also would need to install the Token-Ring Network to handle the heavy data traffic in certain departments. To bridge these two very different networks, IBM has designed the IBM Token-Ring Network/PC Network Interconnect Program. A network workstation on each network serves as a **bridge** running this particular program. These bridge workstations must be dedicated to this function, and every other workstation must possess two network adapter cards, one for each network. The bridge enables a company to share the resources of each network (printers, disk drives, etc.) as well as send information and electronic mail to all users. A NETBIOS name table on each network adapter card must contain the unique names of each workstation on both networks so that they may be located when a message is sent. Each adapter card can contain up to 16 names, a limitation of the PC Network program.

Bridging the Gap Between Linked Token-Ring Networks

Often companies will have several token-ring networks or even a variety of different types of local area networks that need to be linked together; ideally the business end user should never be aware of the processes involved in moving data from location to location. We'll discuss the bridges necessary for moving information between networks in this chapter and consider the **gateways** between networks and mainframe computers and minicomputers in our next chapter.

A bridge acts as a high-speed connection between two token-ring networks. The networks can contain different cabling (data-grade versus telephone wire, for example) and operate at different speeds. The bridge generally is a network workstation that contains two network interface cards, one for each of the networks to be connected, and appropriate cabling from the interface cards to each network.

It is important to realize that each token ring is independent, despite the bridges connecting them. If something happens to disrupt one ring, the other ring continues to operate, since it produces its own tokens and monitors its own operations.

IBM advocates the organization of token-ring users by affinity groups—users who have the greatest need to communicate among themselves. Accounting and other financial users, for example, would most likely need to access the same information and communicate with each other, so they would form a natural affinity group on a token-ring network. Another token-ring network might contain users whose primary tasks revolve around manufacturing. Occasionally they might need to share electronic mail or send data to financial users on the other token-ring network, but for the most part they will communicate with their own affinity group.

Each adapter card on a token-ring network has a unique address. The source and destination fields that are part of the frame transmitted on a token ring have certain bits that indicate the specific ring number, and the address within that ring, of the sending and receiving network workstations. Some network installations may be so complex that additional information will be needed to make sure that the information frame gets to the right ring and workstation in the most efficient manner.

For multiple token-ring networks that require several bridges, it is possible to have a high-speed backbone ring that consists of several bridges to other networks, as illustrated by Figure 4.17. Using a technique called source routing, it is possible to place a list of connecting bridges at the start of the variable-length information field. This header contains a list of bridges that must be traversed by the frame in order to reach the destination station.

A bridge examines this routing information header to see whether or not this information is correct. The bridge may maintain a routing table consisting of the most common destinations it routinely handles. If the routing header contains a destination station whose route is unknown, the bridge can generate a "query" frame to other workstations on its ring or even on adjacent rings to learn the required routing information.

The physical routing decisions by the bridge workstation are handled at the very low media access control (MAC) level of the OSI model, which facilitates the establishment of logical link control connections between source and destination workstations. These are essentially hardware decisions. If a bridge workstation should fail, it is a relatively easy process to provide alternative bridge routing information to the bridges on the backbone network, especially if the bridges are set up in parallel so that there are several alternative paths to the same destination.

Fiber Distributed Data Interface (FDDI)

As we have seen in examining IBM's Token-Ring Network and its ability to link large multiple-ring networks, IEEE 802.5 token-ring networks are designed for heavy data traffic. Yet the speed of these networks (4Mbps) falls well below the standard 10-Mbps speed of Ethernet networks. Even using optic-fiber cabling, which is capable of much faster transmission, does not help; the IEEE 802.5 standards were not designed to take advantage of optic fiber's incredible transmission speed.

Developed by the American National Standards Institute (ANSI) Committee X3T.9, the **Fiber Distributed Data Interface (FDDI)** standard is a counter-rotating

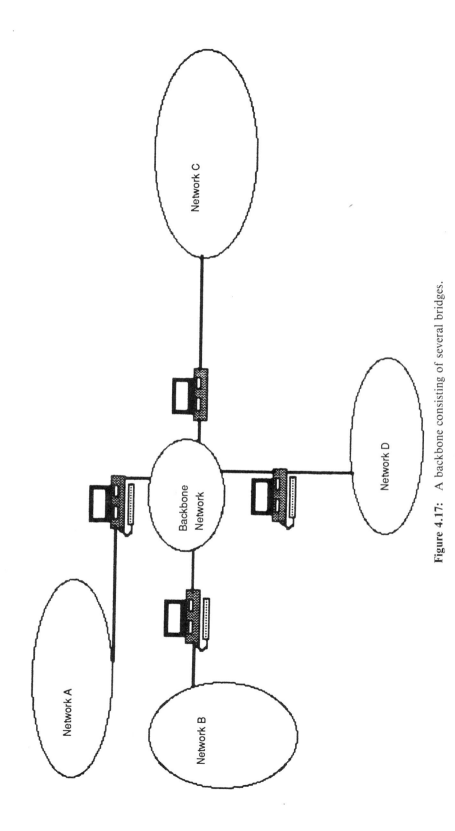

Figure 4.17: A backbone consisting of several bridges.

token ring capable of covering a very large area (200 kilometers) while transmitting data at 100-Mbps speed, a standard that ensures compatibility with IEEE 802.5 token-ring networks by maintaining the same frame fields found in that standard.

The FDDI model is reasonably consistent with OSI model. As Figure 4.18 reveals, the FDDI Physical layer is broken down into a Physical Layer Protocol (PHY) that concerns itself with the actual encoding schemes for data as well as a Physical Medium Dependent (PMD) layer that provides the actual optical specifications. A Media Access Control (MAC) layer handles token-passing protocols as well as packet formation and addressing. Notice that there is also a set of Station Management standards to provide information on such tasks as removal and insertion of workstations, fault isolation and recovery, and collection of network statistical information. SMT utilizes the Connection Management protocol in conjunction with PHY line states to determine whether nodes entering the ring are linked together.

FDDI was used initially primarily for "back-end" applications such as connecting mainframe systems and mass-storage devices and for the backbone network function of connecting different networks together. Today it is ready to take its place along with token bus and token ring as a viable standard for large networks

What makes FDDI so appealing despite its expense are its speed of transmission and its duel-ring approach, the latter offering built-in protection against system failure. One major difference between IEEE 802.5 token-ring networks and an FDDI network is that a token-ring network circulates one token at a time. A sending station transmits its token and then waits until the receiving station returns the token with an acknowledgment that the message has been received. The sending station then passes the token to the next workstation on the ring. In an FDDI network the workstation sending a message passes on the token immediately after transmitting the message frame and *before* receiving an acknowledgment that the message has been received. The result of this procedure is that several message frames can be circulating around the ring at any given time. FDDI networks also use a *restricted token*, an innovation that enhances network speed. With a restricted token, it is

OSI Layer	FIDDI Layer
Data Link Layer	MAC Defines token ring protocols packet formation, CRC
Physical Layer	PHX (Physical Sublayer) Physical layer protocol
	PMD (Physical medium dependent)

Figure 4.18: The FDDI model.

possible to keep other workstations off the network while a time critical task is performed. A timed-token protocol also used in IEEE 802.4 networks ensures that low-priority messages will not clog up a network during peak hours. Timed-Token Protocol uses both synchronous and asynchronous transmission. Workstations utilize a certain amount of transmission bandwidth defined for use as synchronous service, while the remaining bandwidth is used by workstations that transmit signals asynchronously when the token service arrives earlier than expected. They continue to do so until the expected time of token arrival, when they switch to synchronous transmission.

In addition to greater speed of transmission, FDDI networks enjoy a built-in redundancy that protects against system failure: the FDDI's dual-ring characteristic. Figure 4.19 shows a typical FDDI network. Note that if a break in cabling occurs between stations A and B, it would be possible for them to continue to communicate through station C, which is acting as a wiring concentrator. Also, note that it is possible to send data over both sets of cable traveling in opposite directions, so that if there is not a break in cabling, a transmission speed of 200 Mbps is possible.

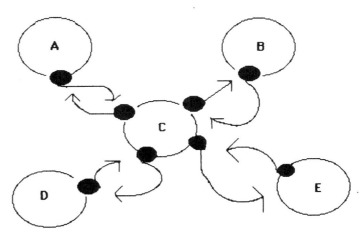

Figure 4.19: FDDI uses dual cabling.

An FDDI Packet

FDDI's MAC layer controls the actual flow of data on the ring. Figure 4.20 describes the actual frame format for an FDDI packet. Sixteen idle control symbols precede packets that begin with Start Delimiter symbols followed by the Frame Control fields that indicates what type of packet is being sent. After these fields come the Source and Destination fields, followed by the Frame Check Sequence field, which is responsible for error checking for all the fields to this point. An End Delimiter field is followed by a Frame Status field, which determines if the packet was received in error or is correct as is.

```
| PR | SD | FC | DA | SA | DA | FCS | ED | FS |
```

PR   Preamble
SD   Start Delimiter
FC   Frame Control
DA   Destination Address
SA   Source Address
DA   Data
FCS Frame Check Sequence
ED   End Delimiter
FS   Frame Status

Figure 4.20: An FDDI packet.

IEEE 802.6 Metropolitan Area Networks

While local area networks can link together departments and even adjacent buildings, how can a manager link together facilities on opposite sides town? The answer is the IEEE 802.6 standard for a metropolitan area network (MAN). While the token-passing routing mechanism for IEEE 802.6 has not been fully developed by the committee, theoretically a MAN will be able to interconnect 256 local area networks. Harvard University has implemented MAN to connect Harvard Observatory, the Harvard Medical School, the Aiken Computational Laboratory, and Massachusetts General Hospital, all of which facilities use local area networks running Ethernet. The Harvard MAN utilizes optic-fiber cabling and several LAN running Ethernet under TCP/IP at 10 Mbps. The system is connected through the telephone company's central offices in Cambridge, Boston, and Brookline.

The Alternative to IBM: Apple's AppleTalk Network

While up to this point our discussion of local area networks has been built around IBM PC and compatible microcomputers. Apple's Macintosh has become a viable alternative for many corporations. In the next chapter we'll take a close look at how unlike networks such as a Macintosh-based AppleTalk LAN can exchange files with an IBM PC–based Token-Ring Network. In this section we'll look at the AppleTalk network itself.

LocalTalk is a serial bus network utilizing a CSMA/CA access method that transmits data at 230 kbps across a shielded twisted-pair wire medium to Macintosh computers. Since the network interface (serial RS-422 interface) is already built in to a Macintosh, the computer workstation requires only a simple network connector and not a network interface card. While Apple indicates that a maximum of 32 workstations may comprise a network zone, zones can be bridged together to form an internet of several hundred workstations. One method of bridging AppleTalk

Figure 4.21: An AppleShare file-server screen. (Courtesy Apple Computer Corp.)

networks is for each network to have a Hayes 2400-baud Smartmodem connected together through a switching device such as Haye's Interbridge.

Every LAN utilizing a file server needs some kind of file-server software. AppleTalk uses AppleShare. As illustrated in Figure 4.21, an AppleShare file server retains the Macintosh's graphic orientation and mouse-driven commands. A user can observe which files are contained in a particular volume simply by moving the mouse cursor to the icon (picture) of the volume and then clicking on it. To copy a file from one volume to another, for example, simply requires using the mouse to select a file's icon and then dragging this symbol from one volume to another.

MAJOR ISSUES IN LOCAL AREA NETWORK MANAGEMENT

System Faulty Tolerance and LANS

Most data processing managers have built successful careers in a mainframe computer environment. For this reason, it has been difficult for them to realize the lack of attention paid until recently to the concept of fault tolerance for mircocomputer-based local area networks. Mainframe computers have several levels of protection to ensure that valuable data will not be lost in the event of a power failure or even the failure of the computer's central processing unit. Tandem Computers' approach illustrates this fact. A special internal fault-tolerant maintenance and diagnostic subsystem known as CHECK continuously monitors the computer system's processors, power supplies, and cabinet environments. A control unit communicates with a fault tolerant application program, the Tandem Maintenance and Diagnostic System (TMDS), running under the direction of the computer's operating system. The TMDS uses artificial intelligence techniques to analyze the data coming to it to determine what action should be taken by the computer operator. The TMDS is even capable of administering stress tests to determine weak components before they actually fail.

In effect, the CHECK system is actually a separate computer based on a Motorola 68000 microprocessing chip. Some of its other capabilities include providing a battery backup clock for setting system time in the event of a power failure. The Tandem system fault tolerance scheme possesses multiple shared power supplies to prevent one power supply failure from "crashing" the entire system. Dual processors ensure that if any board or data path fails, the corresponding alternative device of data path takes over and completes the task. The system even permits users to repair defective components without shutting down the computer.

Since microcomputers until recently were thought of exclusively as single-user/single-tasking machines, there was little thought given to fault tolerance in the past beyond using a surge protector and making periodic tape or floppy disk backups of important data stored on a hard disk. Today it is possible to achieve system fault tolerance with LAN by ensuring protection against failure of the disk system, the file server, individual workstations, and other LAN hardware.

Novell offers a System Fault Tolerant (SFT) version of its popular NetWare LAN software. The SFT will serve as an example of what levels of security are possible with local area networks. In conjunction with normal safety precautions, such as the installation of surge protectors and uninterruptible power supplies (UPSs), the SFT software provides LANs with mainframelike protection.

Novell's SFT NetWare Level I protects against partially damaged media. A file server contains a file allocation table (FAT) that contains the critical information of the locations on disk where different files are stored. If the FAT is partially damaged or destroyed, this information is usually lost, and it is impossible to retrieve files even though the files themselves are undamaged. The closest analogy might be a young man who losses his address book containing the addresses and unlisted phone numbers of several women he wishes to date when he visits New York City. He could begin randomly knocking on doors, but his chances of locating the young women is remote.

Level I duplicates the FAT on other areas of the disk virtually ensuring that at least one FAT will remain undamaged and usable. Another SFT feature at this level is the "hot fix," which deals with the problem of particular areas of a file server becoming unusable. When NetWare tries to write a file to an area that it discovers is unusable, it will write to a different area of the disk and then place the address of the bad area in the Bad Block Table. The user is never aware of this process, but it ensures that data is never written to an area where it cannot be retrieved.

Level II includes all Level I features but adds some additional safeguards against major disk failures. Disk mirroring is a process of using two different hard disks. The second disk is a mirror image of the first and receives identical updates at the same time the first disk is updated. If something happens to the first disk, the second disk takes over immediately without the loss of any data.

One problem with disk mirroring is that both hard disks use the same disk controller. If this controller "crashes," then neither disk is useful. The solution available under Level II as an option is disk duplexing. Disk duplexing consists of two identical disk subsystems including duplicate disk controllers. The duplexed drives contain their own power supplies and cabling as well as separate controllers so that the failure of one system will have absolutely no effect on the other system. One side advantage of this approach is that both disk drive systems can be used for reading and writing purposes, thus access time is generally cut by 50%.

One system fault that has not been addressed as yet is the possible failure of the file server itself. Under Level III of NetWare SFT, in addition to incorporating all SFT features found under levels I and II, there are duplicate file servers connected by a high-speed bus with party checking to make sure that no data is garbled as it travels from file server to file server. If one file server fails, the second one automatically takes over network operation.

One critical SFT problem that NetWare SFT addresses under Level II and III is what happens to a database's integrity if the network fails in the middle of a transaction. The NetWare Transaction Tracking System (TTS) monitors each transaction as it takes place. If the system fails before a transaction is completed, then TTS will roll back the database to its condition prior to that transaction, thus ensuring that the database itself is not corrupted by incorrect data. While Novell's SFT NetWare cannot guarantee that there will not be a network shutdown, its three levels of protection do show that local area networks now enjoy a degree of system fault tolerance heretofore found only with mainframe computers.

NetWare's system fault tolerant version also addresses the issue of security. A network administrator can monitor the system from any workstation and is not limited to just the network console. From any workstation users thus can be given encrypted passwords, limited to certain hours, restricted as to which workstations they may use to log in, and be restricted as to the number of stations they can simultaneously be logged into.

Local Area Network Software

While certain standards have emerged for LAN, including NETBIOS compatibility for IBM PC and compatible networks and AppleTalk Filing Protocol for Macintosh networks, it is still absolutely essential that network managers check to see if a major software package has different network versions (NetWare, 3Com, IBM Local Area Network Programs, etc.).This is especially true with accounting software, since here record-locking is critical; two or more people must be able to access the same accounts receivable file at the same time.

Productivity programs and application programs comprise the two major types of software found on local area networks. Productivity software consists of word processing, spreadsheet, and database management functions. Since most people using word processing and spreadsheet programs create their own files or modify existing templates, record-locking is not critical here. It is critical, though, for a network administrator to require that all network users use the same version of the word processing program or spreadsheet program. Frequently documents and spreadsheets will need to be viewed and perhaps modified by other users. Different network versions of a software program that may even require different versions of DOS can result in fatal system errors that can disrupt network operation for everyone.

Database management programs present a different challenge to the network administrator. Since a LAN often incorporates the entire company's personnel, different users will have different levels of "need to know" and different levels of access to key database management files. Only certain people within a company should be able to see payroll records. Programs such as dBASE IV's network version include password protection that enables network administrators to select which users will have access to which files.

Since a local area network should ensure smooth transfer of data from user to user and not simply serve as a way to share resources such as printers and disk drives, a network administrator may want to examine some of the newer integrated software now available in network versions. A program such as SPI's Open Access II, for example, offers software modules for word processing, spreadsheet, graphics, database, and communications, as well as desktop accessories and utilities. Since all modules are designed to work together, it is easy to move data from one application to another. A user would not have to go through the file conversion process required when moving data from the Lotus 1-2-3 spreadsheet program to the dBase III Plus database management program. One user, for example, might create a spreadsheet, while a second user might want to take this information, graph it, and then incorporate the graphs in a report written with the word processing module. The report thus compiled could be sent via modem to another branch office using the communications module.

Accounting probably represents the major activity found on a LAN, after word processing. It also represents the greatest security threat for a network administrator. Some sophisticated accounting modules, such as those offered by Prologic, permit different levels of access within individual accounting modules. One user might be able to see only the first three menu items under accounts receivable, while a second user with a higher level of access may be able to see the first six menu choices. The accounting software checks each user's request against a table of levels of access. It also provides a clear audit trail of all transactions requested by different individuals. If a user removes an item from inventory, there will be a printed report indicating when this transaction took place and which specific user performed this deed.

CHAPTER SUMMARY

A local area network's components include the workstations, file server(s), adapter cards, media, network software. The IEEE 802 committee has developed a series of standards for bus (802.3), token bus (802.4), token-ring (802.5), and metropolitan (802.6) topologies. The new fiber standard (FDDI) promises to deliver at least 100-

Mbps speed. Networks may use contention or non-contention methods of handling possible data collisions. Some local area networks build levels of system fault tolerance into their hardware and software to ensure that the network will not fail.

KEY TERMS

AppleTalk
Backbone Network
Baseband coaxial cable
Bridge
Broadband coaxial cable
Bus Centralized file server
Contention network
CSMA
CSMA/CD
Datagram
Despooling
Disk server
Distributed file server
Fiber Distributed Data Interface (FDDI)
File allocation table (FAT)
File server
Gateway
Hub architecture
IEEE 802.3
IEEE 802.4
IEEE 802.5

IEEE 802.6
Jam signal
Local area Network
Manufacturing Automated Protocol (MAP)
Mesh topology
Multistation access unit
NETBIOS
Network architecture (topology)
Network interface card
Optic-fiber cable
Pipes
Protected mode
Public volumes
Record locking
Ring topology
Star topology
System fault tolerance
Token
Twisted-pair wire
Volumes
Wire center

REVIEW QUESTIONS

1. Under what conditions would distributed file servers be preferable to a centralized file server?

2. What are the major distinctions between disk servers and file servers?

3. Two of the most common network topologies are the bus and the token ring. Under what conditions would one be preferable to the other?

4. Specifically how does Ethernet differ from the IEEE 802.3 standards? Why can this present problems for a network administrator?

5. Differentiate between a bridge and a gateway on a local area network.
6. What is a backbone on a local area network?
7. What is the FDDI model?
8. What are some advantages and disadvantages to an AppleTalk network?
9. Describe the different levels of system fault tolerance available on large computers and on microcomputers running a local area network.

TOPICS FOR DISCUSSION

1. Explore a number of LAN alternatives, including switch boxes and simple RS-232 cable-to-cable transfer of data. What appears to be some of the key elements that determine whether a company needs a LAN?
2. A number of low-cost networks, such as 10-Net, offer very appealing features such as electronic mail and even an electronic bulletin board for company communications. Look at 10-Net and contrast it with LANs offered by 3Com and IBM. When should a company opt for the higher-cost networks?
3. Many pioneers in the PC movement are not wildly enthusiastic about local area networks, because they feel that LAN require a network manager who then controls personal computing and restricts the user's choice of programs and files. In effect, LANs may represent a movement back toward centralization and away from the decentralization of computing power. Discuss this complex issue from both points of view. Is there any compromise?
4. Examine an office at your university and determine what kind of network topology would appear to be most effective in that environment. What are the major computing functions in that office? What kind of local area network would you recommend?
5. There is no uniformity in network software pricing or licensing. Examine several word processing programs, including WordPerfect, WordStar, and Word. Price a 10-user network and indicate the conditions imposed by the software vendors on such a network.

Bridges, Gateways, and Micro-Mainframe Communications

CHAPTER 5

OBJECTIVES

Upon completion of this chapter, you should be able to:

- compare and contrast a bridge and a gateway
- differentiate between a router and a brouter
- identify some of the major components of a mainframe computer
- discuss the role of each layer of IBM's System Network Architecture
- explain how a local area network can communicate with an IBM mainframe computer
- explain how Apple's Macintosh computers can communicate with the IBM mainframe world

INTRODUCTION

In this chapter we will explore how local area networks composed of microcomputers can be bridged together; we will also examine how such microcomputer-based networks can be linked to mainframe computers using gateways. Finally, we will examine a very real issue for many large companies: the linking together of IBM PC–based networks and Apple Macintosh–based networks. This entire chapter illustrates that communication between different computers is just as difficult and yet just as necessary to achieve as communication between two very different countries.

BRIDGES

What is a Bridge?

In Chapter 4 we observed that many companies may have more than one network. The administratrive functions might be performed by employees with desktop computers linked together by, for example, a 3Com 3+ network that is running Novell NetWare software. The research division might require a large IBM Token-Ring Network that also happens to be running Novell NetWare software.

Often two networks can be managed better as separate entities, because the users are performing different tasks using different software. Sometimes two networks are necessary because the size limitations of one network (260 devices in the case of IBM's Token Ring) have been reached. There are other times when networks have been constructed separately because they utilize different hardware. Separating the two networks reduces traffic congestion on both, but occasionally it is necessary to communicate and send information between the two. The actual physical links required to connect the two homogeneous networks running the same protocol are known as **bridges**.

When information from a user on network 1 is transmitted to a user on network 2, a number of tasks have to be completed to make this process successful. First off, network 1 has to form a frame. This is done by appending to the data to be transmitted a header and a trailer containing specific information about that network to form a frame.

The bridge connecting network 1 to network 2 monitors all frames transmitted on its network. When it encounters a frame containing a station address found on another network, it appends an additional header and trailer to this frame while retaining all current headers and trailers. It then trasmits this frame to the bridge found in network 2. This bridge strips off the linking headers and trailers and then transmits the data that is still encapsulated with headers and trailers to the appropriate network 2 workstation.

Generally bridges on local area network consist of a workstation (often dedicated solely to this purpose) with special network bridge software. A bridge connects networks that run the same network protocol, NetWare in our example. These protocols could consist of NETBIOS, Novell's IPX, or 3Com's own protocol. The address of each workstation on both networks must be unique for a bridge to function correctly. A bridge checks the destination address of each frame and determines if it is destined for another network. If it is, this frame is forwarded to the bridge connecting the next network. In effect, then, each bridge keeps the addresses of its own network workstations and forwards those frames that belong elsewhere. This address checking does delay the forwarding of packets. In addition, the frame size transmitted on each network must be identical.

A relatively recent development has been the **"learning" bridge**, intelligent bridges designed to tie together unlike, as well as like, local area networks, even those offered by different vendors. In order for these learning bridges to function, both networks must support the IEEE 802.2 Logical Link Control protocol. Some of these "smart" bridges are capable of learning locations of different workstations.

They may even update their own routing tables automatically. Since bridges function at the lower levels of the OSI model that we will discuss in Chapter 6, they require that both connected networks support the same higher level protocols, such as XNS, TCP/IP, and so on. Higher-level ISO protocols are supported by Manufacturing Automation Protocol (MAP), a broadband IEEE 802.4 token bus, and Technical Office Protocol (TOP), an Ethernet-based system.

Often an employee would like to be able to use a company's local area network from her or his home or from a customer's site. One possible solution is a **remote bridge,** a bridge that connects a computer at a remote site with a network and enables the caller to use the network's resources as if he or she were at work and physically attached to the network. Novell's NetWare, for example, permits the user calling into the network to create a remote shell under NetWare. This shell intercepts any DOS requests to network devices and makes these devices think that the user is on-site. In effect, the caller "takes over" the network workstation functioning as a bridge, and anything typed remotely goes directly to this workstation. One problem with this approach is that each remote caller requires his or her own bridge workstation that is dedicated to this function as long as the call takes place. While this approach may not be expensive or inefficient after business hours, during business hours it could require a number of bridge workstations.

Novell's NetWare is the market leader in local area network software, but there are other options besides Novell's that might boost a company's productivity as well as its connectivity. Gateway Communications' Universal NetWare Bridge is capable of connecting any NetWare-based network, including Ethernet, ARCnet, IBM's Token Ring, and G/NET. Running on a non-dedicated personal computer and providing transparent operations, this bridge utilizes the synchronous X.25 protocol that consists basically of the first three layers of the OSI model. Unlike Novell's own remote program that runs in slow, asynchronous mode, this bridge can transmit data up to 19.2 kbps. A bridge may contain up to 64 virtual circuits. Since data is routed through the public data networks, networks virtually anywhere around the world can be connected. Figure 5.1 illustrates how several remote networks can be connected using this approach.

ROUTERS

Unlike a bridge, a **router** operates at the network-layer level of the OSI model. A router is not transparent to local network workstations, as bridges are. It is capable of routing packets between networks that utilize different protocols. If one network uses a different-size packet, then the router is capable of segmenting them and later reassembling them. Routers have slower throughput than bridges because it takes them longer to examine a data packet and extract the information it needs.

A good example of a router that is capable of supporting different common protocols is Proteon's p4200 Gateway, an internetwork router which supports TCP/IP, DECnet, and XNS communications protocols. Figure 5.2 illustrates how this router can connect several different microcomputers and minicomputers using different protocols.

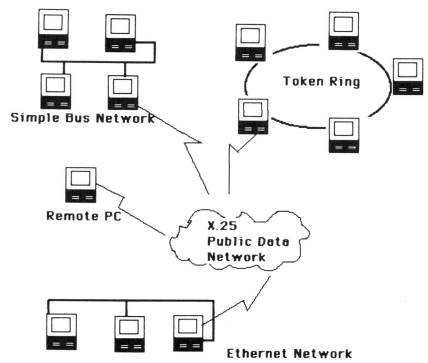

Figure 5.1: A remote bridge utilizing the synchronous X.25 protocol.

A hybrid of a bridge and a router known as a **brouter** recently began to appear on the market. Cisco Systems' HyBridge serves as an example of this technology. It is capable of handling a number of protocols, including TCP/IP, DECnet, and XNS. While it acts as a router providing links between Ethernet-connected devices, it can revert when necessary to the role of a bridge between Ethernet networks. When only a bridging function is required, network managers can suppress the routing function completely and can thus speed up the bridging operations.

THE MAINFRAME ENVIRONMENT

Communicating information from a local area network composed of microcomputers to a mainframe computer is far more complex than simply bridging the gap between two networks. The two environments are so alien to each other that before we can discuss how this communication can take place, we must look at this mainframe world.

IBM's Mainframe Hardware

IBM mainframe computers govern a kingdom of terminals, printers, and storage devices that all speak these mainframe computers' language. Terminals and other

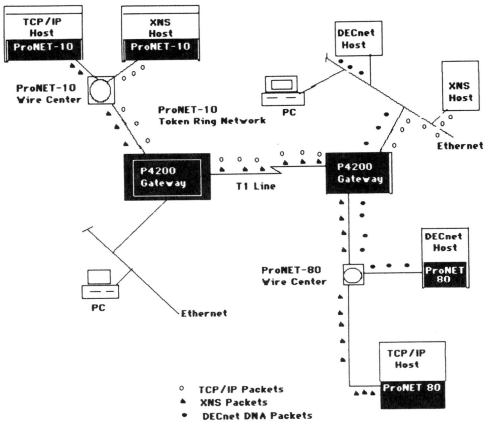

Figure 5.2: An internet router at work.

devices communiate via **cluster controllers,** devices that serve as an interface between the host computer and as many as 32 devices. Information from these devices is multiplexed along coaxial cable that serves as the system's arteries. Information from terminals calling from remote sites via a modem and telephone line is received by a communications controller, a high-speed data "traffic cop" that transmits this information to the mainframe. Figure 5.3 illustrates an IBM 9370 attached to a number of devices, including terminals, printers, cluster controllers, and even a local area network.

Microcomputers store information in an ASCII format in which 7 bits of information represent a character or number while an eighth bit represents a parity bit (a bit that serves as an error-checking mechanism that ensures that information is not lost when transmitted). IBM mainframe computers utilize an EBCDIC 8-bit format for storage that is completely incompatible with ASCII. While many microcomputer-based local area networks use asynchronous transmission, IBM mainframes use synchronous transmission; once again, the two methods are completely incompatible. Finally, while many local area networks' architectures are

Figure 5.3: An IBM S/370-168 mainframe computer. (Courtesy IBM Corp.)

based on the OSI model, IBM's mainframe network architecture is System Network Architecture (SNA), which corresponds only in part with the OSI model. Before we can discuss gateway connections between microcomputer networks and IBM's mainframe computers, we must look at the basic structure of System Network Architecture.

SYSTEM NETWORK ARCHITECTURE (SNA)

SNA Definition

System Network Architecture (SNA) consists of a set of protocols designed to facilitate data communications in an IBM computer system environment. These protocols enable different devices—even when designed by different manufacturers—to communicate with each other. Before we can examine the various functional layers of SNA and note its similarities and dissimilarities to the OSI model, it is necessary to learn a few definitions, a few of the common SNA terms used to describe the relationship between an IBM computer and the devices that communicate with it.

Under SNA any person and any device or application program that needs to use the services of the SNA network is considered to be an SNA user. SNA provides a point of access for its users known as a **logical unit** (**LU**). This LU is like a port for plugging into the SNA system. Since it is a *logical* unit, however, it is not a physical connection. IBM designates LUs from 1 through 6, although at the moment there is no LU 5. LU types 2, 3, and 4 support communication between application programs and different kinds of workstations. LU types 1, 4, 6.1, and 6.2 support communication between two programs. We will return to LU 6.2 later in this chapter, since it offers the hope that in the future microcomputer programs will be able to communicate directly with mainframe programs.

NAUs, Subareas, Domains, Nodes

All components on an SNA network are composed of network addressable units (NAUs) and the path-control network. We will concentrate on the NAUs, since the path control network is more concerned with the physical transmission and routing of data, a subject we will consider later when we look at the layered architecture of SNA. Every component of an SNA network that can either send or receive data is assigned an address and is known as a **network addressable unit (NAU).** We have already discussed logical units. We will see that there are two other types of network addressable units: physical units (PUs) and system service control points (SSCPs). These we will examine below. Regardless of the type of NAU, two NAUs must establish a **session** or linkage in order to communicate on the network.

SNA uses the term **physical unit (PU)** to *represent* a second type of network addressable unit, the actual physical device or communication link found on this computer network. In effect, SNA does not deal directly with a physical device but with the PU that represents the combination of hardware, software, and microcode that defines the services performed by that device. Each device on an SNA network must have a PU. Communication processors, terminals, and cluster controllers all have their services defined in terms of PUs. This PU performs certain functions when the device is activated or deactivated as well as during error recovery when this function is necessary.

The third type of network addressable unit is a system service control point (SSCP), which provides the services necessary to manage a network or a portion of a network. An SSCP activates sessions with physical units and logical units (SSCP-PU) and (SSCP-LU). A logical unit that needs to communicate with another logical unit (LU-LU) must first submit a request for this LU-LU session initiation to an SSCP during an SSCP-LU session. The SSCP resides in the Virtual Telecommunications Access Method (VTAM) control program on the host computer. Not only does the SSCP assume responsibility for activating and deactivating network resources, but it also assumes responsibility for taking action during error recovery and gathering statistics on the devices with which it communicates. The SSCP is assisted in its network management functions by the **Network Control Program (NCP),** which is a physical unit running on a communications controller (3705 or 3725). The NCP connects the VTAM program to the remote portions of the network and controls the flow of information between the VTAM and remote PUs and LUs. Figure 5.4 illustrates a sample network.

Whenever an SSCP activates a resource it becomes the owner of that resource. This means that before NAUs can communicate with each other, their physical and logical connections must be activated by the SSCP. After all the physical devices have been turned on and the network control program is loaded into the 3705 or

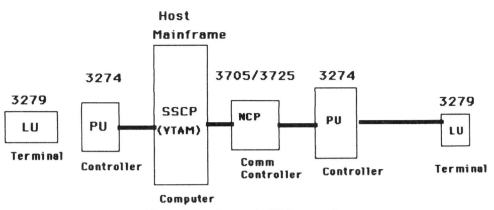

Figure 5.4: A sample SNA network.

3725 and all cabling is in place, the SSCP provides the services desired by each of these resources.

The NCP and SSCP, as well as all network resources physically linked to these, including all lines and links terminating at the 3705/3725, are known as a **subarea.** All application programs running on a single host computer would be located in the same subarea. A systems programmer assigns each subarea a number so that messages can be directed to the right location. The only way to change subarea boundaries is by physically moving a resource to another 3505/3725 or host computer.

In contrast to the physical nature of subareas, a **domain** is logical. Since a specific SSCP activates certain network resources, we say that these resources are in the domain of that particular SSCP. PUs and LUs as well as network links (physical communications connections) can only belong to one domain at a time, but ownership can be changed dynamically when an SSCP deactivates a particular resource, at which time it may be activated and thus owned by another SSCP. Figure 5.5 illustrates a single domain network, while Figure 5.6 shows how more than one SSCP can provide a multiple-domain network.

An SNA node represents one physical unit (PU). If the node has application programs or terminal devices, it might contain one or more logical units (LUs). They also must contain path-control components necessary for network transmission of data. At least one node on a network must have an SSCP. Those that do not have SSCPs have **physical unit control points (PUCPs),** a subset of SSCP functions that enables the node to establish communication with other nodes. There are actually several different types of nodes differentiated by SNA on the basis of their routing capabilities and the presence or absence of different NAU types. Table 5.1 reveals the basic differences between node types 2.0, 2.1, 4, and 5. Notice that node type 5 contains an SSCP.

Resources that want to communicate on an SNA network must do so in a specific sequence. The host computer must be running the SSCP software, and the SSCP-NCP session must have been established. The SSCP must then establish own-

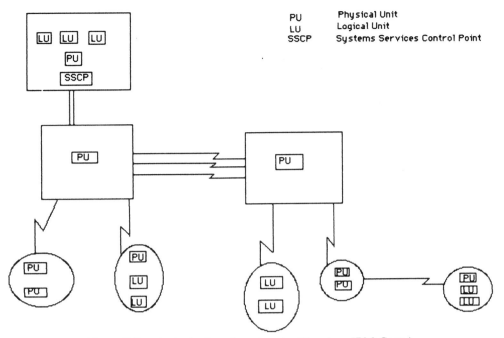

Figure 5.5: A single-domain network. (Courtesy IBM Corp.)

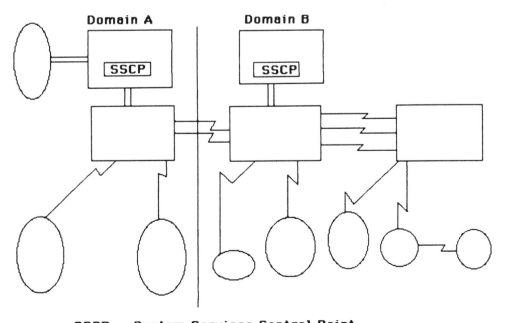

SSCP = System Services Control Point

Figure 5.6: A multiple-domain network. (Courtesy IBM Corp.)

ership of all resources that wish to communicate. An application program must be running on the host computer, and an SSCP-LU session must be established. In the case of a terminal, for example, an SSCP-PU session must be established first, followed by an SSCP-LU session. Figure 5.7 reveals how a local terminal (LU A) can be activated and then run any one of the application programs running on the host computer under VTAM.

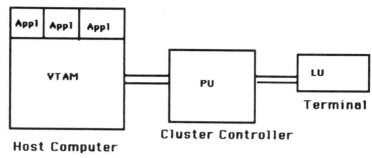

Figure 5.7: A terminal establishes a session with an application.

TABLE 5.1
Architectural Definition of Node Types (Courtesy of IBM Corp.)

Node Type	Architectural Description	Primary Functions
5	Subarea node	Control network resources
	Contains an SSCP	Support application and transaction programs
	Contains a PU type 5	Provide network operators access to the network
	Supports all LU types	Support end-user services
4	Subarea node	Route and control the flow of data through the network
	Contains a PU type 4	
2.1	Peripheral node	Provide end users access to the network
	Contains its own control point, which provides PU service	Provide end-user services
	Supports LU type 1, 2, 3, and 6.2	
	Supports direct link connections to other type 2.1 nodes	
2.0	Peripheral node	Provide end users access to the network
	Contains a PU type 2.0	Provide end-user services
	Supports LU types 2, 3, and 6.2	
	Supports LU type 1 for non-SNA interconnect	

TRANSMITTING DATA UNDER SNA's LAYERED ARCHITECTURE

SNA offers a layered architecture that provides a number of advantages including the ability to modify or revise one layer's services without having to create an entirely new architecture. Similarly, it is possible to substitute a protocol such as the CCITT X.25 protocol for Synchronous Data-Link Control (SDLC) at the lower layers. This substitution enables a non-IBM packet switched public network to communicate with SNA devices. We have just discussed at a system level how data in the form of an application program is sent to a terminal on an SNA network, but to understand how the data physically makes this journey we need to look at the functions governing each of these layers. This information will become important later when we examine how non-SNA networks can communicate through the use of gateways with an SNA network.

As Figure 5.8 shows, SNA may be considered to have a seven-layered architecture. The Physical Control Layer specifies both serial connections between nodes and high speed parallel connections between host computers and front end processors. The Data Link Layer corresponds to OSI layer 2. Synchronous Data Link Control (SDLC) is the major protocol found in this layer, although IBM does provide a high-speed parallel link protocol for its S/370 host, as well as protocols for its Token-Ring architecture, and the international CCITT X.25 standard.

The Path Control Layer is responsible for routing and flow control. It can be divided into three sublayers: Virtual Route Control, Explicit Route Control, and Transmission Group Control. A virtual route is a logical connection between two endpoints along a path or route, while an explicit route is a path that has been defined as a series of sequenced network groups. Finally, a transmission group represents a set of physical links between adjacent nodes along a network. The Virtual Route Control and Explicit Route Control sublayers correspond roughly to the Network Layer found under the OSI model. The subarea addresses found under SNA correspond roughly to the network addresses used in other networks. Note that sequencing and expedited network services are both functions of the Virtual Route Control sublayer, while routing is the primary responsibility of the Explicit Route Control sublayer. This sublayer will also notify the Transmission Control Layer (the next higher layer) of the failure of an explicit route.

The Transmission Group Control sublayer makes the physical links between adjacent nodes within the network *appear* to be a single link to the higher layers. This technique results in greater reliability, since the other layers do not have to concern themselves about addressing specific physical links.

The Transmission Control Layer is concerned primarily with pacing the data exchanges between NAUs so that information is not lost. It also is responsible for the encrypting of information should security require it. This layer adds a request/response header to each message it receives from the Data Flow Control Layer before it forwards the frame to the Path Control Layer. It uses this header to indicate such control information as the pace desired and whether or not encryption is desired.

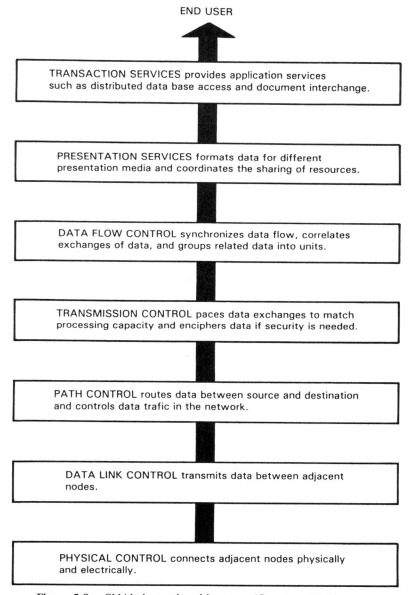

Figure 5.8: SNA's layered architecture. (Courtesy IBM Corp.)

The Data Flow Control Layer is responsible for the establishing of sessions. These sessions can result in the rough equivalent of half-duplex transmission in the form of a one-way transmission across the network (a chain) and the rough equivalent of full-duplex transmission in the form of two-way transmission across the network (a bracket).

The Presentation Services Layer provides format translation as well as sequencing restrictions when required. The Transaction Services Layer corresponds with the OSI model layer 7. At this, roughly the equivalent of the Application Layer, we find SNADS (SNA Distribution Services), which is a way of distributing data asynchronously between distributed application programs. We also find Document Interchange Architecture (DIA), IBM's method of ensuring that documents can be exchanged among different application programs.

How Data Moves Through the SNA Layers

Let's assume that an application program user has produced some data that must be transmitted to another user under SNA. The basic data unit under SNA is the Request Unit (RU). The Transaction and PResentation layers provide this RU with the information needed for interfacing and formatting in order to be understood by the device that will receive this information. The Transmission Control Layer adds a Response/Request Header (RH) to the RU. This RH contains information on bracketing and chaining, flow control, and the type of request or response.

When this RU-RH unit reaches the Path Control Layer, a Transmission Header (TH) is added. The TH contains destination and source addresses, information on segmenting (if necessary), as well as route flow control.

The Data Link Layer adds a Link Header (LH) and a Link Trailer (LT). This information contains the necessary protocols for the data to travel over common carriers. This information packet created under SNA is transmitted to a destination node, where it begins its journey from the lowest SNA layers (Physical Control and Data Link Control) to the highest layer (Transaction Services). During the course of this journey the headers and trailers are stripped off at the layers corresponding to where they have been attached until only the RU remains.

CONNECTING LOCAL AREA NETWORKS TO IBM SNA NETWORKS

There are a number of different ways of connecting an IBM mainframe computer running under SNA with microcomputers linked in one of the local area networks we described in Chapter 4. We will describe terminal emulation, which permits both local and remote gateway connection, and we will discuss IBM's long-term solution to increased connectivity, LU 6.2 and Advanced Program-to-Program Communication (APPC).

A Local LAN Gateway Permitting Terminal Emulation Under SNA

For several years it has been possible to install a 3270 terminal emulation board in a microcomputer and then attach the computer directly to an IBM cluster controller, which in turn conveys information to an IBM mainframe. This rather expensive procedure results in the microcomputer's ability to emulate, or imitate, an IBM dumb terminal. The data from the microcomputer is converted from asynchronous to synchronous form by the terminal emulation board, which then transmits the information via coaxial cable to the mainframe via the cluster controller device. As far as the cluster controller is concerned, it is "talking" directly to an IBM dumb terminal.

It is far more cost-effective for a company that already has a local area network installed to install a LAN gateway to a mainframe. Using this approach, the company would install a network gateway adapter card in a microcomputer, which would run special gateway software. The combination of hardware and software not only converts data from asynchronous to synchronous, but it also contains all necessary protocols IBM's SNA computers require. A network PC acting as a communications server functions as an IBM 3274 or 3276 Terminal Control Unit capable of being connected via synchronous modem to an IBM 3705 ro 3725 or equivalent communications processor.

What makes this network gateway approach so cost-effective is that the one PC acting as a communications server is capable of providing up to 32 concurrent mainframe sessions for the networked PCs on a first-come-first-serve basis. Each workstation on this network runs special 3278/79 emulation software, which enables the workstations to emulate the typical IBM terminal display and use various combinations of keys to match the special keys found on an IBM terminal. This emulation software also permits these workstations to access printers exactly the same way that a real terminal would access a printer. Once mainframe information has been received by a LAN workstation, this data can either be sent to a printer as a real terminal would do or it can be stored on disk where all other LAN workstations can access it.

Remote LAN Gateways to an IBM Mainframe

A remotely connected gateway emulates a cluster controller such as the IBM 3274, which is capable of 32 sessions. This gateway microcomputer is connected to a synchronous modem that transmits the session information over a phone line. At the other end of the phone line a synchronous modem receives this information, translates it back into a digital form, and then forwards it to a front-end processor for transmission to the host mainframe computer.

Remote gateways cannot transmit information as quickly as local gateways, because typically they are limited to 56-kpbs transmission over phone lines, compared to speeds of more than 1 Mbps with a local coaxial cable connection. Because of the heavy traffic generated by tasks such as large file transfers from host to LAN, it is a good policy to have more than one remote gateway with special session seeking software that enables a workstation that wishes to initiate a micro-mainframe session to seek an available session from the closest gateway.

ADVANCED PROGRAM-TO-PROGRAM COMMUNICATIONS (APPC) AND LU 6.2

In 1985 IBM announced an addition to SNA called **Advanced Program-to-Program Communications (APPC)** for the IBM PC. While APPC had been available since 1982 for IBM's System/36 and System/38 minicomputers, its availability for microcomputers was a major breakthrough. It meant that PCs would eventually be able to communicate with minicomputers and mainframe computers on a peer-to-peer basis.

APPC incorporates two new SNA protocols: PU 2.1 and LU 6.2. PU 2.1 permits two processors to communicate on a peer-to-peer basis. While PU 2.1 does permit the connecting of a network node to a mainframe in the conventional hierarchical host-slave relationship, it also permits two intelligent PU 2.1 remote nodes to be connected together in a peer-to-peer relationship.

LU 6.2 permits two application programs to have a peer-to-peer conversation. Ideally, this situation would permit two programs running on different machines (a mainframe and a microcomputer, for example) to exchange information without users needing to even be aware of how to communicate directly with the mainframe computer. LU 6.2 also permits transactions to synchronize themselves across network boundaries. This means that changes are reflected in all related databases even if they are distributed across significant distances.

APPC offers a number of features that data-link protocols such as Token Ring, Ethernet, and Xerox Network Systems (XNS) do not have. It permits the encryption of data so that it cannot be intercepted and interpreted. It is able to start a program located at a remote site, and it is also able to transform user data into a form required by a remote computer. Finally, APPC can correct errors in application processing. This error-correction facility is important, since APPC is concerned with distributing processing between different machines in different environments.

Many non-IBM vendors have seized upon APPC and LU 6.2 as ways of entering the IBM world without becoming completely IBM-compatible. Companies such as Apple, DEC, AT&T, Prime, Honeywell, Unisys, and NCR all have products that support LU 6.2.

APPC is a standard that requires programming tools. IBM's Advanced Program-to-Program Communication for the IBM Personal Computer software contains the tools a programmer needs to include such APPC function calls as "send" and "receive" that computers need to communicate with each other. Once programs share this common language it becomes easy for microcomputers to talk with larger computers.

CONSIDERATIONS IN SELECTING A LAN-IBM-MAINFRAME GATEWAY

1. *How many sessions are available?*
   Not all gateways provide a network with 32 sessions. Some offer 5, 8, or 16 sessions.

2. *Is the LAN gateway dedicated or non-dedicated?*
   A dedicated gateway means that this workstation performs only the gateway function and is not used as an ordinary workstation. Dedicated gateways can handle more traffic, since they do not have to allocate some of their processing time to an application program.

3. *Is adequate security provided for the mainframe?*
   A gateway permits any LAN workstation to access the company mainframe. Since LAN gateways can also be remote, this means that virtually anyone with a PC and a modem can attempt to connect with the LAN remotely and then with the mainframe via the network gateway. Obviously a single password shared by a number of gateway users would not provide adequate security for the mainframe. A way of identifying each LAN workstation uniquely might be required for certain LAN gateway environments.

4. *Will the gateway be used primarily for file transfer or for "conversations" with the mainframe?*
   The answer to this question will go a long way toward helping you determine what kind of file-transfer software will be required. Some gateway software packages come with file-transfer software already.

5. *Is the gateway a full-function gateway?*
   A full-function LAN gateway should be capable of more than the one-way initiation of a session with a host that characterizes an IBM 3270 terminal. A full-function gateway should be able to hold a session initiated by the host computer.

6. *Is the gateway software completely compatible with the LAN operating system?*
   If the gateway software is supplied by a vendor other than the one who supplied the LAN operating system, there is more than a slight chance that the two products may not be 100% compatible.

7. *Are graphics a major consideration?*
   If graphics from a mainframe computer need to be downloaded to a microcomptuer via a LAN gateway, then use gateway software compatible with IBM's new DFT (distributed function terminal) devices that feature APA (all points addressable) or vector graphics. This approach reduces the amount of mainframe processing and downloading required. In addition, there is need to purchase a special programmable symbol set to correspond with that used by the mainframe.

8. *Is the gateway compatible with the API (application program interface)*

that is becoming an industry standard for micro-mainframe communications?

As more and more application programs become API-compatible, there will be a premium placed on this type of compatibility.

9. *Does the gateway support LU 6.2 and APPC?*

Eventually there will be host-independent, application-to-application communications that will exist regardless of the type of processors or even the language in which the programs are written.

DEC'S DIGITAL NETWORK ARCHITECTURE (DNA)

Digital Equipment Corp.'s **Digital Network Architecture (DNA)** as implemented in DECnet is based on the OSI seven-layer model. In September 1987, DEC announced Phase V of DNA. Under this scheme DEC will support both the OSI protocols and its own set of Digital-specific protocols. At the very lowest levels equivalent to the OSI Model's Physical, Data Link, and Network Layers, DEC supports Digital Data Communications Message Protocol (DDCMP), IEEE 802.3, and X.25 protocols.

At the higher levels there will be two distinct methods of communicating with DEC computers. Communication between two DEC computers will follow DEC's own proprietary protocols, since in effect the two computers will be running the same operating system.

Unlike the bit-oriented High-Level Data Link Control (HDLC) protocol, DDCMP is byte-oriented protocol, in which the number of bytes in a message must be conveyed along with the message to ensure that the information remains intact. DDCMP performs the task of framing its messages by placing unique 8-bit patterns in the bit stream.

A header containing a message number is included in each data stream by DDCMP. The station receiving this message checks the message number to make sure that it is correct. It forwards this message to the next higher layer and then sends an acknowledgment message back to the sending station indicating the number of the message received. The station that sent the original message checks the message number received against its own records as to the message number it sent. If the two numbers correspond, the station can release the copy of this message it was keeping for recovery purposes and then proceed with the next message. If there is a negative acknowledgment, the procedure can be repeated. Figure 5.9 illustrates the flow of data through the DNA model as compared to the OSI model.

Can DEC Networks Talk to IBM Networks?

We have pointed out that DDCMP is not compatible at all with HDLC or SDLC. Yet, many companies have a real need for their Digital Equipment Corp. computers and IBM computers to communicate with one another. DEC offers a DECnet/SNA Gateway, which makes the DEC equipment look like a PU 2 node (a cluster controller) to an IBM SNA network. The problem with such an approach is speed. The 56-

The OSI Model	DEC Network Architecture
Application	User Applications
	Network Management
Presentation	Network Applications
Session	DECnet Session Control
Transport	DECnet Transport
Network	DECnet routing
Data Link	Digital Data Communications Protocol
	High Level Data Link Control
Physical	Ethernet

Figure 5.9: Digital network architecture.

kbps speed can support approximately eight IBM 3270 terminal emulation sessions, but proves to be too slow to support large-file transfers.

DEC's approach is to take advantage of IBM's Advanced Program-to-Program Communications (APPC) interface, which, as we have mentioned, consists of Logical Unit (LU) 6.2 providing a program-to-program protocol and Physical Unit (PU) 2.1 that consists of a node-to-node protocol. Unfortunately, DEC's approach supports only serial links, which explains why 56 kbps is a practical limitation. For large companies with major investments in equipment, there is an expensive solution to the problem of fast file transfers as long as both the IBM and DEC equipment reside at the same location. Companies such as Interlink Computer Sciences and Systems Interconnect Operation offer 3711 and 3732 Gateways. A network controller resides on the DEC equipment, while a software portion resides on the IBM mainframe. Interlink's product can support 128 simultaneous sessions, with speeds of up to 800 kbps. A problem with this approach is that since these products do not support IBM's major terminal access method under SNA (i.e., Virtual Telecommunications Access Method, or VTAM), the DECnet equipment is treated by IBM equipment as if they are disk drive. This means that file transfers are possible, but sophisticated network management functions are not available.

While speed is obviously a prime requirements for bulk file transfers, there are other communications functions between the IBM and DEC worlds that place more of a premium on the ability of both sets of equipment to understand high-level protocols. There will be more and more applications in the future that will require electronic mail compatibility (X.400) as well as the ability to share information adhering to MAP and TOP protocols. TCP/IP is available for both worlds, and this protocol may serve as a kind of common protocol until the implementation of OSI protocols is completed.

Still another option for companies who need to connect the DEC and IBM networks worlds is to find a third standard, one that both worlds are willing and able to use. TCP/IP protocols are widely available in both worlds and offer one possible solution.

APPLE NETWORKS LINKED TO THE IBM AND DEC WORLDS

As Apple's Macintosh networks have become more and more sophisticated, users have clamored for a way of connecting these products with the IBM and DEC worlds. There are several companies that offer micro-mainframe connections that

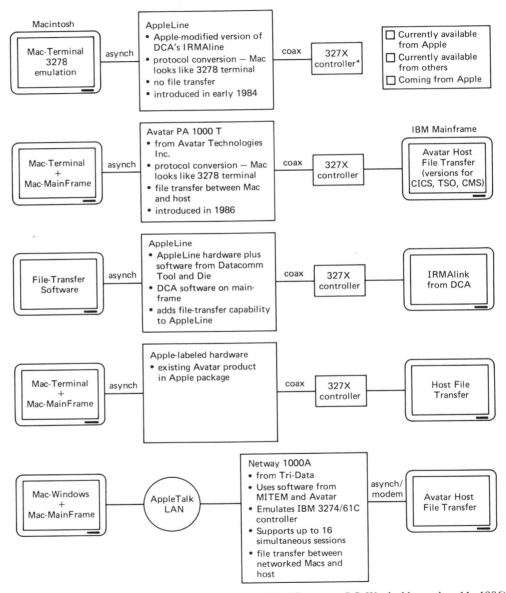

Figure 5.10: 3270 Connections between Apple and IBM worlds. (Courtesy *PC Week,* November 11, 1986)

166   Chapter 5   *Bridges, Gateways, and Micro-Mainframe Communications*

permit Apple Macintosh computers to communicate with IBM mainframe via 3270 terminal emulation. Figure 5.10 illustrates a number of different ways in which this terminal emulation can be achieved.

Connecting Macintosh computers and DEC's computers is far easier than linking with IBM's computers, because DEC uses asynchronous rather than synchronous communications. The result is that an Apple-DEC link can be software-only, while an IBM connection requires expensive hardware. A Macintosh running Apple's MacTerminal communications program need only emulate a DEC VT-100 terminal. A product such as White Pine Software's Mac240 even permits a Macintosh to emulate an advanced DEC VT-240 terminal. Figure 5.11 illustrates the differences between IBM and DEC terminal emulation for a Macintosh.

Many Macintosh users do not want their computers to function as dumb terminals. They are much more concerned with being able to link their Apple networks with mainframe networks so that file transfers can be transparent. They want to let computer hardware and software handle the details of file transfer while their screens still present them with the Macintosh user interface.

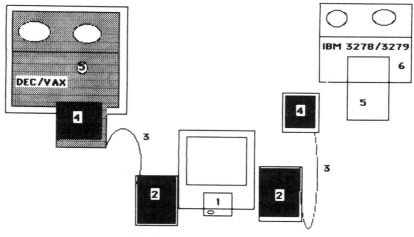

The Mac To DEC Link:
1. Terminal emulator software goes to MAC. 2. Modem. 3. Telephone line. 4. Modem at DEC site. 5. DEC minicomputer

The Mac to IBM Link:
1. Terminal emulator software goes to Mac. 2. Modem. 3. Telephone line. 4. Modem at IBM site. 5. Protocol convertor. 6. IBM mainframe. Note: If Macs are directly wired to host, no modems a used.

Figure 5.11: Connecting Macs to mainframes (Courtesy *MacWEEK,* July 20, 1987, p. 8)

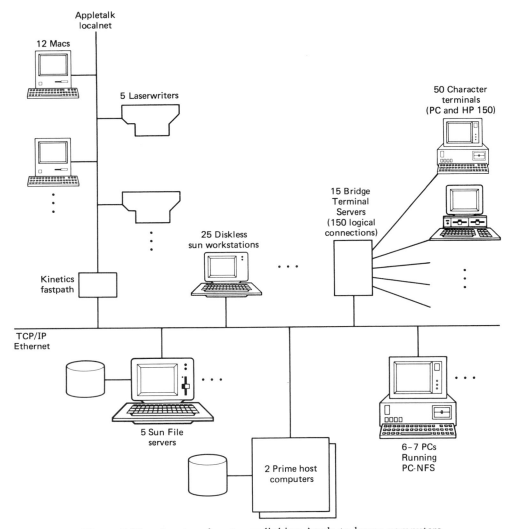

Figure 5.12: A network gateway linking Apple to larger computers.

An Ethernet-to-AppleTalk gateway from Kinetics provides a method of moving information from the Apple world to DEC computers. Figure 5.12 illustrates how a company uses this gateway along with TCP/IP protocol to link these two diverse worlds with mainframes from Prime.

CHAPTER SUMMARY

Bridges connect local area networks that utilize the same media but different protocols. Gateways are used to connect networks that utilize completely different protocols. The IBM mainframe world is completely alien to microcomputers, since transmission tends to be synchronous rather than asynchronous and data tends to be stored in EBCDIC rather than ASCII form. IBM's larger computer systems tend to use System Network Architecture (SNA), a proprietary architecture that is incompatible with that of Digital Equipment Corp. and other major vendors. IBM's additions to SNA of LU 6.2 and PU 2.1 as well as its concept of Advanced Program-to-Program Communications could result in the future of easier peer-to-peer communication among very different computers.

KEY TERMS

Advanced Program-to-Program Communications (APPC)
Bridge
Brouter
Cluster controller
Digital Network Architecture (DNA)
Domain
Emulate
Learning bridge
Logical unit (LU)
LU 6.2
Network addressable unit (NAU)
Network Control Program (NCP)
Physical unit (PU)
Physical unit control point (PUCP)
Remote bridge
Router
Session
Subarea
System Network Architecture (SNA)
System service control point (SSCP)

REVIEW QUESTIONS

1. Explain the movement of data from one network to another across a bridge.
2. Explain how a LAN can communicate with a remote IBM mainframe.
3. Discuss how the Apple Macintosh can communicate directly with an IBM mainframe computer.
4. Trace the journey that information follows under SNA.
5. Differentiate between a brouter and a router.
6. What does LU 6.2 add to SNA?
7. Compare and contrast DNA with SNA.

TOPICS FOR DISCUSSION

1. At one time PC owners wanted terminal emulation software so that they could access mainframe computers. Today many PC users want to download infor-

mation and even add information to a mainframe database. Discuss some of the major security issues raised by PC users' demands for greater access to mainframes from their personal computers and local area networks.

2. A router seems to be a very economical way of moving information among several different computer systems. What disadvantages must you accept when selecting a router?

3. Research the current battle for marketshare between Novell and 3Com. How does Novell address charges that its NetWare is not compatible with the OS/2 LAN Manager? Is it possible to have a bridge or gateway between the two networks?

4. IBM's AS/400 is quickly emerging as one of the leading minicomputers in a crowded field. Investigate what products are available for providing gateways to LANs, particularly IBM's Token-Ring Network.

The OSI Model and Data Communications Protocols

CHAPTER 6

OBJECTIVES

Upon completion of this chapter, you should be able to:

- describe the seven-layer OSI model
- explain how data is encapsulated with headers added as information flows from one DTE to another DTE
- explain why TCP/IP is currently so popular
- describe the XNS model
- explain what GOSIP is
- discuss the advantages and disadvantages of layered protocols
- compare and contrast bit-oriented protocols and character-oriented protocols
- discuss some of the latest trends in electronic mail and EDI

INTRODUCTION

In this chapter we will explore a number of data communications standards that are in the process of increasing the compatibility of different vendors' products. We will examine how these standards determine how information flows from one computer to another computer. We will see that these standards ultimately will make it possible to have a world-wide electronic mail service.

THE OSI MODEL

As the need for communication between heterogeneous computer systems has increased, various international organizations have worked to establish a set of standards so that dissimilar network equipment will be able to send and receive messages from each other. In 1983, after several years of committee development, the International Standards Organization (ISO) along with the CCITT (*Comité Consultatif Internationale Telegraphique et Telephonique,* or International Consultative Committee for Telephone and Telegraph) adapted the Reference Model for Open Systems Interconnection (OSI). This **OSI model** consists of a seven-layer communication architecture with each layer consisting of specific **protocols** or rules for communicating.

Figure 6.1 illustrates the OSI architectural model. Notice that levels 4, 5, 6, and 7 make it possible for two host computers to communicate directly (peer-to-peer). The bottom three levels communicate using CCITT Recommendation X.25, which specifies the network access protocols for a packet-switching network. Chapter 7 will examine packet-switched networks in depth. In this chapter we will take a close look at how these packets are assembled and transmitted from one computer to another through the various layers of this OSI model.

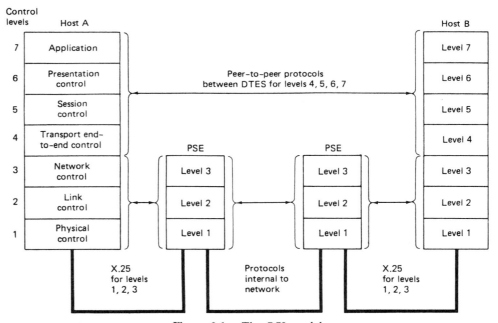

Figure 6.1: The OSI model.

172   Chapter 6   *The OSI Model and Data Communications Protocols*

Why is the OSI Model a Layered Architecture?

There are a number of advantages to using a layered architecture for the OSI model. Different computers are able to communicate at different levels, as Figure 6.1 illustrates. Also, as technology changes, it will be possible to modify one layer's protocols without having to modify all the other layers. Because each layer is essentially independent, different design groups have been assigned to each layer, which has expedited the entire development process. Because of the essential independence of each layer, many of the functions found in the lower layers have been removed entirely from software tasks and replaced with hardware. These advantages more than offset the one major disadvantage of this layered approach, that is, the tremendous amount of overhead including duplication of effort required in adding headers to the information being transmitted through the various layers. Each layer, with the exception of the first layer, adds a header containing key information for its corresponding layer on the machine receiving the data. These headers accumulate and can contain more than 100 bits of information.

Each layer (with the exception of the Physical Layer, which interfaces directly with the physical medium) receives services from the layer below it. Similarly, each layer (with the exception of the application layer, which interfaces directly with an application process) provides services to the layer above it.

If Computer A wishes to send Computer B a report, several different types of negotiations must take place. The corresponding Presentation Layers will need to negotiate the syntax, or form, that the conversation between the two computers will take. Unless the layers can agree upon the way that data will be represented, neither will be able to understand the other.

In a similar way, the two corresponding Session Layers will have to agree upon such issues as how the sessions will be conducted. Will both computers transmit at the same time (full-duplex), or will they take turns (half-duplex)? The corresponding Transport Layers will negotiate such items as the quality of service during the tranmission. Are line conditions bad enough to warrant a high level of error checking and correction procedures? Do both computers want an acknowledgment that their messages were received and understood?

The bottom three layers of the OSI model concern themselves with the actual mechanics of moving the data (at the bit level) from one machine to the other. The Network Layer is concerned with routing a packet of data from Computer A to Computer B. This trip may involve crossing through several networks along the way. The Network Layer must ensure that the packets carry accurate enough addressing information to take them from one network gateway to another until they reach their ultimate destination. Even if the packets must be divided up into a series of smaller segments, the Network Layer must ensure that these segments can be assembled in the appropriate order at the receiving end.

The Data Link Layer is concerned with bits at the frame level. It must ensure proper flow control so that information is not lost during transmission. Finally, the Physical Layer is concerned with the actual physical medium and lives in a world of voltages and cable.

Data Assumes Different Forms in the OSI Model

The OSI architectural model handles five types of data units, which are transmitted by its different layers. At the very top layer, users communicate directly by sending and receiving user messages and data. The details at this layer are still be worked out by various international standards committees. At a slightly lower level (layers 4 through 6) data is transmitted in the form of session messages. In layer 3 the messages are placed in packets with headers that contain information on the routing of this information. In layer 2 this packet is encapsulated in a *frame* that contains both a header and a trailer, both of which contain information needed to transmit this information over a physical link. Finally, layer 1 transmits the frames as a stream of bits. In reality, it is this lowest layer, where signals are actually transmitted, that communication between two computers actually takes place.

Figure 6.2 illustrates how each layer that services the layer above it adds a header containing key control information to the actual data. The corresponding layer at the receiving end of a transmission removes the header, which provides it with specific directions.

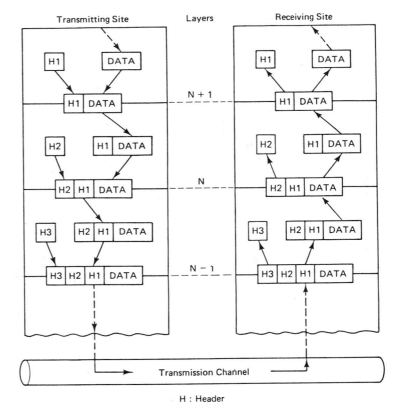

Figure 6.2: Each OSI layer provides control information.

THE PHYSICAL LAYER

At the **Physical Layer,** information is transmitted as *bits* across a physical medium connecting two or more computers. At this level we are concerned with generating and detecting voltages. The Physical Layer is responsible for activating, maintaining, and then finally de-activating the physical link between a computer and its transmission medium. At this very basic level the OSI model refers to two devices, data terminal equipment (DTE) and data circuit-terminating equipment (DCE). The data terminal equipment usually refers to terminals and computers, while data circuit-terminating equipment usually refers to communication devices such as modems.

Two computer networks connected by modems in Figure 6.3 illustrate how DTE-DCE pairs must work closely together. Bits in the form of voltages are sent from a DTE to its modem (DCE), which transmits this data in a form that can be interpreted by the receiving DCE, which, in turn, sends this information to its DTE. At this physical level, it is critical that devices have the same electrical characteristics, including the same coding schemes for data, the same voltage levels, and the same duration of signals. Their timing must be synchronized.

The OSI model spells out very specifically what kinds of pluggable connectors are acceptable (25-pin connectors for RS-232C, 34-pin connectors for CCITT V.35 wide-band modems, 15-pin connectors for public data network interfaces found in CCITT Recommendations X.20, X.21, and X.22, etc.). It also spells out what electrical characteristics are acceptable, including RS-232C, RS-449, RS-410, and CCITT V.35. The Physical Layer can handle half-duplex and full-duplex transmission, as well as point-to-point and multi-point connections. It is also able to handle both synchronous and asynchronous transmission. The Physical Layer is able to interface with analog signals through a modem using the RS-232 or RS-449 standard. It can also handle synchronous transmission using the CCITT X.21 recommended standard.

The Physical Layer provides a number of services to the Data Link Layer, the OSI layer directly above it. It transmits the stream of bits that it receives over the physical medium, notifies the Data Link Layer of any fault conditions in transmission, and ensures that the bits are transmitted in proper sequence. The Physical Layer also defines the type of data that it is handling, that is, asynchronous or synchronous. It establishes a number of quality-of-service parameters, including transmission rate, error rate, and service availability.

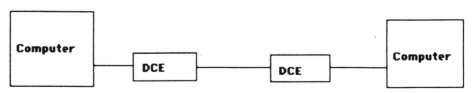

Figure 6.3: Network connections at the physical layer level.

Summary of Physical Layer Services and Functions

- Physically establishes a connection when requested to do so by the Data Link Layer
- Physically transmits data as bits
- Concerns itself with establishing synchronization of bit flow—duplex, half-duplex, point-to-point, multi-point, asynchronous, or synchronous transmission
- Defines quality-of-service parameters

THE DATA LINK LAYER

The Data Link Layer is responsible for making sense out of the flow of bits coming from the Physical Layer. It must establish where a transmission block starts and where it ends, as well as detect whether there are transmission errors. If it does detect errors, this layer is responsible for initiating action to recover from lost data, garbled data, and even duplicated data. Several data links may exist simultaneously and function independently between computer systems. The Data Link Layer is also responsible for ensuring that these transmissions are not overlapping and that data does not become garbled.

One of the first tasks of the Data Link Layer is to initialize a link with the corresponding Data Link Layer on the data terminal equipment with which it plans to communicate. Usually this task consists of sending a few signals that establish that both parties are ready to receive and to transmit information. Part of this process consists of determining how to decode the synchronized bit stream being sent by the Physical Layer. Obviously, the sender's encoding scheme and the receiver's decoding scheme must be in harmony for communication to take place.

Once the data link between two devices over a physical medium has been synchronized, the Data Link Layer can concern itself with dividing the stream of bits it receives into frames with defined fields containing specific types of information. The segmenting of this bit stream into definable blocks makes it easier for the Data Link Layer to detect errors, since it knows what type of bit pattern should exist in a particular field within a frame.

In addition to transforming the raw bit stream it receives into frames that can be interpreted by the Network Layer directly above it in the OSI model, the Data Link Layer assumes responsibility both for the smooth flow of these frames through its layer and for the management function of maintaining statistics as it monitors this information flow. Upon the completion of a user's transfer of information, the Data Link Layer assumes responsibility for determining that all data was received correctly before terminating the link. This termination of a logical connection does not necessarily mean that the physical link established by the corresponding Physical layers is broken.

Summary of Data Link Layer Services and Functions

- Segments bit stream into frame
- Coordinates data flow split over multiple physical connections
- Responsible for error detection and correction
- Monitors flow of data frames and compiles statistics
- Ensures that data frames are in sequence

DATA LINK LAYER PROTOCOLS

There are a number of protocols that co-exist within the Data Link layer. Among them are the character-oriented protocol ANSI X3.28 (Bisync), and the bit-oriented protocols Advanced Data Communication Control Procedures (ADCCP), High-Level Data Link Control (HDLC), Lap B of CCITT Recommendation X.25, and CCITT Recommendation X.75.

Character-Oriented Protocols

Binary Synchronous Protocol (BSC), usually referred to as "Bisync," was developed by IBM as a synchronous, half-duplex protocol; this means that it transmits in one direction at a time. One advantage of half-duplex transmission is that since a single block of information is transmitted in one direction and is followed by a positive acknowledgement transmitted in the opposite direction prior to the next information block transmission, only two buffers are needed. One buffer is required at the central site for transmission of information, and a second buffer is required at the remote site where the information is received. These buffers store information until it is possible to send or receive it. BSC is a **character-oriented protocol** because ASCII or EBCDIC character sets are used for data link control functions. Figure 6.4 illustrates the format of a BSC frame.

The sending station transmits two packet-assembler/disassembler (PAD) characters to establish bit synchronization with the receiving stations. It then transmits a sequence of two SYN characters to establish character synchronization with the receiving stations. Since we are dealing with synchronous transmission, the sending station transmits a start of text code (STX) to indicate that it is beginning to transmits frames. After transmitting the text portion of this frame, which consists of whatever message needs to be sent, the sending station transmits an ETX character which indicates the end of transmission of this frame. Finally, the sending station transmits a block check character sequence (BCC) that consists of two characters—this is done to permit the receiving station to verity that it received the frame correctly. Two PAD characters complete this BSC frame transmission. Binary Synchronous Protocol provides for error checking only and not for error correction. If errors are located, the entire frame is retransmitted. Figure 6.5 illustrates a typical sequence when two stations communicate using BSC.

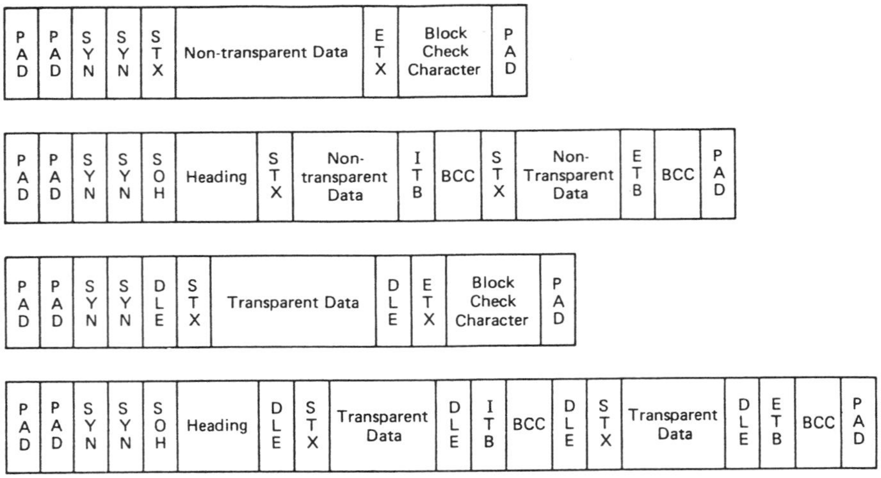

| P A D | P A D | S Y N | S Y N | S T X | Non-transparent Data | E T X | Block Check Character | P A D |

| P A D | P A D | S Y N | S Y N | S O H | Heading | S T X | Non-transparent Data | I T B | BCC | S T X | Non-Transparent Data | E T B | BCC | P A D |

| P A D | P A D | S Y N | S Y N | D L E | S T X | Transparent Data | D L E | E T X | Block Check Character | P A D |

| P A D | P A D | S Y N | S Y N | S O H | Heading | D L E | S T X | Transparent Data | D L E | I T B | BCC | D L E | S T X | Transparent Data | D L E | E T B | BCC | P A D |

BSC Control Character:

Character	Function
SYN	Synchronous idle (keeps channel active)
PAD	Frame pad (time fill between transmissions)
DLE	Data link escape (used to achieve code transparency)
ENQ	Enquiry (used with polls/selects and bids)
SOH	Start of heading
STX	Start of text (puts line in text mode)
ITB	End of intermediate block
ETB	End of transmission block
ETX	End of text
EOT	End of transmission (puts line in control mode)
BCC	Block check count

Figure 6.4: The BSC frame format.

Two stations using BSC that are linked by a point-to-point connection must initiate communication, since the line's default condition is an idle state. Station 1 transmits an enquiry code (EN), which essentially asks if Station 2 is ready to receive a message. Station 2 replies with an even acknowledge (ACK0) signal indicating it is ready to receive.

Station 2 receives the message in the frame form we have described earlier in this section. Assuming that its error checking routine finds that the message has been received correctly, Station 2 then sends a positive acknowledgment, in this case an odd acknowledge (ACKI) signal, since it used an even acknowledge signal earlier.

Station 1 now has the option of sending a second message. This process continues until Station 1 sends an end of transmission code (EOT), which causes the line to return to an idle condition, where it remains until an enquiry signal is sent once again.

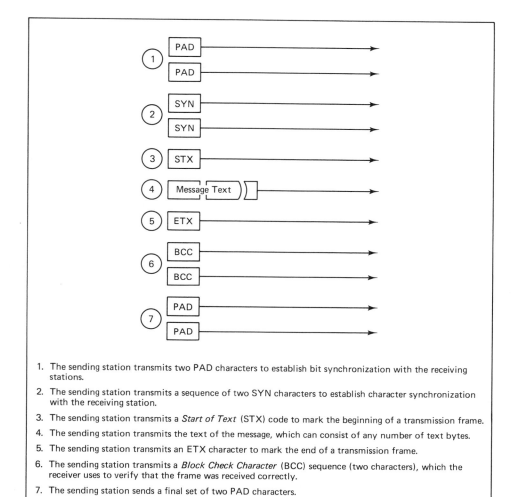

Figure 6.5: Two stations communicate using BSC protocol.

Bit-Oriented Protocols

Bit-oriented protocols do not use ASCII, EBCDIC, or any other information coded characters for data link control, since they perceive communication as a continuous bit stream. Unlike BSC, which operates only in half-duplex mode, bit-oriented protocols can operate in full-duplex mode as well, which permits simultaneous communication in both directions. Bit-oriented protocols have sophisticated error checking procedures unlike character oriented protocols that are limited to simple format checking. Another major advantage of bit-oriented protocols is that they were developed to prevent **aliasing,** a condition in which a bad message "looks" like a good message. Still a third major advantage of bit-oriented protocols is that the set of control messages can be expanded easily. Under BSC, the character-code set determines which control characters are available. To add new control messages in the future, it would be necessary to string together longer and longer sequences of control characters. With bit-oriented protocols, the mere position of a bit in a field makes it a control code; therefore, it is possible for a bit-oriented protocol with an 8-b control field to have 2^8 different control codes. We will look at IBM's Synchronous Data Link Control (SDLC) and at a superset of this protocol adopted by the ISO called High-Level Data Link Control (HLDC).

Synchronous Data Link Control (SDLC)

Under IBM's **Synchronous Data Link Control (SDLC),** stations have a primary-secondary relationship with each other. A station with primary status is responsible for data communications, while the secondary stations simply respond to its commands. It is also possible for stations to have equal status with each other and to function in **Asynchronous Balanced Response Mode,** in which both stations have primary and secondary functions. Finally, it is possible for the primary station to function in full-duplex mode, while secondary stations function in half-duplex mode. This means that the primary station can transmit to one secondary station while simultaneously receiving from a different secondary station.

Figure 6.6 illustrates the various fields found in an SDLC frame. Notice that the first and last fields contain the identical flag, a sequence of six straight 1-bits. Because it is critical that information sent within the frame not be mistaken for these flags, SDLC uses a technique known as **bit stuffing,** or zero insertion, to ensure that only the flags will have this pattern. The communications controller transmitting these frames automatically inserts a 0-bit after the fifth 1-bit. The receiving communications controller will interpret six straight 1-bits as a flag. It also will assume that five 1-bits followed by a 0-bit probably was the result of bit stuffing and it will remove the 0-bit to reconstruct the correct message.

The Address field is an 8-bit pattern used to identify the secondary station. When the primary station sends a message to the secondary station, it does not identify itself with its address, since the secondary station already knows where its messages are coming from. Similarly, if a secondary station sends a message to the primary station, the address field still contains the secondary station's address.

The Control field is 1 byte in length, and it is used to identify what type of

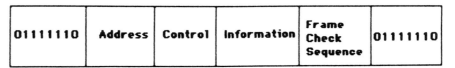

Figure 6.6: An SDLC frame.

information the frame will be carrying. It is possible to have information frames and command/response frames. Information frames contain a 0-bit in the first position of this field and indicate that the frame will carry a message. Command/response frames contain a 1-bit in the first position of this field. To prevent one station from dominating communications in half-duplex mode, a maximum of seven information frames can be sent unacknowledged before a positive acknowledgment is required. Primary stations generate commands in this field, while secondary stations generate responses. This type of frame is used to help control the flow of data along a data link, and the frames can be either Supervisory frames or Unnumbered frames. Supervisory frames are used for such functions as acknowledging information frames and requesting retransmission of information frames in which errors were detected. Unnumbered frames have 5 bits in the Control field that can be used to provide up to 32 command functions and 32 response functions. A typical function of the unnumbered frames is to set the operating mode of a secondary station.

Under SDLC, secondary stations are in normal response mode, initialization mode, or disconnect mode. Secondary stations generally can be set to initialization mode, normal disconnected mode, or normal response mode. Initialization mode is simply the state a secondary station assumes to perform a hardware-specific function such as receiving downloaded program code from a primary station. Normal disconnected mode is a state in which a secondary station is logically and/or physically disconnected from the data link. This type of situation might occur after a power failure. Finally, **normal response mode** is just what it implies, a situation in which a secondary station assumes the position of responding to a signal from a primary station.

It is possible for a secondary station to respond to a primary station with a request for on-line status (ROL) or even a request for initialization (RQI). Under most conditions the secondary stations await their turns to be polled by the primary station. In effect, the primary station asks each of these stations if they have any information to transmit.

When a primary station transmits to a secondary station, it expects a response within a certain period of time. There are various reasons why the secondary station might not respond, including that it might be in an idle condition or that the message received may have been garbled. After the response time period has expired ("timed out"), a station may retry a transmission to resume communication. Stations are authorized to automatically retry a specified number of times. If communication is still not possible, help comes from protocols at a higher level than the Data Link Level.

High-Level Data Link Control (HLDC)

In 1975, the ISO adopted **High-Level Data Link Control (HLDC)** procedures as a standard. HLDC can be used on point-to-point and on multidrop lines, and like SDLC can function in full-duplex or half-duplex mode. As a superset of SDLC, HDLC offers a number of features not found under IBM's protocol. An extended Control field of 16 bits under HDLC provides additional supervisory and unnumbered commands. A major advantage for high-speed communication links and satellite transmission in HDLC's ability to transmit up to 127 information frames before positive acknowledgment, in contrast SDLC's seven frames. An extension to the Address field permits multi-byte addresses under HDLC without the 8-bits field restriction found under SDLC. While information fields can be of unlimited lengths under both protocols, under SDLC the Information field must be a multiple of 8 bits. In addition to normal response, normal disconnect, and initialization modes found under SDLC, HDLC offers asynchronous response mode (ARM). Under asynchronous response mode, secondary stations can send unsolicited response frames and do not have to wait to be polled. Finally, HDLC uses a slightly different method for error checking, although the differences are not really relevant to our discussion.

THE NETWORK LAYER

The **Network Layer** receives a packet, a frame stripped of a Data Link Layer header and trailer. As a switching station of sorts, responsible for routing and relaying of information, this OSI layer must decide which line to use to forward this packet. It may multiplex several network connections to a single data link connection for added efficiency. It also assumes responsibility for maintaining a designated quality of service, regardless of the different networks it routes packets through. The CCITT has devised a three-layer protocol of its own, known as X.25, specifically designed for packet-switched networks. Public long-haul networks is a subject of sufficient importance that we will consider it in a chapter of its own later in this book. At that time we'll look at the actual format of the fields found within an X.25 packet. In this chapter, though, we will examine the Network Layer as the ISO envisioned it within the framework of the OSI seven-layer model.

The Network Layer performs a number of services for the Transport Layer immediately above it in the OSI model. It notifies the Transport Layer when unrecoverable errors have occurred. It also maintains flow control and helps avoid congestion by stopping the transfer of packets whenever requested to do so by the Transport Layer.

Of primary importance to the Transport Layer and all layers above it, though, is the Network Layer's ability to create virtual circuits that are transparent. In effect, the Network Layer handles routing of information through several physical connections that may change from time to time. The upper layers of the OSI model do not have to know through which physical connection the packets are being routed. Similarly, it is possible for these packets to arrive out of sequence, since they may have been routed along different routes. The Network Layer takes responsibility for ensuring that these packets are reassembled in proper sequence for the Transport Layer. Upon request of the Transport Layer, the Network Layer will even monitor the receipt of information by the corresponding Transport Layer and when it receives an acknowledgment from this layer that the packets were received in good order, it will notify its own Transport Layer that everything has been received correctly.

The Network Layer must deal with several problems associated with internetwork connectivity. It even has a special protocol (Internetwork Protocol) designed to help it in this area. It must be able to provide addressing and routing for information that may have to go through several different networks to arrive at its ultimate destination. Since different networks may have different-size packets that are permissable, the Network Layer must be able to ensure that segmented information can be reassembled in the correct order. Since information may take different paths over public and private data networks, segments may not arrive in sequential order. The Network Layer must ensure that the information can be reassembled correctly. Finally, it is possible that a packet's time in its network travels may expire ("time out") and an error message may cause a duplicate packet to be transmitted just as the errant packet arrives. The Network Layer must have means to recognize and then discard duplicate packets.

Summary of Network Layer Services and Functions

- Establishes virtual circuits and provides higher OSI layers with interface to it
- Routes packets
- Establishes connections between Transport Layers on two computers that will communicate with each other
- Control flow: stops the flow of packets upon request from the Transport Layer
- Maintains correct sequence of packets
- Provides Transport Layer with acknowledgment that packet has been received correctly
- Maintains quality of service
- Responsible for detecting and correcting errors
- Responsible for multiplexing several network connections to a single data-link connection for maximum productivity
- Handles internetworking, the movement of data from one network to another network

THE TRANSPORT LAYER

The **Transport Layer** is concerned with the transporting of data between two Transport users. In effect, the bottom three OSI layers support the Transport Layer in performing this function. The key to differentiating the role of this layer from the bottom three layers is to realize that it is not concerned with the specific details of how data is routed through a network, only with whether the actual transfer process took place smoothly and efficiently. Within this layer, data is encapsulated into a packet known as a **transport protocol data unit (TPDU)**, which contains control data including the destination address as well as the user information.

The Transport Layer is concerned with monitoring the quality of service found in a transport connection. The Session Layer residing just above the Transport Layer in the OSI model provides instructions on the quality of service desired. The ISO and CCITT have defined five classes of transport protocol:

```
Class 0    Simple
Class 1    Basic error recovery
Class 2    Multiplexing
Class 3    Error recovery and multiplexing
Class 4    Error detection and error recovery
```

A network manager would select the appropriate transport protocol class to use in conjunction with the type of network to be serviced. The ISO defines types A, B, and C networks. A type A network has an acceptable residual error rate and an acceptable rate of signaled failures, while a type B network has an acceptable residual error rate but an unacceptable rate of signaled failures. Finally, a type C network has a residual error rate that is not acceptable to a Transport Layer user.

The Class 0 transport protocol is compatible with CCITT T.70, the telex transport protocol. This most simple of transport protocols simply *assumes* a reliable, error-free environment and supplies only the information needed for establishing a connection. The Network Layer assumes responsibility for flow control and error checking when this transport protocol is used.

Class 1 protocol was developed primarily to run on X.25 networks. Its basic error recovery consists of numbering its packets so that lost packets may be retransmitted. It is not unusual for packets to be lost after an X.25 reset. After a network's virtual circuit is reset, the transport entity receiving the packets notifies the transport entity sending packets what the last numbered packet received was. Any packets sent with numbers greater than this number are assumed to be lost and are retransmitted.

Class 2 includes flow control, since it has the capability of multiplexing multiple transport connections. Class 3 includes the capabilities of classes 1 and 2, including flow control, multiplexing, and the ability to retransmit lost packets after a network reset. Finally, class 4 provides all the capabilities of the other classes plus the ability to detect errors. Its basic assumption is that the network is unreliable, and so the emphasis is on error detection and recovery.

Classes 0 and 2 are used with type A networks, while classes 1 and 3 are used with type B networks. Finally, class 4 is used with type C networks.

Summary of Transport Layer Services and Functions

- Ensures that TDPUs arrive in proper order
- Specifies grade of service, including acceptable error rates
- Monitors status of connection and TDPUs
- Establishes connection with another Transport Layer entity
- Detects and recovers errors

THE SESSION LAYER

The purpose of the **Session Layer** is to help users organize and synchronize their "conversation" with each other. These conversations can be organized as one-way, two-way alternate, or the most common, two-way simultaneous (full-duplex). For half-duplex service, the Session Layer provides a number of different tokens. A data token is used to manage the half-duplex connection, while a synchronize-minor token is used to set minor synchronization points. A major/activity token is used to establish major synchronization points, while a release token is used to release a connection. These tokens provide a user with the right to transmit, a right only one user can have at a time under half-duplex mode. The Session services synchronize conversations by establishing certain fixed points. These points must be acknowledged by the other user before data is permitted to be transferred.

Session services use their Transport Layer connections to support flow control. In the event of a failure, the session services re-establish transport connections. It is important to note at this point that consecutive session connections can be established and then later terminated over one continuous transport connection. A single continuous session connection can also be maintained over several different transport connections that are established and then terminated sequentially. Before a connection can be released, however, the Session entity has the final word. Only it can give the order that releases the transport connection. This is the way the OSI model ensures that no data is lost. This ability of the Session Layer to synchronize services is essential within the wider OSI model framework that includes the CCITT X.400 message-handling facilities protocols. We will return to the issue of electronic mail and the prospect of universal standards later in this chapter.

How does the Session protocol work? Perhaps an example will clarify how this OSI layer functions. A user (user 1) makes a request for a connection. The user's request results in the creation of a **session protocol data unit (SPDU)**. This SPDU contains key information, including connection identification, serial numbers, token selections, and parameters required for connection and quality of service.

This information is transmitted via an SPDU to the session protocol entity for user 2, who replies with a response message. If the parameters are acceptable, user 1's session entity replies with an Accept SPDU, which results (hopefully) in user 2 replying with a Confirm message.

The Accept SPDU is not automatic. A session entity may refuse a connection because of congestion or the unavailability of the Application requested. If such is the case, then the session protocol will reply with a Refuse SPDU, explaining the reason for the refusal, which will result in a Confirm message from the other Session entity.

Data is transmitted encapsulated within a session service data unit (SSDU), which consists of the data and a Session header. If the data units exceed the size that the Transport Layer can handle, then the Session Layer will segment these SDDUs and enclose a number within the SPDU specifying when the last SPDU is received.

Session Protocol Formats

An SPDU consists of four fields. The first field is the SPDU identifier which can specify one of 34 different SPDUs. An SPDU can be identified as a Connect, Accept, Refuse, Finish, Disconnect, Not Finished, Abort, Minor Sync Point, Major Sync Point, and so on. The second field is the length indicator, which indicates the length of the header. The third field contains the parameters, if there are any, and the fourth field contains user information. The third and fourth fields can be broken down further into a number of subfields, but this information is well beyond the scope of this book.

Summary of Session Layer Functions and Services

- Establishes and maintains transport connection with a certain designated quality of service
- Organizes and synchronizes dialogs between Presentation entities
- Ensures the reliable transfer of data
- Uses flow control of Transport Layer to prevent Presentation entity from becoming overloaded
- Re-establishes transport connection if there is a transport connection failure
- Expedites data transfer for high-priority items

THE PRESENTATION LAYER

The **Presentation Layer** is responsible for representing information to communicating Application Layer entities in such a way as to resolve any syntax differences that might exist. The classic analogy often used to explain this function is the difficulty that two people would have if each spoke a language that was unintelligible to the other. The Presentation Layer ensures that the two Application entities use a common language, or *representation*.

The two Presentation Layer entities negotiating the syntax that will be used for Application Layer entities to communicate have a number of transfer syntax options that can be used for encoding information. The two entities must agree on a common syntax for the exchange of information. Once the syntax is negotiated, the Presentation Layer entities can exchange protocol data units (PDUs) consisting of headers, data, and control information. These PDUs are very similar in structure to those found in the Session Layer PDUs. The exact contents of the fields composing these protocol data units is beyond the scope of this discussion.

While the Presentation Layer is concerned primarily with syntax, it also must deal with the Session Layer. It requests the establishment of a session, negotiates the syntax to be used in this particular session, and actually transforms and formats data to fit within this specified syntax. Finally, the Presentation Layer is responsible for requesting the termination of a session.

Summary of Presentation Layer Functions and Services

- Selects the syntax to be used
- Negotiates syntax with the corresponding Presentation Layer entity
- Requests the establishment of a session
- Transforms and formats data to match specified syntax
- Encapsulates data in protocol data units

THE APPLICATION LAYER

The **Application Layer** does not provide services to any higher layer, but to application processes that reside within specific applications. These applications can vary widely to include actual programs, terminals, and electronic mail. The Application Layer protocol must establish how communicating terminals will handle common control functions. Among the other tasks of this top layer in the OSI model is providing the user directly with file service, directory service, print service, mail service; virtual terminal emulation; and other types of emulation, including NETBIOS emulation.

FTAM, X.500, and X.400 Protocols

The File Transfer Access and Management (FTAM) protocol has developed slowly. At present it is possible to exchange files between systems. It is even possible to exchange these files even if they have different data formats; EBCDIC and ASCII files can be exchanged and translated into the appropriate form. Sometime in the future, however, it will be possible to access individual records within a file and to provide remote file service.

The CCITT introduced an X.500 proposal in 1985 that, if and when it finally becomes accepted as a standard, will provide a universal directory that will cover not merely electronic mail but also telephone, telex, and other data. Theoretically, users from different networks will be able to send information to each other, since the networks will be able to use the directory to route data to its appropriate address.

To expedite the development of such a difficult standard, the committee envisioned X.500 not as a single database but as a standard to link together several autonomous databases already residing within existing electronic mail networks.

The CCITT has defined two key entities in its X.500 standard. The directory user agent (DUA) is what a customer will use to find an address and the directory server agent (DSA) will supply the correct address from the appropriate network. At this point there is still some unfinished pieces to this universal directory puzzle, including precisely how these different directory server agents representing different network directories of addresses can be linked together.

There are other problems still unresolved with this universal directory. The CCITT has expressed belief in the principle that those users who do not want to be listed in an electronic mail directory should not have to be listed. Somehow X.500 will have to be capable of ignoring requests for users who have "unlisted addresses."

The CCITT has established a number of standards for electronic mail consistent with the OSI model. Under the X.400 standard its message handling system has two major elements, a user agent (UA) and a message transfer agent (MTA). The UA element is used to help a user submit and retrieve messages. MTAs hold the messages and convert them to a form the user can understand. This task is made even more difficult by the fact that so many major computer operating systems have incompatible electronic mail formats.

The message handling system envisioned by the CCITT offers a number of electronic mail services, including non-delivery notification, submission and delivery time stamps, grade of delivery service, deferred delivery, and even delivery notification. The X.400 standard is closely linked to the Session Layer of the OSI model as well as the Application Layer, since it is the Session Layer that provides reliable transfer service. The Session Layer uses half-duplex dialog for electronic mail, while the Transport Layer is involved because it is required to provide class 0 protocol service with a class 1 service option. Finally, the Network Layer provides the addressing service so that neither rain nor sleet nor incompatible network addressing schemes can keep the electronic mail from getting through to its destination.

The CCITT X.400 standard is not simply theory, since there are a number of vendors actually offering this service today. Recently Telenet announced it will debut an X.400-compliant version of its Telemail service. The new Telemail service provides interfaces to international public mail services and private mail systems using the X.400 standard. Telenet signed agreements with state-run telephone companies in the United Kingdom, Italy, Chile, Australia, Belgium, and Sweden as well as with Canada, Japan, and Taiwan. It also has signed agreements with such corporate giants as Hewlett-Packard, Honeywell-Bull, and Wang Laboratories. Similary, IBM recently released its Open Systems Message Exchange program offerings, which provide X.400 service permitting IBM computer system users to send and receive electronic mail from other X.400-compatible electronic mail systems.

Electronic Data Interchange (EDI)

While this book was being written, the CCITT has been working on linking electronic data interchange (EDI) with its X.400 electronic mail standard. Electronic data interchange is a set of standard formats that allows organizations to transmit common business documents electronically from computer to computer, thereby eliminating many labor-intensive paper-based transactions. Ultimately, this marriage between two related functions will permit companies to exchange business documents with other companies as well as distribute copies of the documents to a number of different departments and divisions within their own organizations.

Since EDI is estimated to become a $2 billion industry by 1992 and since it is already used by such corporate giants as General Motors, Hewlett-Packard, Texas Instruments, and General Foods, it makes a good deal of sense for it to be consistent and compatible with electronic mail standards.

The Virtual Terminal Protocol

The **Virtual Terminal (VT) protocol** resolves differences between terminals by defining a generic type of terminal in terms of the number of characters per line, lines per screen, methods of underlining and boldface, and so on. This protocol enables vendors to write programs that run on a certain computer but can then be accessed by a number of different types of terminals. One problem that must be resolved in the future is that Virtual Terminal protocol is a bare-bones basic method of communication. It is not able yet to handle all the variations of graphics and windows that different programs often use. In the future, this protocol will have to be revised periodically to reflect the increasingly complex user interfaces found in most microcomputer application programs.

Summary of Application Layer Functions and Services

- Establishes authority to communicate
- Identifies intended communication partners
- Agrees upon level of privacy
- Determines whether resources are adequate for communication
- Decides upon an acceptable quality of service
- Agrees upon responsibility for error recovery
- Agrees upon responsibility for data integrity

FASCIMILE MACHINES ILLUSTRATE THE OSI MODEL IN ACTION

One of the major growth areas in telecommunications the past few years has been the facsimile machine market. The CCITT issued the first facsimile machine standards two decades ago, and it has been forced to add additional standards to keep up with the changing technology. The first facsimile machines were analog models that took 6 minutes to send one page of text. Today's digital Group 3 models can transmit an entire page in under 1 minute. The Group 4 standard adopted in 1984 covers digital machines that are capable of transmitting an entire page in under 10 seconds.

The CCITT Study Group VIII has drafted Recommendation T.a, entitled "Apparatus for Use in the Group 4 Facsimile Service." This plan for facsimile machines follows the seven layer OSI model. At the Application Layer of the OSI model the facsimile machine uses special software to scan a document that is to be faxed somewhere. At the Presentation Layer the facsimile machine is concerned with resolution conversion and the compressing or decompressing of text. Facsimile machines compress text for transmission to increase speed and also to reduce the memory required to store the information.

The Session Layer is concerned primarily with the structure of the facsimile machine dialog sessions. Will communication be simplex, half-duplex, or full-duplex? The Transport Layer concerns itself with providing end-to-end transport service. The Network Layer formats data into packets and then establishes a virtual circuit using the public telephone system. The Data Link Layer provides the appropriate data link protocol as well as error control procedures. Finally, the Physical Layer contains the X.21 bis protocol for the new high-speed V series of modems. It also performs the appropriate signaling and modem functions. In the near future when facsimile machines and computer systems are all OSI-compatible, it will be a very elementary process to send a document around the world, modify it in any one of dozens of different computers, and then transmit the revised version back.

PROTOCOLS NOT INCLUDED IN THE OSI MODEL

TCP/IP = Origins and Operation

The Defense Advanced Research Project Agency (DARPA) decided in the early 1970s that it needed a set of protocols that would expand the communications ability of its ARPANET packet-switched network. **Transmission Control Protocol/Internet Protocol (TCP/IP)** was developed by Vinton G. Cerf and Robert E. Kahn under a grant from the Department of Defense expressly to increase interconnectivity among heterogeneous computer systems. The Department of Defense needed a common protocol that would permit the thousands of different computers running operating systems as diverse as Unix, VMS, VM, IX/370, to be connected together on its network. Under TCP/IP, IBM PCS running PC-DOS, DEC VAX minicomputers running VMS, and even the Cray supercomputers can communicate with each other. Because TCP/IP has a built-in multiplexing function, it is possible for this protocol to co-exist on a network with a host computer's own operating system communication protocol. TCP/IP's popularity can be explained in part by its acceptance by government agencies and also by its incorporation by the University of California at Berkeley in the Berkeley Unix version 4.2 operating system. Because TCP/IP is part of the Unix 4.2 kernel, it has an interface to the world of local area networks because the Unix kernel also includes an Address Resolution Protocol (ARP) that maps TCP/IP addresses to Ethernet IEEE 802.3 addresses. Transmission Control Protocol (TCP) corresponds to OSI's Transport Layer (layer 4), while Internet Protocol (IP) corresponds to OSI's Network Layer (layer 3).

Perhaps the key to understanding how TCP/IP works is to observe its protocol hierarchies, seen in Figure 6.7. Notice that the data reaches the IP in the form of blocks of data, known as datagrams. The IP places its own header on these datagrams for routing to a local network or even a global network. Before taking a close look at the TCP and IP structures, we will briefly examine the upper-level application protocols linked to them. These protocols form a "utility layer," since they each have specific, practical utilitarian tasks to perform.

Telenet is an application level protocol that permits a terminal tied to one computer network to link to another computer network and appear to it as a local terminal. This protocol supports EBCDIC mode for IBM 3270 terminal emulation as well as ASCII TTY mode. This upper level protocol makes it possible for pro-

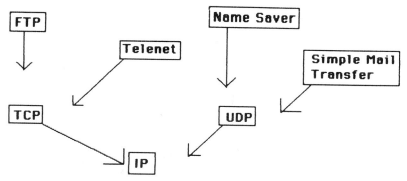

Figure 6.7: TCP/IP protocol hierarchies.

grammers to ignore the special features of various terminals, and it means that they do not have to write special terminal "drivers" for every terminal on the market.

Simple Mail Transfer Protocol (SMTP) provides electronic mail that can travel in both directions. Messages can be sent in batch mode to individuals and to distribution lists. File Transfer Protocol (FTP) permits the transfer of files between networks. The data can take the form of ASCII, EBCDIC, or even a binary format. This particular protocol is much more primitive than the FTAM protocol found in the Application Layer of the OSI model. Finally, the Name Server Protocol (NSP) provides the translation of host and server names into internet addresses. This procedure eliminates the need to keep long lists of tables for names and addresses.

Higher-level application protocols establish TCP sessions for transmitting data. The TCP program creates a datagram, a finite-length packet containing both a header section and a section for data, which contains the source port (the station calling the TCP program) and destination port (the station to be called) numbers in its header. The datagram is encapsulated at the IP layer to include the IP header. At this point, the datagram is ready for routing to another station in the same network or a station on a completely different network. We will see that the IP header includes the information a gateway needs to route the datagram through several different networks until it reaches its ultimate destination.

TCP is a protocol that uses its Windows and Checksum fields to make sure that data arrives correctly. User Datagram Protocol (UDP), on the other hand, is designed when such careful monitoring is not needed. A user may simply wish to send a message ("I am here") to another user on a different network. The UDP is more than adequate for this type of message.

The Structure of the TCP Header

The TCP layer receives data and encapsulates it in a datagram with its own 32-bit header. Figure 6.8 illustrates the TCP datagram structure. Briefly, the numbers found in the source and destination ports fields represent the station calling the TCP and the station being called. The Sequence Number keeps track of the order of the

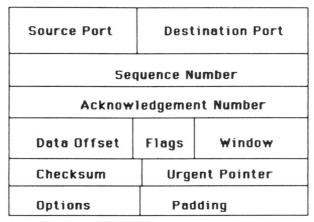

Figure 6.8: A TCP datagram structure.

datagrams to be received so that data is received in the correct order. The Acknowledgment Number field indicates the next byte sequence that is expected.

The Data Offset field tells us the number of 32-bit words in the TCP header; this is so that we do not confuse the information in the header with the data field. The Flag section gets more technical than we want to at this point, but among the functions represented here are an indication of how urgent this datagram is as well as notice of a session's setting up or closing down.

The Window field indicates the number of octets of information the sender is willling to accept or receive. The Checksum field is a 16-bit number that indicates whether the datagram is received intact. The Urgent Pointer contains the number of octets of from the beginning of the TCP segment to the first octet following any urgent data. The Options field can contain a variety of information, including the maximum TCP size. Finally, the TCP header includes some zeroes to pad the datagram to the next multiple of 32 bits.

The Structure of the IP Header

The IP header is illustrated in Figure 6.9. The Version field indicates which version of IP is being used and, therefore, the Internet header's format. The Internet header length field is simply its length in 32-b words. The Type of Service field indicates the quality of service desired. The parameters for this field provide information on the datagram's precedence (importance), its intervals (will there be a steady stream of these datagrams at regular intervals in the future?), the level of reliability desired (are mistakes critical?), the importance of speed of delivery of this datagram, and an indication of the relative importance of speed versus reliability in the event of a conflict between the two.

The Quality of Service field is followed by the Total Length field, which is similar to the same field found in the TCP header. It represents the total length of the datagram including its IP header. The Identification field contains a unique

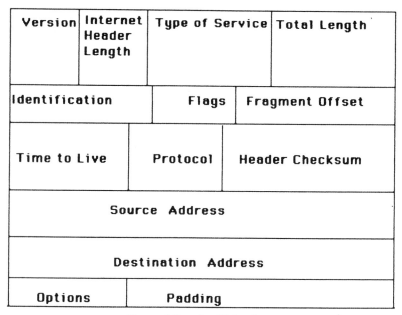

Figure 6.9: An IP header.

number for identifying this particular datagram, while the Flags and Fragment Offset fields include information on whether a datagram is fragmented, whether more fragments are still coming, and the number of octets from the beginning of the datagram for a specific fragment in 64-bit units.

The Time to Live (TTL) field records how long the datagram should be permitted to exist in the network. This number is decremented each time the datagram is processed along its journey through the network, and it is discarded when the number reaches 0. The maximum number permitted is 255 seconds; the point of having a TTL field is to preserve the resources of the network and avoid time delays due to a large number of undelivered datagrams.

The Protocol field following the TTL field indicates which protocol follows the IP. In Figure 6.10, we see that the datagram encapsulated by the IP includes a TCP header. This protocol would be reflected by the number in this Protocol field. The functions of the Header Checksum, Source Address, and Destination Address

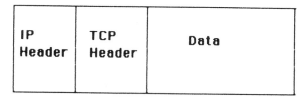

Figure 6.10: A TCP/IP header.

Protocols not Included in the OSI Model

fields are self-explanatory, and they serve the same purposes as their counterparts in the TCP header, which we have already discussed. Finally, the Options field can contain information on such tasks as routing specifications.

Why TCP/IP is Still So Popular

TCP/IP has a number of weaknesses. Its performance suffers because it is byte-oriented rather than packet-oriented. Also, it has mandatory checksums, an error checking procedure that slows up transmission. Despite these limitations, TCP/IP has flourished because it offers a standardized protocol that can co-exist with a number of operating systems and provides the capability for terminal emulation and file transfer. It is possible for a PC using the TCP/IP protocol's Telnet program and running under Xenix to emulate a VT 100 terminal and log onto a DEC VAX minicomputer. Using TCP/IP's File Transfer Protocol (FTP), the PC user can download a key file from the VAX, modify the contents, and then upload the same information to the minicomputer. TCP/IP is standardized at all layers. This means that different vendors' implementations of this protocol can co-exist within the same medium and even communicate with each other across this medium.

Xerox Network Systems (XNS)

Xerox Network Systems (XNS) is a protocol developed by Xerox for office automation environments. As seen in Figure 6.11, XNS's lower layers correspond to the OSI model's Data Link and Physical layers.

The real strength of the Xerox Network System is its emphasis on connectivity with all types of office systems. Since office automation users will call directly for

OSI Layer	XNS Layer
Application	Clearing House (Network Director)
Presentation	Courier
Session	None
Transport	Internet Datagram Protocol
Network	Ethernet
Data Link	
Physical	

Figure 6.11: Xerox network systems compared to the OSI model.

such services as electronic mail and printing, they are found in XNS's Application Layer of its architectural model. Figure 6.12 shows how the services provided in the Application Layer are built on a set of protocols. The Information Format and Encoding Standards area provides the Interpress Electronic Printing Standard, a universal method of describing two-dimensional text and graphics for printing in a network environment. The Interpress standard makes sure that documents will look the same, regardless of the device used for printing.

The Clearinghouse Service (CHS) is a directory of network users and network services and resources. It uses a standard syntax for naming consisting of a network name, domain, and organization. The Mail Service (MS) provides efficient electronic mail; messages are available from any workstation and can be forwarded.

The Services area of XNS's Application Layer also includes asynchronous communication protocol software that supports TTY emulation. This means that virtually any asynchronous terminal, including the very common DEC VT 100 terminals, can communicate. The Services area also supports IBM terminal emulation in the form of 3270 BSC Communication Protocol and 3270 SNA/SDLC Communication Protocol support.

The Courier Layer of XNS corresponds roughly to OSI layers 5 and 6. It provides a single format for handling requests and replies between two different computers. The Courier Layer can translate any number of higher-level protocols into a form that can be understood by a computer located remotely. Also associated with the Courier Layer is Bulk Data Transfer Protocol, a method for transferring large data files between XNS devices and non-XNS devices without having to translate this information.

The Internet Transport Layer corresponds to layers 3 and 4 of the OSI model. There are several protocols involved in this layer. The Echo Protocol is a way of determining that a device is operating correctly and that the communication path is working. Echo packets are sent from one system to another and then returned to establish that everything is working smoothly. The Sequenced Packet Protocol keeps

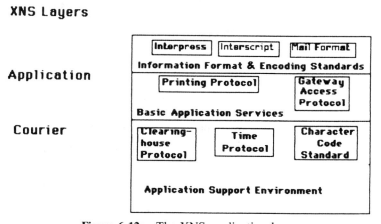

Figure 6.12: The XNS application layer.

track of large packets that have been divided into smaller packets for transmission. These packets are sequentially numbered so that they can be kept in order and reassembled without needless duplication. The Error Protocol indicates how a receiving system should respond when it perceives an error in transmission. If the type of error can be identified, this information along with the defective portion of the packet are returned to the originating computer.

Another protocol that increases transmission reliability is Packet Exchange Protocol. This protocol is designed for single packet exchanges in which the request resides in one packet and the reply in another packet. The Internetwork Datagram Protocol is exactly what it sounds like, a protocol for routing packets to their destination by providing addresses and routing instructions.

As Figure 6.11 illustrates, the lower two layers of XNS correspond roughly with the OSI model's layers. The Ethernet protocol at this layer makes it possible for XNS to establish communication among a variety of non-XNS networks through the common Ethernet format.

THE GOVERNMENT OPEN SYSTEMS INTERCONNECTION PROFILE (GOSIP)

The Government Open Systems Interconnection Profile (GOSIP) represents a set of U.S. government specifications based on agreements reached at the National Bureau

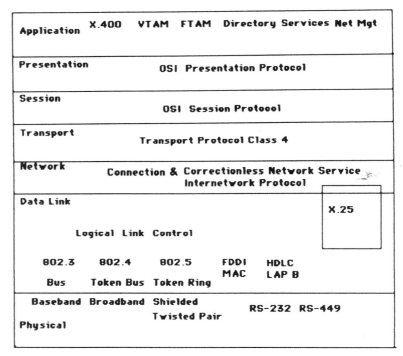

Figure 6.13: GOSIP's protocols.

of Standards' Workshop for Implementors of OSI. Because the OSI model consists merely of a framework of protocols and not a complete set of detailed specifications, the government has been concerned that different vendors' computer systems may not be compatible even though they may claim to be OSI products. The government issued a draft document of GOSIP in December 1986. This document consists of numerous subsets and options of existing OSI standards. GOSIP includes a number of Application Layer protocols, including File Transfer, Access, and Management (FTAM) and X.400's Message Handling System. This document also supports X.25 packet switching as well as a number of local area networks we discussed in Chapter 4, including bus networks (IEEE 802.3), token bus networks (IEEE 802.4), and token-ring networks (IEEE 802.5). In the future, additions to GOSIP are expected to include Integrated Services Digital Network (ISDN). Figure 6.13 illustrates the grand plan for this new protocol.

CHAPTER SUMMARY

The ISO OSI model is a seven-layer set of protocols for helping to ensure compatibility in a multi-vendor environment. Layers perform services for the layers above them as well as perform their own functions. TCP/IP is a very popular protocol today for companies that do not want to wait for the OSI model to be implemented. The XNS protocol is particularly appropriate for companies in which office automation is a prime concern. The Government OSI Profile (GOSIP) is becoming a standard that government contractors will have to meet.

KEY TERMS

Aliasing
Application Layer
Asynchronous Balanced Response Mode
Binary Synchronous Protocol (BSC)
Bit-oriented protocol
Bit stuffing
Character-oriented protocol
Data Link Layer
Government Open Systems Interconnection Profile (GOSIP)
High-Level Data Link Control (HDLC)
Network Layer
Normal response mode
OSI Model

Physical Layer
Presentation Layer
Protocol
Session Layer
Session protocol data unit (SPDU)
Simple Mail Transfer Protocol (SMTP)
System Network Architecture (SNA)
Synchronous Data Link Control (SDLC)
Transmission Control Protocol/Internet Protocol (TCP/IP)
Transport Layer
Transport protocol data unit (TPDU)
Virtual Terminal protocol
Xerox Network Systems (XNS)

REVIEW QUESTIONS

1. What is GOSIP? What impact will it have?
2. Why has TCP/IP become so popular?
3. What is a character-oriented protocol?
4. What is a bit-oriented protocol?
5. What are the essential differences between HDLC and SDLC?
6. What are the different classes found in the Session Layer?
7. What are some of the more common standards found in the Physical Layer?

TOPICS FOR DISCUSSION

1. What impact do you think the ISO set of standards will have on medium-size businesses? There are a number of industry experts who feel that it simply is not worth the time, expense, and effort for most businesses to become ISO-compatible.
2. Talk to an IBM dealer or representative and learn some facts about IBM's OSI/File Services program. Does it support FTAM? Does it permit files to be deleted or copied remotely?
3. 3Com has an X.400 gateway for its 3+ LANS that utilizes an X.25 router. Since there are security problems with X.25 transmission, what kinds of security safeguards does X.400 offer? You may need to have a computer search performed on recent literature to find the answer to this question.
4. Examine a number of electronic mail packages. What are some of the features that you feel are absolutely essential for a large company? Does the X.400 standard support these features?

Packet-Switched Networks

CHAPTER 7

OBJECTIVES

Upon the completion of this chapter, you should be able to:

- explain the difference between a circuit-switched network and a packet-switched network
- determine when using packet networks is advantageous
- explain how information flows through a packet network
- describe a virtual circuit
- name the most commonly used packet network protocol
- name two commercial packet networks

INTRODUCTION

Packet-switched networks are in widespread use in the United States, Canada, Europe, and many other parts of the world. As common as they are, they are not well known. This chapter takes a look at packet networks and examines their importance and how they are used.

SWITCHED CIRCUIT VERSUS PACKET CIRCUIT

Packet-switched networks are a new phenomenon, compared to circuit-switched networks. Circuit-switched networks have been around since the earliest days of the telephone and have become the foundation of the Public Switched Network. Packet networks, on the other hand, did not come into use until the early 1970s.

Each type of network performs the same function: they move information from one location to another. The way they accomplish this function, however, differs radically. Circuit-switched networks establish a physical path for every call that is made. The path is dedicated and remains intact until completion of the call. At that point, the circuit is taken down and the legs comprising it can be used in other circuits.

Packet-switched networks do not set up a physical path for each call. Instead, a virtual path, or **virtual circuit,** is established. The virtual circuit acts as if it is a dedicated physical path though in fact the physical path it uses is shared with many other calls.

Sharing a physical path allows a packet network to use its facilities more completely, and therefore more efficiently, than a circuit-switched network. For this reason, it costs less to move information through a packet-switched network than a circuit-switched network. Why, then, do we not see circuit-switched networks being phased out in favor of packet networks? The answer is that packet networks lend themselves far better to data transmission than to voice transmission, and voice still accounts for the majority of communication traffic.

Two reasons explain why: (1) packet networks require digitized information—data transmissions are always digital, while voice transmissions often are analog; and (2) parts of a transmission move at different rates through the network (see more detailed explanation below), which is not a problem for data, but can cause a noticeable delay on a voice call.

HOW A PACKET NETWORK WORKS

Packet networks deliver messages through a network by sending them as small bundles, or **packets.** All packets are a uniform size, creating a standardized environment for information to flow. If a message is very small, it will fit into one packet. The packet is then sent through the network and delivered at the distant end. Most messages are too large to fit in a single packet. They are divided into segments, and the segments are assigned to packets. The packets are sent through the network separately and are then reassembled into a complete message at the receiving end.

To help understand how this works, an analogy using the U.S. Postal Service will help. Letters are sent to a destination by being placed in an envelope. The envelope can be placed in any mailbox and it will be delivered.

The envelope must conform to certain rules for the postal service to be able to handle it. It must have an address with a name, street address, city, state, postal code, and perhaps a country name. There is usually a return address allowing the letter to be returned in case something goes wrong. Finally, the envelope must have proper postage.

A standard format is used for placing information on the envelope. Among other things, it is important to help distinguish the mailing address from the return address. The mailing address should be toward the lower right corner, the return address should be at the upper left corner, and postage should be at the upper right corner. If all these things are done, the letter should get delivered without difficulty.

Note that the structure and rules apply to the envelope, but there is very little restriction on its contents. The correspondence inside can be a short, one-page note, or a lengthy, multipage document. It can be a handwritten letter or a birthday card or a photograph. It can even be a computer diskette or a painting. So long as it is reasonably appropriate, the postal service will deliver it.

Suppose an article is too big or has too many pages to fit in a single envelope. It must be split among two or more of them. They will be mailed separately and then put back together at their destination.

These same requirements apply to packet networks as well. A summary of them shows:

1. Correspondence must be sent in an envelope
2. The envelope must have proper delivery information and should follow a standard format
3. The contents of the envelope are relatively unrestricted
4. Mail may enter the system at any point and it will be delivered to any known location.

These points will be discussed in more detail as they apply to packet switching.

Packetizing, Routing, Delivery, and Cost

The "envelope" used in a packet network is the packet. It contains the message that is to be transmitted. Packets are relatively small, having a maximum size of 2,048 bytes, and messages usually must be broken into many pieces before they can be sent through the network. The process of breaking them down is called **packetizing.** Special protocols define the procedures used. They will be discussed later in the chapter. The procedures are handled either by software running on a computer or by a separate piece of hardware. Figure 7.1 shows an example of a packet-switched transmission.

As can be seen in Figure 7.1 there are multiple paths through the network. The path an individual packets takes is determined by the switching equipment, and no two packets will take the same route.

The packets are likely to pass through many nodes on the way to their destination. At each node, the validity of the packet data is checked before it is routed on. If the node receiving a packet detects an error, it requests the sending node to retransmit. In this way, only good packets of information are allowed, and the infor-

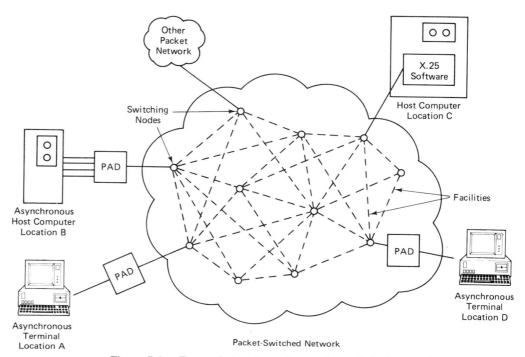

Figure 7.1: Transmission through a packet-switched network.

mation coming out of network is assured to be the same as the information that went into it.

The method used to route packets gives users the freedom to send from any location. Messages can enter the network at any point and, so long as they are properly addressed, they will be delivered to the correct location.

Final delivery is dependent upon proper reassembly of packets into a message. The **packet assembler/disassembler (PAD)** (of which more later) is responsible for accomplishing this. It accepts packets as they arrive from the network and holds them until all are received. The various routes that packets take make it likely that they will arrive out of sequence. For instance, packet 6 may arrive before packet 4. Once all have been received, the PAD strips out the routing information and merges them back into a single message. The message is then presented to the terminal or computer.

The cost of using a packet network is based on the volume of information transmitted and not on the distance between location. Most networks use 1,000 packets as a unit of measurement. There is a charge of about $2.00 for every 1,000 packets of information sent. This, again, is similar to the methods used by the postal service, where payment is made per letter, not by the distance to the destination.

PACKET NETWORK STANDARDS

The first packet network was developed at the request of the U.S. government. The General Accounting Office found that some government data processing centers were overloaded, while others were idle a great deal of the time. It was looking for a way to shift some of the processing from the heavily burdened centers to the lightly used ones. To achieve this goal, the government funded the development of a network known as ARPANET. Packet-switching theories and techniques were developed and tested on this network.

In the 1970s, commercial companies implemented packet-switching technologies and began their own networks. The use of packet switching spread from the United States to Canada and Europe. With the proliferation of networks came the need for standards. Each network had its own protocols and network interface hardware. As a result, none of the networks could communicate with one another. In 1976, the countries of North America and Europe agreed to accept recommendations of the CCITT. There are a whole series of recommendations, the most important being Recommendation X.25. Table 7.1 summarizes the recommendations.

X.25

X.25 is formally known as "Interface Between Data Terminal Equipment (DTE) and Data Circuit Terminating Equipment (DCE) for Terminals Operating in the Packet Mode on Public Data Networks." It establishes precise guidelines for computers and terminals (DTE) to communicate with packet-switch nodes (DCE). Its importance is that it defines how non-packet equipment (DTE) can connect to the packet-switching world (through a packet-switch node, the DCE).

The benefit of commercial networks supporting the X.25 standard is that information can be exchanged between them. To date, networks based on X.25 are operating in Europe, Australia, Singapore, Hong Kong, Japan, the Soviet Union, South Africa, Canada, the United States, and Mexico. These networks are not totally compatible, but many of them can interchange data with relative ease. The incompatibilities result from the fact that, in addition to a core set of recommendations, X.25 has a series of extensions and optional features.

X.25 as a standard predates the Open Systems Interconnect (OSI) Reference Model of the International Standards Organization (ISO). Nonetheless, X.25 is intended to be an open standard in the same spirit as the OSI Reference Model. As such, it is divided into three layers that largely conform to the OSI definitions. OSI layer 1, the Physical Layer, corresponds to X.25 level 1, the Physical Level. OSI layer 2, the Data Link Layer, corresponds to X.25 level 2, known as the Link Level or Frame Level. OSI layer 3, the Network Layer, corresponds to X.25 level 3, the Packet Level. Figure 7.2 shows the relationship between X.25 levels and the OSI Reference Model.

The upper four levels of the OSI Reference Model, the Transport Layer, the Session Layer, the Presentation Layer, and the Application Layer, are not defined for X.25. This is typical of network protocols. They may be defined in the future.

TABLE 7.1

A Summary of CCITT Interface Recommendations

Name	Formal Title	Description
Overall		
X.25	Interface between Data Terminal Equipment (DTE) and Data Circuit Terminating Equipment (DCE) for terminals operating in the packet mode on public data networks	Standards for connecting to a node in a packet-switched network
X.75	Terminal and transmit call control procedures and data transfer system on international circuits between packet-switched data networks	Standards for communication between two packet-switched networks
Level 1	Physical interface	
X.21	Interface between Data Terminal Equipment (DTE) and Data Circuit Terminating Equipment (DCE) for synchronous operation on public data networks	Physical interface standard between DCE and DTE for synchronous operation
X.21 bis	Same as CCITT Standard V.24	An alternate physical interface standard to X.21. It is compatible with EIA RS-232C
X.26	Same as CCITT Standard V.10	Electrical interface standard for DTE operating synchronously at rates up to 9600 bps
X.27	Same as CCITT Standard V.11	Electrical interface standard for DTE and DCE operating synchronously at rates above 9600 bps
Level 2	Link/Frame level	
X.1	International user classes of service in public data networks	Standards for user operating environments
HDLC	High Level Data Link control procedure	ISO-based standard for link level procedures
Others		
Interactive Terminal Interface (ITI)		A combination of three recommendations defining interface standards for low speed, asynchronous terminals and devices
X.3	Packet Assembly/Disassembly Facility (PAD) in a public data network	Definition of functions to be performed by a Packet Assembler/Disassembler (PAD)
X.28	DTE/DCE interface for a start/stop mode data terminal equipment accessing the Packet Assembly/Disassembly Facility (PAD) in a public data network in the same country	Standards for an asynchronous terminal to interface with a Packet Assembler/Disassembler (PAD)

| X.29 | Procedures for the exchange of control information and user data between a Packet Assembly/Disassembly Facility (PAD) and a packet mode DTE or another PAD | Definition of procedures used by a Packet Assembler/Disassembler (PAD) to communicate with another Packet Assembler/Disassembler (PAD) or with a packet mode DTE |

At present, however, they must be dealt with by higher-level software, such as an application program.

Level 1

The Physical Level defines the physical hardware and electrical connections that will be used. For the X.25 standard, these are defined in Recommendation X.21.

X.21 differs from most other physical standards because it does not define functions in terms of the physical connection. Other interfaces, such as EIA RS-232C, define a specific function for each pin in the interface.

Instead, X.21 defines a coded string of characters for each function. There are advantages to this approach. First of all, there is no limit to the number of functions the interface can have. Adding functions is as easy as defining new coded strings. The EIA RS-232C interface, in contrast, is limited to the 25 pins it has available. Another advantage is that X.21 has built-in dialing capabilities. This allows for quick and easy reporting of trouble when a connection cannot be made.

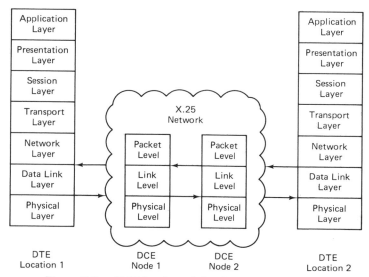

Figure 7.2: X.25 levels and the OSI reference model.

The flexibility of using coded strings opens X.21 for use in types of communications besides data. It could be adapted as an interface for digitized voice or digital image transmission. Should this happen, X.21 could become a universal interface standard supporting voice, data, and image through a single physical connection.

X.21 has four different operating phases. They are the Quiescent phase, the Call Control phase, the Data Transfer phase, and the Clearing phase. Each one puts the DCE and DTE in a specific state. Figure 7.3 shows the possible transitions between phases.

The Quiescent phase is an inactive phase in which either the DCE or the DTE can initiate a call to the other. The DCE can be in either a Ready or Not Ready state. The DTE can be in a Ready, Controlled Not Ready, or Uncontrolled Not Ready state. (Controlled Not Ready indicates a temporary type of interruption such as buffers full or equipment not in interactive mode; Uncontrolled Not Ready indicates a more serious problem, such as equipment malfunction or failure.) If both ends are in a ready state, service requests can proceed. If either end is not ready, service requests must wait. From the Quiescent phase, equipment can transition to the Call Control phase or the Data Transfer phase.

The Call Control phase handles many aspects of call processing. It includes normal call control procedures such as Call Request, Connection In Progress, and Ready For Data. It also includes abnormal procedures such as Unsuccessful Call and Call Collision. From this phase, equipment enters the Data Transfer phase.

In the Data Transfer phase, packets of information flow between the DCE and the DTE. Transmission is full-duplex, allowing for independent transfers in both directions simultaneously. Data transfer is terminated by entering the Clearing phase.

The Clearing phase serves to terminate the Data Transfer phase. Either the DCE or the DTE can initiate it. The DCE issues a Clear Indication to signal its intent. The DTE responds with a Clear Confirmation. When the DTE initiates clearing, it sends a Clear Request. The DCE acknowledges with a Clear Confirmation followed by a Ready signal.

The electrical interface used with X.21 is defined by two other recommenda-

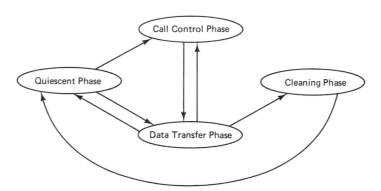

Figure 7.3: Phases of operation defined by recommendation X.21.

tions. Recommendation X.26 relates to synchronous transmission at speeds up to 9,600 bps. Above this speed, Recommendation X.27 is used. The hardware interface used with either recommendation is a 15-pin connector.

While X.21 is the stated level 1 standard for X.25, the interface is not yet fully accepted. In the United States particularly, there has been resistance to its implementation. The principal reason for the reluctance is the current widespread use of the EIA RS-232C interface standard. The EIA RS-232C interface is available on virtually every type of data equipment. Changing to a new standard would be a major undertaking. Furthermore, EIA RS- 232C is a less complex interface. It requires only certain electrical voltages and pins in a specified layout. X.21, in addition to this, requires intelligence in the interface to interpret the coded character strings. This makes it more expensive than EIA RS-232C.

For these reasons, the move to the X.21 interface has been slow. In recognition of this, CCITT allows for an alternative interface called X.21 bis. X.21 bis is equivalent to CCITT standard V.24, which in turn is compatible with EIA RS-232C. This allows the vast number of EIA RS-232C devices to connect to packet networks immediately, without any modification to existing interfaces.

X.21 is likely to become more popular during the coming years. Its backing by CCITT and its flexible format should bring this about. It is currently closer to acceptance in Europe and elsewhere than in the United States. Over time, however, it is likely to find increasing acceptance in the U.S. as well. In any case, X.21 bis appears to be an intelligent interim solution for providing packet network access.

Level 2

Level 2, the Link Level or Frame Level, has several important tasks to perform. It is responsible for maintaining control of the transmission, for conducting error-checking, for adding control information to the transmission, and for stripping off control information before delivering a packet to level 3.

Software programs are used to accomplish the tasks. The programs run on both DCE and DTE. The procedures used come from CCITT Recommendation X.1 and the ISO High-level Data Link Control (HDLC) protocol.

Recommendation X.1 is a document defining user classes of service. X.25 specifies the use of Class B (2,400 bps), Class 9 (4,800 bps), Class 10 (9,600 bps) and Class 11 (48,000 bps). Also defined is the requirement to provide full-duplex transmission.

HDLC is used to implement two procedures required of the level 2 software. The Link Access Procedure Balanced (LAPB) is based on HDLC's Asynchronous Balanced Mode (ABM). This will be discussed in more detail below. The other procedure is the **Multilink Procedure (MLP),** which allows for one or more concurrent data transmissions. MLP can be thought of as an extension of LAPB to multiple channels.

HDLC recognizes two types of stations, **primary stations** and **secondary stations.** During normal operations, primary stations control the data flow and handle level 2 error-recovery routines. Secondary stations are any other stations connected to the link.

In the **normal response mode (NRM),** secondary stations cannot transmit without permission from a primary station. In this way, the primary station maintains control of activity on the link. Another option is the **asynchronous response mode (ARM).** Under ARM, a secondary station can transmit at will, without permission from a primary station. ARM can only work on a point-to-point connection where there is one primary station and one secondary station. NRM works also in multipoint environments where there may be many secondary stations.

X.25 works in the point-to-point mode. Originally, it supported ARM. This has been extended to encompass **asynchronous balanced response mode (ABRM).** With ABRM, the stations on the link are **combined stations.** Combined stations can act as both a primary and a secondary station; they have equal capabilities, and either one can initiate or terminate transmission activity. In this way, both stations on the link share the management and error-recovery responsibilities.

HDLC defines a structure called a **frame.** It is a combination of fields in a precise order that allows for control and error-checking of information passing through the network. Figure 7.4 shows the makeup of a frame.

The frame is composed of a packet plus additional fields containing level 2 control and addressing information. The contents of the packet are never actually touched, but fields are appended in front of and behind it. It is frames, and not just the packets themselves, that are sent through the network.

At each end of the frame are Flag fields. They mark the separation from one frame to the next. Flags are always the same sequence of bits: 0 1 1 1 1 1 1 0. DCE and DTE are designed to continually scan for this sequence. They keep track of the datastream from flag to flag and, in this way, identify each frame. When frames are sent consecutively, a single flag may serve as the end of one frame and the beginning of the next. This feature helps minimize the overhead in the network.

Following the frame field is the address field. Normally, it is an 8-bit field giving a total of 256 addresses. Recognizing that this is often insufficient, HDLC has a provision for extended addressing. Under this scheme, the first bit of the address frame is set to 0 if extended addressing is needed. If a 0 is found, the remaining bits are taken as the first part of the address and the next field is interpreted as an extension of the address field, not as the control field. The address field can be extended again and again, as far as necessary. Successive fields will be interpreted as part of the address until a 1 is encountered as the first bit.

The control field follows the address field. It is 8 bits long and serves to identify

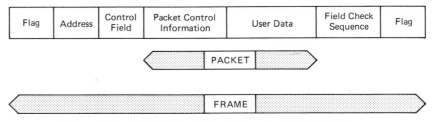

Figure 7.4: A high-level data link control (HDLC) frame.

the type of frame that is being sent. There are three possibilities: it may be an Information frame (I-Frame), a Supervisory frame (S-Frame), or an Unnumbered frame (U-Frame). An I-Frame contains a packet with user information. S-Frames and U-Frames contain no user packet and are used to transmit network commands. The distinction between S-Frames and U-Frames is that S-Frames are limited to supervisory functions, such as acknowledging I-Frames or requesting retransmission of I-Frames. They also affect the sequence counters at the sending and receiving stations. U-Frames serve to transmit non-supervisory functions; they do not increment sequence counters. An important U-Frame command is to initialize Asynchronous Response Mode (ARM) in the DCE and DTE. This allows either piece of equipment to institute transmission requests.

Next within the frame comes the packet. It contains two fields of its own. The first is the Packet Control Information field, and the second is the User Data field. The Packet Control information field contains a Send Sequence Count, a Receive Sequence Count, and a Poll/Final bit. The Send Sequence Count gives the sequence number of the frame being sent. The Receive Sequence Count indicates the expected sequence number of the next frame that will be received. The Poll/Final bit determines whether the frame must be acknowledged. (HDLC allows from 7 to 127 frames to be outstanding before an acknowledgment must be sent; the exact number allowed varies from network to network.) The User Data field also varies in size from network to network. It has a maximum limit of 2,048 bytes.

The Frame Check Sequence (FCS) is a 16-bit field used to detect errors in transmission. Before a frame is sent, a number is calculated based on the binary value of the bits in the frame. This number is placed in the FCS field and transmitted with the frame. At the receiving end, the same calculation is done and the result is compared to the contents of the FCS field. If they match, the frame is deemed good. If they do not match, the frame is rejected and a request is sent to have it retransmitted.

When frames are received without error, the level 2 software strips off the fields surrounding the packet. The packet is then delivered to level 3 software for further processing.

Level 3

Level 3 of the X.25 Recommendation is called the Packet Level. It defines the procedures for packet handling between the DCE and the DTE. These include call initiation, data transfer, and packet-level error handling. There are several types of packets in addition to user information packets. They are used for administrative purposes.

Call initiation encompasses the procedures for establishing a call. The DTE begins by sending a Call Request packet. This requests that a channel be set up for communication. The channel is a logical channel, not a physical one. Remember that the virtual circuits of the packet network will route packets over many different paths, but they all end up at the same place. Figure 7.5 shows the steps involved in call initiation.

The Call Request packet names a logical channel to be used and gives the

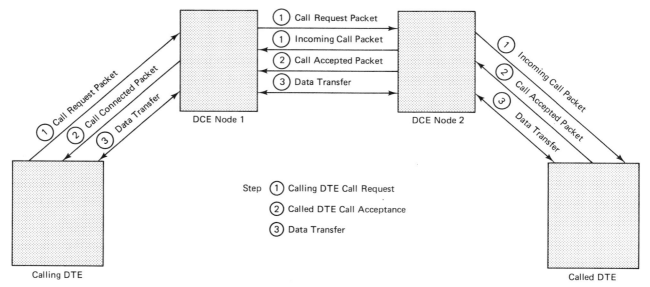

Figure 7.5: X.25 level 3 call initiation and data transfer procedures.

addresses of the calling DTE and the called DTE. The network retains this information for the duration of the call. Subsequently transmitted packets need not carry addressing information, but only logical channel information. The network can derive the addresses to identify where the packet originated and where it is going.

The Call Request packet is transmitted by the calling DTE to the DCE associated with it (we will call it DCE Node 1). DCE Node 1 passes the request through the network to DCE Node 2 (the DCE connected to the called DTE). DCE Node 2 sends an Incoming Call packet to the called DTE. The called DTE responds with a Call Accepted packet if it is able to receive the call. DCE Node 1 alerts the calling DTE that the call has been accepted by sending a Call Connected packet. The logical channel is then in the Data Transfer state and user information can begin to flow.

User information and error recovery packets can be sent while a channel is in the Data Transfer state. User information is contained in Data Packets, while error recovery is handled by Interrupt packets, Reset packets, Clear packets, and Restart packets.

In normal operation, Data packets flow back and forth transmitting user information. Interrupt packets serve to halt the data flow. They can be generated by the DTE at either end.

The channel can be placed back in the Data Transfer state with a Reset packet. Both DTE and DCE can issue them. DTE generates a Reset Request packet, while DCE generates a Reset Indication packet. The channel is actually reinitialized by a reset, and any Data packets or Interrupt packets in the channel are cleared. Resets are associated with specific, relatively minor error conditions, such as network congestion or local/remote procedure errors.

Clear packets are generated by the DCE when more serious errors occur. These might be a host computer crash or loss of a link to a host. When this happens, the entire session is cleared and must be re-established.

Restart packets affect an entire link. There may be several logical channels, and all of them will be disrupted by a restart. It clears all calls and reinitializes the link.

Other Packet Network Standards

X.25 establishes the standards for DTE to communicate with a DCE node in a packet-switched network. While there are many computers and intelligent workstations that can accommodate these requirements, there are also many types of equipment that cannot. To allow more equipment the opportunity to access a network, other standards have been presented.

The first standards to be recognized and implemented were standards for asynchronous terminals. There are an abundance of them in use and, for the most part, they lack the intelligence to handle a packet switch interface. Three standards are involved. Recommendation X.3 defines the requirements of an interface between an asynchronous terminal and the X.25 network; it is known as a packet assembler/disassembler, or PAD. Recommendation X.28 defines procedures used between the asynchronous terminal and the PAD. Recommendation X.29 defines procedures to be used between two PADs or between a PAD and packet mode DTE. Collectively, Recommendations X.3, X.28, and X.29 are called the **Interactive Terminal Interface (ITI).**

Recommendation X.3 sets forth a list of 22 parameters that a device must handle to act as an interface to the network. The device, called a packet assembler/disassembler (PAD) facility, is normally a separate piece of hardware placed between a terminal and the packet network.

The functions it performs can be categorized as follows:

1. creation of packets from user data (packetizing)
2. packet disassembly
3. virtual call setup
4. error recovery
5. asynchronous terminal command interpretation
6. operational parameter setting (data rate, parity, etc).

The second standard for asynchronous terminals, Recommendation X.28, describes the procedures allowing a PAD to interact with an asynchronous terminal. There are four different categories defined:

1. procedures to establish an access path between an asynchronous terminal and a PAD
2. procedures for character interchange and service initialization

3. procedures for the exchange of control information between an asynchronous terminal and a PAD
4. procedures for the exchange of user data between an asynchronous terminal and a PAD.

Access paths are usually provided by leased line facilities or public dial-up circuits. Modems are often used on these circuits, and the standards for them are set out in this recommendation. The CCITT V-series specifications are used. The exact specification depends on the type of circuit and the speed of the modem. X-series interfaces are also sometimes used with these circuits, and these are defined in Recommendations X.20 and X.21 bis.

Character interchange and service initialization refers to the bit codes that will be used to represent different characters. X.28 requires that they conform to International Alphabet No. 5 (IA5). Furthermore it defines that bit strings for characters be 8 bits long, with the eighth bit used for parity checking. Some of the 22 parameters set out in Recommendation X.3 are further defined here. For example, parameter 21, parity treatment, is described. If it is set to 0, parity is ignored. If it is set to 1, parity will be checked between the terminal and the PAD. If it is set to 2, the PAD will set the parity bit on traffic coming from the network to the terminal. If it is set to 3, the PAD does both a parity check on characters from the terminal and sets parity on characters going to the terminal (1 and 2 combined).

Control information includes Command Signals, Service Signals, the Break Signal, and the Prompt PAD Service Signal. Command Signals always go from the DTE to the PAD. They include commands to request call set up, clear a call, send an interrupt, request a reset, request PAD parameters, request circuit status, and select a PAD profile (defines the operating parameters of the terminal for the PAD). Service Signals always go from the PAD to the DTE and include call progress signals, command signal acknowledgments and operating information. The Break Signal can be transmitted by either the PAD or the DTE. It allows for signaling without the loss of character transparency. The Prompt PAD Service Signal is sent from the PAD to the DTE and indicates it is ready to receive a command.

Procedures regarding user data are also defined by X.28. For instance, the specification sets forth that data to the DTE will be delivered in octets (groups of eight bits) preceded by a start bit and followed by a stop bit. Data from the DTE is expected in the same format, and the PAD will buffer, or store, octets until one of the following occurs: (1) enough octets have been received to fill a packet, (2) a predetermined time threshold has elapsed, (3) a data forwarding character is received, or (4) a Break Signal is received.

The third standard for asynchronous terminals, Recommendation X.29, deals with communication between two PADs or between a PAD and a packet mode DTE. It specifies that the data fields defined in Recommendation X.25 be used. Also that the procedures and physical interfaces conform to it.

X.29 also defines a PAD message, which is used to transmit control information. A PAD message consists of a control identifier field, a message code field, and optionally, a parameter field. Three PAD messages are also specified: Set, Read, and Set and Read. *Set* allows PAD parameters to be changed. *Read* causes the PAD to transmit a list of current parameter values. *Set and Read* causes a parameter list to be sent and allows changes.

A fourth standard, Recommendation X.75, defines the procedures for multiple packet networks to exchange information with one another. In many respects, it mirrors X.25. There are some differences at the Physical Level. These relate to the definition of links A1 and G1 and are set out in Recommendation X.92.

The Link Level conforms in terminology and function to HDLC. Frames are used for data transfer. Their format is virtually the same as X.25. A Single Link Procedure (SLP) and a Multilink Procedure (MLP) are defined. The SLP is used to transmit data over a single physical circuit. It can also be used on multiple parallel circuits when each one is operated independently. The MLP is used when there are multiple links. It serves to distribute packets over the available links.

The Packet Level procedures are also very similar in function to those of X.25. The conformity between X.25 standards and X.75 standards is a great advantage. It simplifies the process of transferring data and minimizes the cost of designing equipment to work with both of them.

PUBLIC DATA PACKET NETWORKS

Although commercial data packet networks are less than 20 years old, they now comprise a market with hundreds of thousands of users. Since the early 1980s, the market has grown on an average of 30% per year. This growth is due to several factors, among them the increasing use of personal computers in the corporate world and a growing need for dissimilar equipment to be able to share information.

Another factor that will spur growth and influence the market is the likely entry of the Bell operating companies as service providers. With the breakup of the Bell System, the now-independent operating companies are looking for growth opportunities. Several of them have signaled intentions to compete in the packet-switched network market.

Trends within the industry show a declining demand for the packet network's original role, providing economical transport of user data. Demand has shifted to value-added services such as custom applications and database retrieval. Some companies, such as General Electric Information Services and CompuServe have geared their offerings to this area for several years. They see it as their best opportunity for growth and revenue. Other, more traditional companies, such as Telenet and Tymnet, recognize the changes and are gearing their futures in this direction also.

Together Telenet and Tymnet make up 85% of the U.S. packet data network market. The balance is divided among a dozen or so competitors, including such companies as IBM, ITT, and AT&T.

Telenet

Telenet was one of the first commercial public switched data networks. It was the first in the United States to implement the X.25 standard.

It is the largest domestic provider with a market share over 45%. There are over 400 cities in North America with access to Telenet. In addition, there are over 50 overseas locations.

The network supports user equipment at speeds from 110 bps to 56 k bps. Asynchronous terminals can be used as well as synchronous terminals operating with Bisync, HDLC, 3270 or 2780/3780 protocol.

Tymnet

The "other" major public data network provider is Tymnet. It has a market share of about 38%. Over 600 U.S. cities are served, with approximately 70 non-U.S. locations in addition.

Tymnet uses a variation of standard X.25. The company maintains that its variation provides faster throughput across the network. Instead of dedicating packets to a single user's information, data from several users is multiplexed together in a single packet. This, according to Tymnet, reduces the transmission time delay of data waiting for a packet.

Tymnet's pricing differs from other companies as well. Normally, users are charged a set fee for every 1,000 packets of information transmitted. Since each Tymnet packet has data from several users, this approach is unfeasible. Instead of charging per packet, Tymnet charges per character. Their basic charge is for transmission of 1,000 characters.

Equipment operating at speeds from 110 bps to 56 k bps can connect to the network. Asynchronous communication as well as the following synchronous protocols are supported: 3270 Bisync, SDLC, 2780, 3780, HASP, or Bisync.

General Electric Information Services Co.

General Electric Information Services Co., or GEISCO, has been a provider of time-sharing computer services for many years. It is seeing that market decline as personal computers and departmental computers bring the cost of computing down to affordable levels for virtually all businesses. As a strategic move, GEISCO chose to enter new markets to shore up its diminishing revenue base. In 1984, it introduced its value-added packet-switched network, called Mark*Net. Over a year of field testing and $30 million were spent before its unveiling.

Mark*Net offers a wider range of features than much of its competition. For instance, it supports shared applications, batch data transfer, interactive timesharing, distributed data processing, office communications, and integration of microcomputers. In addition, customized software can be used.

Asynchronous communication from 110 to 19,200 bps is available. Synchronous communication from 2,400 bps to 56 k bps are also available. Synchronous protocols supported include Bisync and SDLC. Over 600 cities are served in the U.S. and Canada.

CompuServe

Like General Electric, CompuServe's strategy is to target value-added services for users. Toward this end, CompuServe offers an extensive database service, electronic mail, and protocol conversion. It also offers point-of-sale credit card verification for Visa and other financial institutions.

The network provides access from 110 to 9,600 bps asynchronously and up to 9,600 bps synchronously. The Bisync protocol is supported. Service is available in approximately 300 cities in the U.S.

CHAPTER SUMMARY

Packet-switched networks provide an efficient means of transporting information from one location to another. Information is bundled in packets. Packets move through the network, routed according to addresses they carry. There are several physical paths linking locations. Each packet takes the most direct one available at the moment of transmission. A logical path, as opposed to a physical path, then, links origination and destination points. The logical path is known as a virtual circuit.

Efficiency in the network is derived from packets sharing physical paths. Sharing allows the paths to be in use a high percentage of the time. Sharing also means that costs can be kept low because the cost of the path can be split among the packets using it.

The nature of packet switching lends itself to data transmission rather than voice transmission. Information must be in digital form to be carried through a packet network. Data transmissions are digital, while voice transmissions are often analog.

Packet networking standards have been established by the CCITT. Chief among them is the X.25 standard, which defines interfaces between computers and terminals (DTE) and packet-switching nodes (DCE). X.25 incorporates three levels, comprised of a physical level, a data link level, and a packet level. They correspond to the ISO Open Systems Interconnect Reference Model layers 1, 2, and 3. Each level is also defined by other CCITT recommendations.

KEY TERMS

Asynchronous Balanced Response Mode
Asynchronous Response Mode
Combined station

Frame
Interactive Terminal Interface (ITI)
Multilink Procedure
Normal response mode

Packet
Packet assembler/disassembler (PAD)
Packetizing
Primary station
Secondary station
Virtual circuit

REVIEW QUESTIONS

1. What is a PAD?
2. Give examples of DTE.
3. Give examples of DCE.
4. What is the purpose of zero insertion (bit stuffing)?
5. How does a frame differ from a packet?
6. What is a virtual circuit?
7. Make a diagram showing DCE and DTE and identify the procedure or CCITT recommendation used at each juncture. (Use Figure 7.1 as a guide.)

TOPICS FOR DISCUSSION

1. Security of data is a potential weakness of packet-switched networks. What is data security? How do different commercial packet networks deal with it?
2. Discuss the likelihood that packet-switched networks will eventually replace circuit-switched networks. What are the strengths and weaknesses of each?
3. Alternate standards, such as X.21 and X.21 bis, point out the difficulty of setting a single standard. Select a country currently using packet networks and research the country's position on standards for levels 1, 2, and 3. Take part in a discussion with others advocating the viewpoint of your country.

ISDN

CHAPTER 8

OBJECTIVES

Upon completion of this chapter, you should be able to:

- explain the forces leading to the development of ISDN
- describe the services ISDN provides
- define the function of layers 1, 2 and 3 within ISDN

INTRODUCTION

ISDN stands for **Integrated Services Digital Network.** It is a cumbersome name, but describes well the network's function. ISDN will provide a range of services, including voice, data, and image, which will be integrated, making it easy to select the desired ones. All communications will travel digitally through the network.

ISDN is a concept born of user demand for more control of their communications networks. Users want a simple way to reach the people they want to reach and to retrieve the information they want to retrieve. They want to change their network connections at will to meet changing requirements. The communications industry has responded by determining how to meet these needs and by establishing standards that all will adhere to.

ISDN is in its infancy. The past few years have seen telephone companies and select users conduct tests of new ISDN technology. Gradually, as the industry and

Figure 8.1: An Integrated Voice/Image/Data (IVID) terminal.

users determine the best use of different technologies, acceptance of ISDN will increase. Widespread use is not expected until the mid-1990s.

THE PROMISE OF ISDN

Some of the benefits end users will experience are (1) preindication of the calling party—the network will display the telephone number, and perhaps the name, of the calling party before the call is answered; (2) simultaneous calls—using a single connection, a person can carry on a voice call while simultaneously viewing video images or retrieving information from a computer; and (3) dynamic facility allocation—a person can select and change, for example, whether connections are to a private voice network or a public data network.

Today, different interfaces are needed for voice communications, data communications, and image communications (video). Analog interfaces, digital interfaces, and video interfaces are all used. Each provides a single service, and none are interchangeable. Voice communication normally uses an analog telephone, data communications uses a data terminal, and video communications requires a video camera and monitor.

Under ISDN, all forms of communication, voice, data, and image, will pass digitally through the network. End users will have a uniform interface for all services. A PBX, for example, will receive voice, data, and image through a single ISDN interface circuit pack. Individual end users will likely use an **integrated voice/image/data (IVID) terminal** that can handle all forms of communication. Even if a user did not have an IVID terminal, the network would be flexible enough to deliver information in an appropriate form. For example, if a user wanted to pick up a data message while standing at a pay telephone, the network would convert it from data format to digitized voice and deliver it verbally.

The standard interface allows a single connection to provide access to a theoretically unlimited pool of resources. For instance, through a single connection, a user could reach the Public Switched Network and any number of other networks. They may be switched-circuit networks, virtual-circuit networks, packet networks, local area networks, wide-area networks, and so on. The single interface simplifies both the hardware required and the methods for accessing each network.

Once past the end-user interface and into the network, a call may be routed to a switched-circuit network, a packet network, a private network, or any other type of facility. See Figure 8.2 for an example of the changes ISDN will bring. Eventually, all networks may have uniform interfaces. However, with hundreds of millions of dollars invested in existing hardware, changes toward a single interface for all networks will come very slowly.

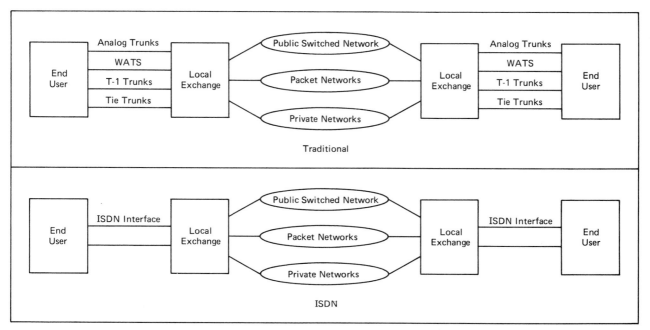

Figure 8.2: Network changes ISDN will bring.

ISDN STANDARDS

Standards are a key component of ISDN. They provide that all vendors of ISDN equipment will work with the same set of rules. Some standards have been established, while others continue to be developed. As new technologies become available in the future, they too will be incorporated into standards. In this way, ISDN will provide a stable yet evolving environment for telecommunications.

Setting the standards is the responsibility of the *Comité Consultatif Internationale Telegraphique et Telephonique,* that is, the International Consultative Committee for Telephone and Telegraph, or CCITT. The CCITT is an international group of telecommunications companies that meets to discuss and resolve telecommunications issues. Formal meetings are held once every four years. During the interim, working groups meet to come up with recommendations for specific problems. The recommendations are taken to the formal meeting for approval.

Standards for ISDN fall into CCITT's I Series Recommendations. Figure 8.3 shows the series' groupings.

The committee's recommendations adhere to the International Standards Organization's Open Systems Interconnect (OSI) Reference Model. Only layers 1, 2, and 3 have been defined for ISDN. The upper layers, 4, 5, 6 and 7, are beyond the scope of a network such as ISDN (see Figure 8.4).

Also among the committee's recommendations is a list of standard terms to refer to specific types of equipment or specific locations within the network. Figure 8.5 is an illustration of the interfaces and functional equipment types found on the list.

The lettered reference points R, S, T, U, and V correspond to demarcations that separate functionally different parts of the network. The boxes labeled TE1, TE2, and so on correspond to types of equipment that perform a particular function within the network.

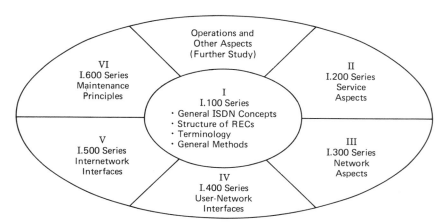

Figure 8.3: CCITT I series groups.

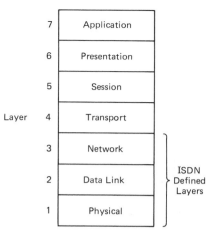

Figure 8.4: Open Systems Interconnect (OSI) reference model.

Functional Equipment Types

Different categories of equipment have been defined according to the function they perform in the network. TE1 and TE2 refer to Terminal Equipment. This is the equipment an end user must have to access the network. It may be a telephone, data terminal, video monitor, or IVID terminal. TE1 equipment is ISDN-compatible and can be directly connected to the network. TE2 equipment is not ISDN-compatible and requires an interface device before it can be connected to the network. The interface device is a TA, or Terminal Adapter. The TA provides conversion from a

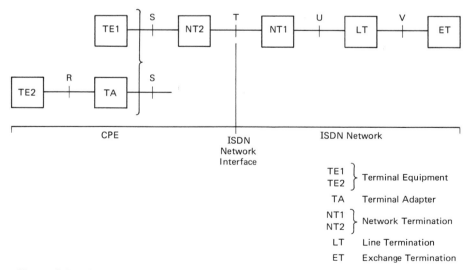

Figure 8.5: Sample ISDN network with interfaces and functional equipment types.

ISDN Standards 221

non-ISDN standard interface, such RS-232C, to the ISDN Basic Rate Interface. The fact that non-ISDN-compatible equipment is accommodated by the network reflects a realistic view that the transition to ISDN will be gradual and lengthy. Users' will collectively save millions of dollars by keeping and implementing their existing equipment rather than discarding it when they make the move to ISDN.

The next catagory of equipment is NT, or Network Termination equipment. This is the equipment that provides the physical interface to the TE and some basic network functions including timing and synchronization and bit-stream multiplexing. Two types are defined: NT1 and NT2. NT1's place in the network is analogous to a Public Switched Network demarcation device such as a **termination block** or **registered jack**. NT1, however, will be an intelligent, electronic device because of the functions it must perform. NT2 is customer-owned switching equipment, such as a PBX or LAN. NT2 can provide additional capabilities beyond NT1's—for example, call switching and concentration.

LT, or Line Termination, equipment is the next catagory. It is located within the local exchange company's or common carrier's network in situations where lines must be extended beyond the normal range of the central office.

ET is Exchange Termination equipment. It terminates the **Digital Subscriber Line (DSL)** or **Extended Digital Subscriber Line (EDSL)** in the local exchange. It may be thought of as central office equipment.

Network Interfaces

The T interface separates customer premises equipment (CPE) from the network (owned by local exchange companies or common carriers). It is based on two-pair wiring: one pair for transmitting and one pair for receiving. It has a maximum distance limitation of 3,300 feet. This presents a problem, since most users are not within 3,300 feet of the local exchange. The U and V interfaces can extend the distance, with the U interface having a range of 9,000 to 22,000 feet and the V interface having a range up to 100 miles.

The U and V interfaces, however, have no set standard; they may vary from country to country. In fact, they may vary within a country from one type of network to the next. The lack of a standard arises from the fact that these interfaces are within a local exchange company's or common carrier's network. They own the equipment on both sides of the interface. As such, they have the responsibility to ensure the equipment's compatibility, and there is little reason to impose an additional standard.

The Federal Communications Commission (FCC) in the United States has taken the position that the demarcation between CPE and the network should be the U interface, not the T interface. The position is based on the T interface's 3,300-foot distance limitation being too short to be practical. If the FCC is successful in establishing the U interface as the point of demarcation in the United States, then a standard for it will have to be reached. In addition, NT1 equipment will become CPE, not network-owned equipment.

Unequivocally on the CPE side are the R and S interfaces. The S interface connects an ISDN-compatible terminal to an ISDN network termination device. The

TABLE 8.1
ISDN Terminology

	Functional Equipment Types		Interfaces
ET	Exchange Termination Terminates the DSL/EDSL in the local exchange	R	Interface between non-ISDN compatible equipment and terminal adapter equipment
LT	Line Termination Extends a line beyond the normal range of central office equipment	S	Interface between ISDN compatible equipment or terminal adapter equipment and network termination equipment
NT1	Network Termination 1 Provides layer 1 functions: Timing Synchronization D Channel sharing B, D, & H channel multiplexing	T	Interface between CPE and network
NT2	Network Termination 2 Provides layers 1, 2 & 3 Functions: Timing Synchronization D channel sharing B, D, & H channel multiplexing Switching of calls Concentration	U	Interface between network termination equipment and exchange termination equipment or line termination equipment, if used
TA	Terminal Adapter Converts non-ISDN interfaces to ISDN standard	V	Interface between exchange termination equipment and line termination equipment
TE1	Terminal Equipment 1 User equipment with an ISDN standard interface		
TE2	Terminal Equipment 2 User equipment without an ISDN standard interface		

R interface connects a non-ISDN-compatible terminal to an ISDN terminal adapter. With these two interfaces, virtually all types of customer equipment can be connected to the network.

Layer 1

Two recommendations are made for layer 1, the Physical Interface Layer. The first is the **Basic Rate Interface (BRI)**, and the second is the **Primary Rate Interface (PRI).**

The Basic Rate interface defines a single access point into ISDN. It consists of two **bearer channels** and one **data channel.** Each bearer channel operates at 64 kbps and carries user information, either voice, data, or image. The data channel operates at 16 kbps and is used for signaling and control information. This is used for setting up and taking down calls and for passing call status information. The technique of using a separate channel for signaling, called **out-of-band signaling,** leaves the bearer channels completely free for user information. The bearer channels are said to be **clear channels,** since they have no restriction on the format or type of information that passes through them.

Bearer channels are often referred to as **B channels** and data channels as **D channels.** The arrangement of two bearer channels and one data channel is termed "2B + D" (read 2 B plus D). It is also called the Digital Subscriber Line (DSL). Figure 8.6 shows a representation of the Basic Rate Interface.

A second interface defined by the CCITT is the Primary Rate Interface. It is used to connect multiple users to ISDN. It is also known as the Extended Digital Subscriber Line (EDSL). The most common application will be connection to a PBX, although a LAN or other multiuser switching device could be used as well. Figure 8.7 shows how a PBX will interface with the Primary Rate Interface.

There are two standards for the Primary Rate Interface: the North American standard and the European standard. The North American standard is followed by the United States, Canada, Mexico, Japan, and South Korea. Western European countries have solidified behind the European standard.

The North American standard contains 23 B channels and 1 D channel. Both the B and D channels are 64 kbps. The aggregate capacity has been set at 1544 kbps, or as it is more commonly shown, 1.544 Mbps. This is the same bandwidth as a T-1 facility, and it is no accident that they are the same. T-1 is intended to be the chief facility used with the North American standard Primary Rate Interface.

The second Primary Rate Interface standard is the European standard. It differs from the North American standard in that 30 B channels and 1 D channel are used. Its aggregate capacity has been set at 2.048 Mbps.

The increased bandwidth of the D channel allows it to provide signaling for 23 B channels under the North American standard or 30 B channels under the European standard. These may be considered the minimum number of channels that can be supported. In fact, the bandwidth required for signaling will vary depending on

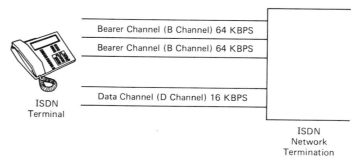

Figure 8.6: The ISDN basic rate interface.

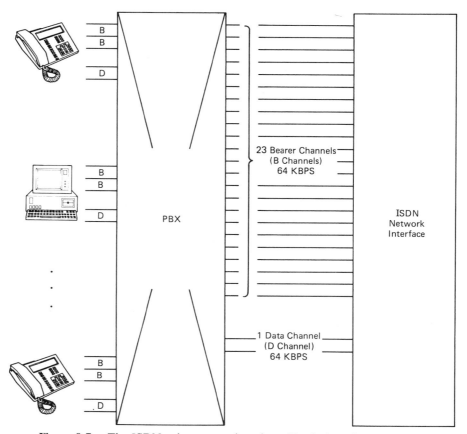

Figure 8.7: The ISDN primary rate interface, North American standard.

the type of calls and the degree to which the B channels are utilized. There is likely to be spare capacity on the D channel. This capacity can be put to use in one of two ways. First, it can handle signaling for B channels in another facility. That means the other facility would not need its own D channel and could have 24 B channels. The second way the spare capacity can be used is for transmitting user information. Some types of user information, such as low-speed data or alarm circuit monitoring, do not require a clear channel and these could be placed on the D channel.

The greater capacity of the Primary Rate Interface over the Basic Rate Interface allows it to support an additional type of channel, the H channel. There are three H channels, each with a different bandwidth. They are:

- The H0 channel (384 kbps)
- the H11 channel (1536 kbps)
- the H12 channel (1920 kbps)

ISDN Standards **225**

The North American standard incorporates the H0 and H11 channels; the European standard incorporates the H0 and H12 channels. All three are clear channels allowing any type of information to pass. Table 8.2 shows a comparison of B, D, and H channels and their expected uses.

Layer 2

OSI layer 2, the Data Link Layer, is responsible for the successful transport of information between the network termination and the user terminal equipment. While the user's information is carried on the B channel, the D channel carries network-generated information that tracks the transfer process. Within this layer, the error checking, addressing, and sequencing are defined. Also defined is the structure for handling these functions. The structure is called the **frame.** A frame may be thought of as a freight train. Each of the boxcars on the train is a field within the frame. Just as there are different types of boxcars, there are different types of fields, each carrying a particular type of information. Some boxcars will always be located in the same place, others will be added or deleted as needed. In the same way, most fields are in a fixed position within the frame. The information field, however, is variable, depending upon the amount of information to be carried. The other fields within the frame are the address field, the control field, and the CRC field. Flags are found at the beginning and end of each frame and act as markers to separate one frame from the next. See the Data Link Layer 2 portion of Figure 8.8 for a picture of the frame.

The address field contains the address of the frame's destination. The control field identifies the type of frame: supervisory frame, information frame, and so on.

TABLE 8.2

Comparison of ISDN Channel Types

B Channel	D Channel	H Channels
Clear channel	Not clear channel	Clear channel
64 Kbps	16 Kbps (BRI) or 64Kbps (PRI)	H0 384 Kbps (NA & E)* H11 1536 Kbps (NA)* H12 1920 Kbps (E)*
Carries user information	Carries signaling information	Carry user information
Such as: Voice Standard data Standard facsimile Slow scan video	Such as: Call setup Call routing Call teardown Frame acknowledgements	Such as: High quality audio High speed data Fast facsimile Full motion video

*NA = North American
E = European

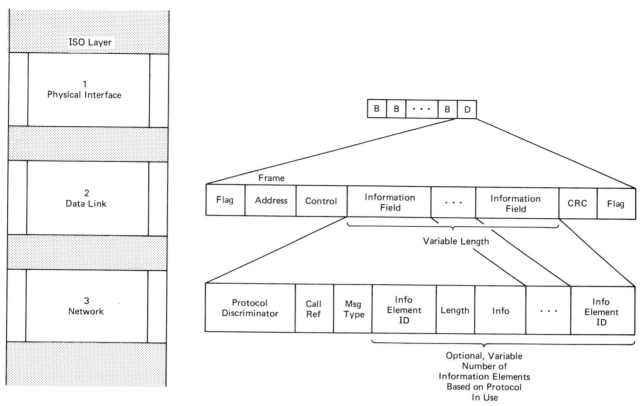

Figure 8.8: Basic Digital Subscriber Line (DSL) frame structure.

The information fields are handled by layer 3 and are, therefore, not examined by layer 2. The CRC field contains a cyclic redundancy check character that is generated when the frame is first created. At the destination, layer 2 generates another cyclic redundancy check character and compares it with the character in the CRC field. If the characters match, the user information is valid; if they do not match, the user information is corrupted and must be discarded. Another copy must be sent from the originating end.

Layer 2 uses the Link Access Protocol—D channel (**LAP-D**) protocol to handle the flow of frames through the D channel. It is a variation of the **LAP-B** protocol used on the B channel. LAP-B allows only one logical link across an interface. LAP-D allows multiple logical links by permitting the address field to change from frame to frame. The flexibility to change addresses explains why a single D channel can control many B channels. Both the LAP-B and LAP-D protocols are based on the High-Level Data Link Control (HDLC), which the International Standards Organization recommends as the standard for level 2 protocols.

Layer 2 has two modes of operation, the Unacknowledged Information Transfer mode and the Multiple Frame Acknowledged Information Transfer mode. The

Unacknowledged Information Transfer mode is unsuitable for most user information transactions, because frames are not acknowledged—that is, once sent, there is no confirmation that the frame is delivered and there is no error recovery. This mode is primarily used for broadcasting and management functions.

The Multiple Frame Acknowledged Information Transfer is the more commonly used mode. In this mode, frames are acknowledged upon receipt and bad frames are retransmitted. Frames are sequentially numbered, and the sequence is checked before the frames are delivered. One feature of this mode is that the originating end may send several frames before receiving any acknowledgments. Both ends keep track of the frames sent and received. For example, if the originating end gets acknowledgments for frames 1, 2, 3, 4, and 6, but not for frame 5, it will check whether frame 5 was received and will retransmit if necessary. The ability to send multiple frames increases network throughput.

Several frames are used with the Multiple Frame Acknowledged Information Transfer mode. Each frame is identified by a message type. The messages are:

- Set Asynchronous Balanced Mode (SABM—Sets mode to allow multiple frame transmission before receipt of acknowledgment
- Disconnect (DISC)—Terminates operation
- Receiver Ready (RR)— Indicates ready to receive a frame
- Receiver Not Ready (RNR)—Indicates temporarily not ready to receive a frame
- Information (I)—Contains information; referred to as an "I" frame
- Unnumbered Acknowledgment (UA)—Acknowledges receipt of unnumbered frames such as SABM
- Disconnected Mode (DM)—Indicates operations cannot be performed
- Reject (REJ)—Requests retransmission of I frames
- Frame Reject (FRMR)—Indicates an error that will not be corrected by retransmission of the identical frame.

Layer 3

Layer 3, the Network Layer, controls the establishment, maintenance, and termination of calls through the network. Establishment involves setting up the call, selecting the type of service, and routing the call. Maintenance requires monitoring the call to ensure that it is not dropped or disrupted before normal termination. Termination is the orderly disconnection of the call.

The information field of the frame discussed above carries data used by layer 3 in performing its functions (see Figure 8.8). Subfields within the information field define its purpose. The first one is a protocol discriminator field, which designates the protocol to be used. Next is a call reference field, which correlates the frame with the correct call. Then comes the message type field, which identifies the frame as a setup frame, an information frame, an acknowledgment frame, and so on. The remainder of the field contains information elements. Information elements consist of an information element identifier, a length indicator, and the information. The information element identifier is a sequence number used to assemble information elements in proper order. The length indicator determines the amount of information to follow.

Some of the messages defined for layer 3 functions are:

- Alerting (ALERT)—Indicates a call is received; triggers ringing or other indication that a call has arrived
- Call Proceeding (CALL PRO)—Call setup in progress
- Connect (CONN)—Indicates the terminating end has connected
- Connect Acknowledge (CONN ACK)—Acknowledgment of connect message
- Setup (SETUP)—Initial request to establish a call
- Setup Acknowledge (SETUP ACK)—Acknowledgment of setup request
- Disconnect (DIS)—Request to terminate a call
- Release (REL)—Indicates release of a facility (e.g., a B channel)
- Release Complete (REL COM)—Verification of the release of a facility

Calling Sequence

A better picture of how layer 3 functions can be seen by examining the sequence a call takes in the ISDN environment. Figure 8.9 shows a sample message flow among the originating end, the ISDN network and the terminating end.

The call setup sequence occurs over the D channel and begins with the originating end sending a Setup message to the ISDN local exchange. The exchange returns a Call Proceeding message to the originating end and forwards the Setup message through the network to the terminating end. The terminating end sends its own Call Proceeding message to its local exchange when the Setup message is received. It may appear that duplicate sets of commands are being passed. In fact, the ISDN network keeps track of both ends of the call and sets up each end independently. This makes for faster overall setup times. The network has sufficient intelligence to tally the message flows at each end and ensure that nothing gets lost.

It is possible that, as each end is being set up, additional messages need to be passed back and forth between the network termination and the local exchange. These messages are optional and are only used in situations such as when the originating end does not initially supply sufficient information for a call to be set up, or when the terminating end has to negotiate for a B channel to carry the call. In such cases, a Setup Acknowledge message will be returned in response to the Setup message, and information messages will pass back and forth until the problem is cleared. Then the Call Proceeding message will be returned as stated above.

Following the Call Proceeding message, the terminating end sends an Alerting message and then a Connect message. The Alert message causes ringing or some other form of alerting to be initiated at the originating end. The Connect message indicates that network connections are complete, and the flow of information can begin.

The flow of information—voice, data, or image—occurs on the B channel. It

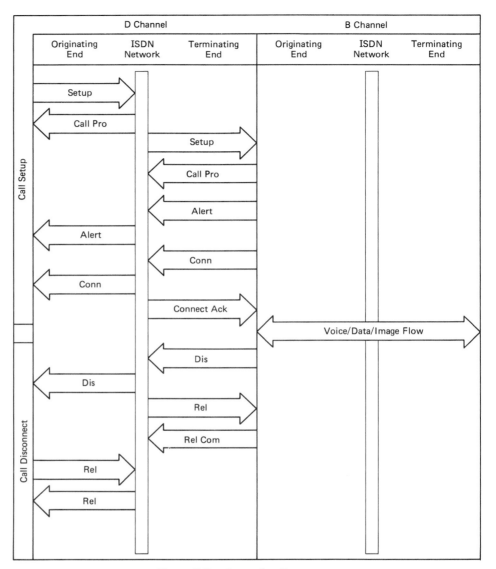

Figure 8.9: Layer 3 call sequence.

continues as long as necessary. At the end of the call, the disconnect sequence takes over and tears down the connection.

The call disconnect sequence again occurs on the D channel. A Disconnect message is sent from the terminating end to its local exchange. The exchange forwards the message to the originating end and sends a Release message to the terminating end. The terminating end responds with a Release Complete message. At this point, the call is disconnected at the terminating end. When the originating end receives the Disconnect message, it sends a Release message to its local exchange, and the exchange responds with a Release Complete message. At this point, the call is disconnected from the originating end.

SIGNALING SYSTEM 7

The implementation of ISDN requires readying the Public Switched Network (PSN) with greater intelligence and uniformity of operation. The method used to accomplish it is called *Signaling System 7,* or SS7. SS7 is a series of recommendations that include the format and content of signaling messages and network design parameters for transferring signaling information. The recommendations are set forth by CCITT and have been accepted as a standard in both the United States and Europe. A key feature of the recommendation is a requirement that out-of-band signaling be used to carry network control information. This conforms to ISDN's use of the D channel for signaling.

SS7 comprises three elements: Service Switch Points (SSPs), Signaling Transfer Points (STPs) and Service Control Points (SCPs). The functions provided by the elements are not new. What is new is the ability to send and receive SS7 messages over a separate signaling channel.

Service Switch Points. Service Switch Points (SSPs) are class 5 central office switches. They connect end users to the rest of the PSN and are usually owned by local exchange companies (LECs). In the SS7 environment, they receive call routing and handling instructions from a Service Control Point. The SSP may access the database itself or may request the information from a Service Control Point.

Signaling Transfer Points. Signaling Transfer Points (STPs) are packet switches that transfer information between network nodes. For example, it would connect calls from a LEC network to a common carrier network. Packet switching is expected to take on increasing importance in the 1990s. That is because the X.25 standard, upon which packet switching is based, is due to be redefined by 1992 to make it more useful in an ISDN environment.

Service Control Points. Service Control Points (SCPS) are nodes that contain computerized database records, provide control information to other nodes in the network, and act as an intermediary to pass information from one node to another. SCPs contain the most intelligence and information of any SS7 elements. The database records will differ from one SCP to the next. One SCP may contain records of the services to which each user subscribes. The records would be used to determine if and how a call can be routed. Another SCP may contain a list of toll-free (800) telephone numbers. Control information would be passed back to the appropriate node.

Centralized versus Decentralized

The description of SS7 elements outlined above sets out a network with centralized control. It is also possible to design a more decentralized network, where database records and routing and control information reside at the SSP node. There are many more SSP nodes than SCP nodes. That means that one SCP node responds to many SSP nodes. Putting information at the SSP node has the advantage of making access to it more direct. However, it also has the disadvantage of requiring that every SSP node be changed whenever information is to be updated. Updates are much simpler in a centralized network because information is stored only at SCP nodes (meaning that fewer locations must be changed). The decision whether to centralize or decentralize is being decided now and will continue through the 1990s. Its impact will mold the type of PSN that will bring us into the 21st century.

THE FUTURE OF ISDN

ISDN is an effort to provide greater end-user control and flexibility through cooperation and collaboration among the major telecommunications companies of the world. The success or failure of ISDN depends largely on the ability of these companies to work in harmony.

An interesting view of ISDN and its future comes from AT&T. The viewpoint stems from AT&T's history as a monopoly. One of its mandates has been to provide universal telephone service. That means making service available to virtually everyone regardless of where they live.

At a United States Telecommunications Assn. (USTA) meeting in 1985, AT&T representatives spoke of ISDN as a stepping stone from universal telephone service to universal information service (see Figure 8.10). The integrated services of ISDN expand the capabilities of existing telephone service. In the same way, ISDN's goals of providing a single mode of access to voice/data/image can be expanded to provide full and free access from virtually any device to virtually any type of information. Getting information as easily as placing a telephone call is the goal of universal information service.

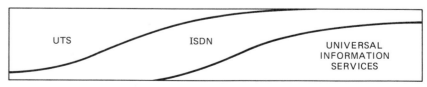

Figure 8.10: Evolution from UTS and UIS.

CHAPTER SUMMARY

ISDN is a plan for establishing a uniform method of transmission for voice, image, and data. This is accomplished by defining standardized network interfaces and by defining the characteristics of transmission between points in the network.

CCITT is the organization with responsibility for setting standards. Its recommendations conform to the ISO Open Systems Interconnect Reference Model. ISDN standards will continue to evolve as user requirements evolve and as new technologies become available.

KEY TERMS

Basic Rate Interface (BRI)
Bearer channel
Clear channel
Data channel
Digital Subscriber Line (DSL)
Extended Digital Subscriber Line (EDSL)
Frame
Integrated Services Digital Network (ISDN)
Integrated voice/image/data (IVID) terminal
Out-of-band signaling
Primary Rate Interface (PRI)

REVIEW QUESTIONS

1. List two features of ISDN that are not available in the Public Switched Network today.
2. What is a B channel? What is a D channel?
3. What is clear channel signaling?
4. Describe Basic Rate Interface.
5. Describe Primary Rate Interface.
6. What is the distinction between TE1 and TE2?
7. Where does the Federal Communications Commission want to locate the ISDN network demarcation point? Explain the reasoning behind its position.

TOPICS FOR DISCUSSION

1. Why are standards important to the success of ISDN?
2. An important factor determining whether ISDN will be a success is the cost of implementing it. What are current cost estimates? Discuss the likelihood of ISDN's success based on cost.

3. Devise a use for some of ISDN's capabilities that would improve the way a business functions.
4. The question whether to implement a centralized or decentralized SS7 design is being debated by LECs and common carriers. Read arguments on both sides in current telecommunications publications and decide which is the better approach. Defend your decision in class.

AT&T PBX Systems

CHAPTER 9

OBJECTIVES

Upon completion of this chapter you should be able to:

- distinguish AT&T PBXs by line capacity
- list features common to all AT&T PBXs
- describe differences in architecture among AT&T PBXs
- name and describe AT&T's proprietary protocol
- name two types of peripheral equipment that can provide enhanced features

INTRODUCTION

AT&T is, by most industry yardsticks, the largest PBX manufacturer in the world. It has more systems installed and sells more systems than any other vendor.

Many have been installed as part of providing overall telephone service. In the 1970s, AT&T introduced a very popular PBX called Dimension. It was a stored program control switch and used a hybrid of digital and analog technologies. It was the first switch by AT&T to offer a host of "central office" type features. These were features that were generally available only on large-size systems, such as central office switches. They included call forwarding, conferencing, and Station Message

Detail Recording (SMDR). These features and many others have become standard on subsequent PBXs.

In 1983, AT&T introduced its first fully digital PBX, the System 85. Since that time other digital PBXs have been introduced, expanding the range of line capacities that can be handled.

Prior to February, 1989, three different PBXs served as the mainstay of AT&T's product line. They were the System 25, the System 75 and the System 85. They differed in size and, to some extent, in architecture. On February 6, 1989, AT&T announced a new PBX called Definity. It represents an evolution of the product line by merging the System 75 and System 85 into a single PBX. We will look at the distinct products first and then follow up with a discussion of Definity and its impact.

SYSTEM 85

Until the announcement of Definity, System 85 was AT&T's premier PBX offering. It is designed to compete in the large PBX market, the most lucrative arena for PBX vendors. It can range in size from a few hundred lines to 32,000 lines, large enough to handle a small city. Most switches are sized far smaller than the maximum, the exact design depending on traffic calling patterns and blockage levels. Private networks often use the System 85 as a node switch because of its ability to handle many types of network interfaces and its ability to tandem.

System Architecture

System 85's architecture is built upon the module. A module contains its own switching matrix and can support up to 254 simultaneous connections. **Carriers** support line **circuit packs,** trunk circuit packs, and related equipment. A module is capable of switching calls to any other line or trunk within the module with minimal intervention from the main processor. Theoretically, 1,536 ports are available, although connecting that many devices could create unacceptable blocking levels, since only 254 could be active at one time.

Increasing the capacity of the switch is done by adding more modules. Each additional module adds another switching matrix. Up to 31 modules can be configured, giving a total capacity of 7,874 simultaneous connections.

For a call to pass from one module to another, it must go through the Time-Multiplexed Switch (TMS). The TMS links all modules together and controls traffic among them. Figure 9.1 shows how the TMS integrates with the modules.

Not only does the TMS connect to the modules, it also connects to the common control, where the main processor resides. The main processor coordinates all activities within the switch. It instructs the TMS how to make connections among the modules. It also receives status reports directly from the modules and issues call processing instructions to them. The channels connecting the common control, the TMS, and the modules operate at 4 megahertz, allowing for rapid handling of multi-

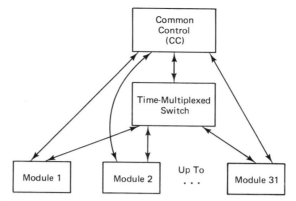

Figure 9.1: System 85 and Definity Generic 2 architecture.

ple calls. The common control is designed to process up to 20,000 calls per hour. That is about 5.5 calls per second!

Sometimes it is desirable to locate a module remotely from the common equipment. This occurs most often in a campus environment, where there are many buildings spread over a large area. A remote module can be used to serve a building or group of buildings. A particular problem with remote modules is providing the high-speed channels necessary for speedy call processing. System 85 solves this problem by routing the channels over fiber-optic cable. The fiber provides a secure, high-capacity connection that allows a module to be located up to 2½ miles from the common equipment.

System 85 also allows placing a single carrier in a remote location. It is called a *remote group* and is linked to the common equipment via a T-1 facility. It can be located up to 100 miles away and can serve up to 23 telephones or data devices. Figure 9.2 shows how modules and groups can be remotely located.

System Administration

An important part of System 85 is the equipment used to administer and maintain it. For example, system maintenance equipment can add new users, change telephone locations, and activate trunks. There are hundreds of tasks that can be performed, each one identified by a numbered procedure, or **proc**. Procs directly alter the settings within the switch's stored program. Procs can also be used to take readings of registers for checking system status or identifying problems.

Two types of equipment are used: a System Management Terminal (SMT) and a Maintenance and Administration Panel (MAAP). The SMT is intended for use by customers. It consists of a light-emitting diode (LED) display, a flipchart for interpreting characters on the display, and an array of buttons to display and execute different procs. Not all procs are accessible through the SMT. Those that assign

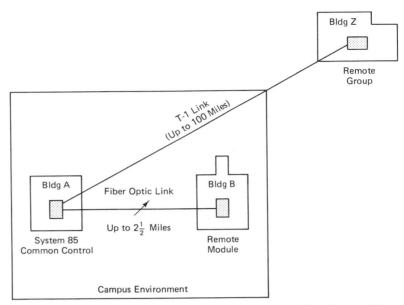

Figure 9.2: Remote modules and remote groups with a System 85.

features, rearrange telephones, and set network routing patterns are available, while those that install new equipment or facilities, or that perform maintenance routines, are not.

These other functions must be performed on a MAAP. A MAAP looks very similar to a SMT, with LED display, flip chart, and button array. However, it is intended for use only by trained service technicians, because it has access to procs that can start a switch or keep a switch from running.

Another way of accomplishing the same thing is with Centralized System Management (CSM). CSM uses a computer to make management tasks easier to perform. Its use and design will be discussed in more detail in the COMMON FEATURES AND EQUIPMENT section later in this chapter.

SYSTEM 75

System 75 is a medium-size PBX with a capacity ranging from 300 to 800 lines. Since its introduction in 1984, it has become AT&T's best-selling machine, with over 10,000 in use. Figure 9.3 shows a picture of the System 75.

It has been well received because it offers a wide range of capabilities at a reasonable price. In addition to handling a standard office, it can be configured for the specialized environment of a hotel, where telephones must be activated and deactivated daily as people check in and check out of rooms. It can also be used as an Automatic Call Distributor (ACD) to handle heavy call volume (see the COM-

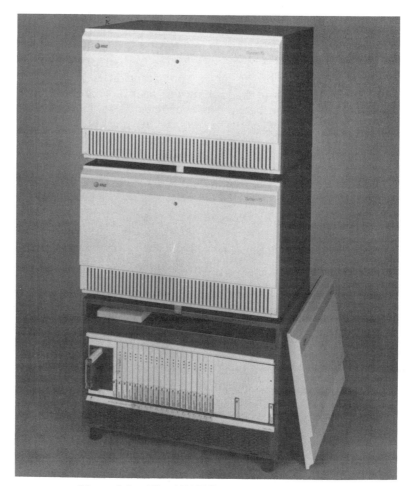

Figure 9.3: The AT&T System 75 Digital PBX.

MON FEATURES AND EQUIPMENT section below). It can carry data traffic and act as a data switch. It can also be a node in a private network.

System Architecture

System 75's architecture is the chief reason for its range of capabilities. It combines stored program control with **distributed intelligence.** Stored program control provides software programmable memory for flexibility in performing many different types of tasks. Distributed intelligence means that each circuit pack has its own microprocessor. It controls the ports on the pack by regulating the flow of information between trunks or terminal devices (telephones and data equipment) and the switch bus. Without distributed intelligence, these functions would have to be performed by the main processor. Spreading them among many processors speeds switch operation and improves overall reliability.

System 75 is also the first AT&T switch designed with **universal carriers.** The slots of the carrier can accept different types of circuit packs (station-line packs, trunk packs, or data-line packs). This is in contrast to a switch like the System 85, where there are different carriers for lines, trunks, and T-1 connections. If a System 85 line carrier is full, another must be ordered, even if there is available space in a trunk carrier. Universal carriers reduce costs by allowing carriers to be completely filled with different types of circuit packs before another is needed.

System 75 Model XE

A smaller version of the System 75 called the System 75 XE, was introduced. It has a size range of 40 to 600 stations and comes in modular cabinets like the System 25 (see below). The Model XE is for users who do not need the full capacity of the System 75, but who want its functionality.

The modular cabinets house a single carrier. A customer can start a system with one cabinet and expand to four, adding only as many as required. This gives a great deal of control over the size, and therefore the cost, of the switch.

System Administration

System administration is performed through a System Administration Terminal (SAT). The SAT is a an AT&T Model 513 BCT, a standard computer terminal. It is much easier to use than the System 85 MAAP and SMT because there are no numbered procs; all instructions are in English and are grouped according to type of function. Screens are displayed in a pleasing format; the layouts and English language commands are built in to the System 75. Both customers and service technicians use the SAT. Access to administrative functions requires a log-in and password—just as a person would access a computer. As in the System 85, the capabilities allowed the customer are more limited than those allowed the technicians. Customers and technicians use different log-ins and passwords, and this is how the system distinguishes the user's capabilities.

SYSTEM 25

The System 25 is the newest and smallest digital PBX offered by AT&T. It is sized for small- to medium-size businesses, accommodating from 30 to 150 users, and is intended for users who need primarily a voice system, but who also want the capability of adding data functionality if desired.

Analog and hybrid telephone sets can be connected to the System 25. A variety is supported, from standard analog sets to Multibutton Electronic Telephone (MET) sets to 7300 series hybrid sets. MET sets are supported on the System 25 as an incentive for customers with Horizon PBXs (AT&T's older, small-sized PBX) to upgrade to newer equipment. The 7300 series sets can be used with the Merlin, a smaller system based on electronic key design, as well as with System 75 and System 85 PBXs. The interchangeability of sets between different systems is an important part of AT&T's marketing strategy. Allowing the same telephone sets to be used with different common equipment makes it easy for customers to change from one AT&T system to another. This helps keep customers loyal to AT&T products and minimizes the expense of changing from one system to another.

System Architecture

Hardware is designed around relatively small cabinets; each one holds a single carrier. The cabinets stack, one on top of another up to a maximum of three cabinets (see Figure 9.4). This approach allows customers to start with a single cabinet and expand to additional cabinets as they are needed.

Most of the slots in the carriers are universal. The first carrier has 10 universal slots, while the second and third cabinets have 12. The difference stems from the Control Circuit Pack, a combined call processor and memory circuit pack, which resides only in the first carrier. It is the master controller for the switch.

The System 25 also utilizes distributed intelligence. It fits well with the switch's building-block approach to adding carrier capacity. Intelligence to handle the in-

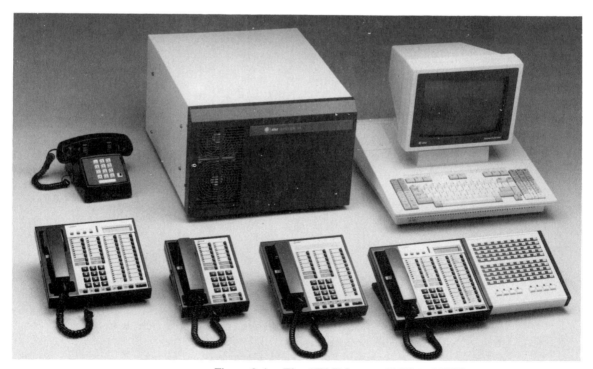

Figure 9.4: The AT&T System 25 Digital PBX.

creased capacity is added with each new circuit pack. By adding it incrementally, costs are not incurred until the capacity actually is used.

The bus uses time division multiplexing (TDM) and has the capacity to handle 115 simultaneous conversations. Once information is on the bus, it is available to all circuit packs. The Control Circuit Pack regulates the flow of information to ensure it gets to the proper destination.

System 25 is the only AT&T PBX that does not use Digital Communications Protocol (DCP). Analog calls come to the common equipment and are converted to pulse-code-modulated (PCM) signals by Station Line Circuit Packs. Digital calls pass from the terminal equipment to the common equipment through **Asynchronous Data Units (ADUs).** As their name implies, these units support only asynchronous transmission. Speeds of up to 19,200 bps can be handled, making them adequate for terminal-to-host interactive sessions. The ADUs essentially boost signal levels so that they travel farther than they could otherwise. ADUs also control voltage levels so that the signals do not interfere with other calls that may be running through the same cable.

An advantage of ADUs is that they require no special wiring. They connect data devices to the common equipment using the same twisted-pair wiring as telephones. Asynchronous data terminals usually have an RS-232C interface. This is where the ADU connects and receives the terminal's digital signals. The signals are then transmitted to a Data Line Circuit Pack in the common equipment. The Data Line Circuit Pack has an ADU built in to it which receives the incoming signals. They are then passed on to the switching matrix where the call is routed to another terminal, a host computer, or a modem. Figure 9.5 shows a diagram of an ADU connection.

System Administration

Administration on the System 25 is performed through a special software program running on a personal computer. The computer is connected to the maintenance port of the switch and provides a menu-driven, easy-to-follow method of administration. The software is called Advance Administration Software (AAS) and allows a customer to add, change, or remove telephone and data users, change available features, or change trunk groups. In this system, the customer has access to all functions.

DEFINITY

Definity is AT&T's newest PBX. It combines the most popular features of the System 75 and System 85 into a single product. For instance, it incorporates universal carriers, single carrier cabinets and SAT-type screen layouts from the System 75. From the System 85 come remote modules, a rich array of features and several networking options.

It comes in two models, Generic 1 and Generic 2. Generic 1 is similar in size range to the System 75 and Generic 2 is similar in range to the System 85.

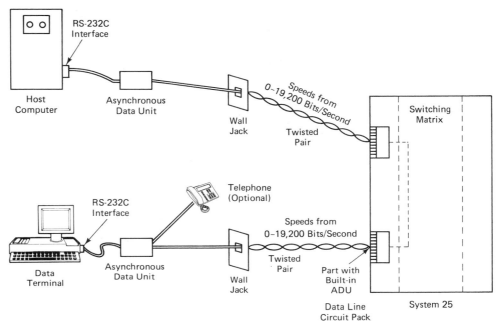

Figure 9.5: Data terminal connection to System 25 with Asynchronous Data Units (ADUs).

Generic 1

Generic 1 is available with modular cabinets (like the System 75 XE) that contain one carrier or standard cabinets that contain up to five carriers. The choice of carriers makes it feasible to offer a line size range from 40 to 1,600 lines.

In is smallest configuration, the Generic 1 is composed of a single modular cabinet equipped with common control equipment and line and trunk ports. When capacity is needed for more users, additional modular cabinets can be added.

At its largest, the Generic 1 contains two standard cabinets. The first cabinet, called the Processor Port Network (PPN), houses the common control equipment and ports for lines and trunks. The second cabinet, called the Expansion Port Network (EPN), is equipped with an expansion interface circuit pack in addition to ports for lines and trunks. The two cabinets are linked by a fiber optic cable. It connects the common control of the PPN cabinet to the expansion interface circuit pack of the EPN cabinet. Fiber provides two benefits, it allows high speed sharing of large amounts of control information and it allows the cabinets to be separated from one another. Normally, all common equipment for PBXs of this size must be situated in one spot. The fiber connection allows the EPN to be located up to 1.8 miles away from the PPN. This capability is similar to the remote modules of the System 85 described above.

Also available with Generic 1 is an ISDN Primary Rate Interface. Recall that

Definity 243

the PRI provides 23 ISDN bearer channels and 1 data channel. Further ISDN enhancements, currently available only on Generic 2, can be expected in the future. These include a Basic Rate Interface and ISDN compatible telephones.

Generic 2

Generic 2 starts with the basic architecture of the System 85: a common control and time-multiplexed switch routing calls to multiple modules. Two types of modules can be accomodated: traditional modules and universal modules. Traditional modules are like those used with System 85. They can contain up to four full size cabinets and support up to 1,536 users. Universal modules are composed of a single full size cabinet which supports up to 1,440 users. This allows nearly as many users to be supported in one-fourth the space. In addition, universal modules accept all types of circuit packs, including new ones for ISDN telephone sets. That means Generic 2 can provide ISDN capabilities all the way to the user's desktop.

System Administration

AT&T announced five new system administration products when it announced the Definity. Called Manager I, Manager II, Manager III, Manager IV and Monitor I, they offer a variety of choices for performing system administration.

Manager I is designed for use with Definity Generic 1. It is virtually identical to the System Administration Terminal (SAT) used with System 75. It allows changes to telephones and facilities, checking the health of system hardware components, and monitoring of system performance. It is the only system administration option available for Generic 1.

The other packages, Manager II, Manager III and Manager IV and Monitor V, are for use with Generic 2. Managers II, III and IV are for different sized systems and offer different levels of sophistication for performing system administration functions.

Manager II is intended for small to medium size systems. It is an MS-DOS based software program that runs on a personal computer. When changes are entered in the computer they are immediately passed to the switch; there is no provision for scheduling (see below). Three different modes of operation are available: basic mode, enhanced mode, and task mode. Basic mode displays proc information from the switch in raw form on the computer screen. This requires the use of separate flip charts to interpret results. While more difficult to use than the other modes, it is included because it can administer System 85 and Dimension PBXs in addition to Definity. The enhanced mode is far easier to use because it displays labels on screen with the proc information. Task mode takes things one step further by grouping commonly performed tasks together. For example, there may be a task for adding new telephones. While there might be ten steps involved in completing the procedure, the task mode would allow information to be entered in one or two screens. Information from the screens then updates all required procs automatically.

Manager III incorporates the enhanced and task modes of the Manager II. In addition, it provides scheduling, modeling, and report generation. Scheduling is the ability to perform a procedure at a later time, for example, in the middle of the night when no users would be disturbed. Modeling allows for standard telephone configurations to be designed and used again and again. They are combinations of features and restrictions used with a particular telephone set. For example, a customer could have one model for office workers, another model for secretaries and another model for managers. Reports allow useful switch information to be printed on a printer. For instance, a report could list the telephone numbers of all users while another could list all the trunks connected to a switch.

Manager III is designed for medium-sized switches and can be used to administer two or more locations. It is a Unix based software program that runs on a super-micro computer.

Manager IV is designed for large switches and for users with multiple locations. It incorporates the features of Manager II and Manager III and has some additional capabilities as well. For example, in addition to handling telephone and facility changes, scheduling, modeling, and report generation are also available. Definity Generic 1, System 75, System 85 and Dimension PBXs can all be administered as well. Manager IV is a Unix-based software program derived from the Facilities Management and Terminal Change Management modules of AT&T's Centralized System Management (CSM) product, described later in this chapter.

Monitor I is a Unix based software program that provides system performance and traffic usage information as an enhancement to the capabilities of Managers II, III or IV. The information is produced in the form of reports which can be used to evaluate if system resources are being used efficiently. Monitor I runs on a super-micro computer and its capabilities match those found in the Traffic Management module of AT&T's CSM product. Like CSM, Monitor I can report on System 75, System 85, and Dimension PBXs in addition to Definity Generic 1 and Generic 2.

COMMON FEATURES AND EQUIPMENT

All of AT&T's PBXs share certain design similarities. Circuit packs slide into slots on a carrier (see Figure 9.6). There are from one to five carriers in a **cabinet,** and there may be from one to dozens of cabinets in the system. The circuit packs have **ports,** which allow the connection of equipment such as telephones and trunks to the system. There are many different types of circuit packs. One may be for analog telephone sets, another for digital telephone sets, and still others for central office trunks or tie trunks. The carriers and cabinets serve to link together the various telephone lines and trunks so that they can be interconnected through the switching matrix.

Other types of equipment and features are used by AT&T's systems as well. The more important ones are: telephone sets, Digital Communications Protocol, Automatic Call Distribution, Station Message Detail Recording, Centralized System Management, Unified Messaging, and networking. They will be discussed below.

Figure 9.6: AT&T PBX's common equipment design characteristics.

Telephone Sets

AT&T's PBXs share the same telephone sets. They fall into three catagories: analog, hybrid, and digital. Analog sets are the 2500 set (TouchTone desk set), the 2554 set (TouchTone wall set), the 500 set (rotary dial desk set) and the 7100 series sets (a variety of single-line, multibutton sets that are no longer actively sold by AT&T). Hybrid sets include the 7200 series sets (an older series that has been replaced by the 7300 series), 7300 series (includes sets used with AT&T's Merlin Electronic Key System), and Multibutton Electronic Telephone (MET) sets (a carryover from an early AT&T PBX, the Horizon). Digital sets are the 7400 series (several models including some with built-in RS-232C interfaces for connection of a data terminal) and the 7500 series. 7500 series sets are only supported by the Definity PBX. Figure 9.7 shows examples of the different types of sets.

Two types of functions are performed with a telephone set: voice conversations (or data transmissions), and activation of system features. An analog set is the most basic type and has a single channel used alternately for voice and features. Features are activated by sending tones from the dialpad, which are interpreted by the common equipment.

Hybrid sets have two channels: an analog voice channel, and a digital feature channel. The digital feature channel is used only for features and not for user data transmissions. Buttons on the set provide access to the features. Pressing one sends

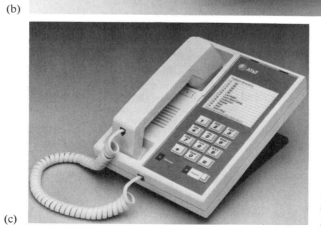

Figure 9.7: Analog (a), hybrid (b), and digital (c) telephones. (Photo courtesy of AT&T Archives)

a digital code along the feature channel to the common equipment, activating the feature.

7400 series digital sets use a proprietary protocol called Digital Communications Protocol (DCP). DCP allows simultaneous transmission of voice and data over a single connection. A detailed discussion appears below. 7500 series digital sets use ISDN's Basic Rate Interface (BRI). This has the same functionality as the 7400 series sets, but is not proprietary and is completely compatible with CCITT standards. (See Chapter 8 for a full discussion of ISDN.)

Digital Communications Protocol

Digital Communications Protocol (DCP) uses a time-division multiplexing scheme between the common equipment and voice or data devices to establish two full-duplex information channels and one signaling channel. One information channel is used for digitized voice, and one is used for data, allowing for simultaneous transmission of voice and data. The signaling channel is used for feature activation, ringing the set, and detecting when a user goes off-hook.

The information and signaling channels are packaged in frames of 20 bits: 8 bits for the voice channel, 8 bits for the data channel, 1 bits for the signaling channel, and 3 bits for framing (overhead, not part of a channel). Frames are sent 8,000 times per second, creating a bandwidth of 160 kbps. 64 kbps are available for the voice channel, 64 kbps for the data channel, and 8 kbps for the signaling channel. Figure 9.8 shows an example of transmission using DCP.

You may notice a similarity between the DCP and the ISDN Basic Rate Interface. The DCP's information channels are similar to ISDN's bearer channels, both consist of two 64-kbps channels that can simultaneously carry voice and data traffic. The DCP's signaling channel performs the same function as ISDN's data channel.

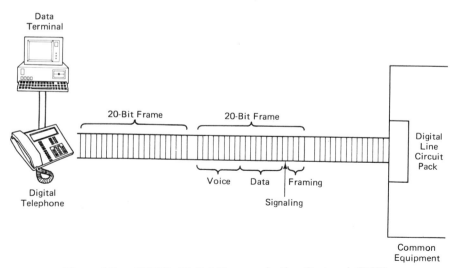

Figure 9.8: AT&T's Digital Communication Protocol (DCP).

This similarity is no accident. AT&T has been an active participant in the formulation of ISDN standards. Incorporating a similar structure into DCP makes the two relatively compatible and simplifies the transfer of information between an AT&T system and an ISDN network.

AT&T allows data devices as well as telephones to use the DCP. Special equipment, called Modular Processor Data Modules (MPDMs) and Modular Trunk Data Modules (MTDMs), convert signals to DCP format. MPDMs connect directly to a data terminal or computer, and MTDMs connect to modems or multiplexers. Both kinds of equipment are about 12 in. long, 8 in. wide, and 3 in. high. A cable is used to connect them to the data device, and standard twisted-pair wiring connects the other end to the switch's common equipment.

Automatic Call Distribution

Automatic Call Distribution (ACD) is a feature shared by all three of AT&T's PBXs. It is a method for handling extremely heavy call volumes. Airline reservation desks, customer service departments, and even telephone operators are often put into ACD groups. The members of the group are known as agents, and they all perform the same job—that is, any agent in the group can answer a caller's questions equally well. Callers dial a telephone number, and the switch routes the call to the ACD. If an agent is available, the call will ring at the free agent's telephone. It there are several agents free, the agent who has been idle the longest will get the call. If all agents are busy, the call is put into a **queue.** A queue is a holding area where calls wait for an agent to become free. The call is then completed.

A PBX may have more than one ACD group. Each group is called a **split** and is independent of any others. Different telephone numbers may be used to direct calls to the appropriate split. For example, an insurance company may have a split for claims, a split for policyholder questions, and a split for general customer information. A different telephone number will be published in the telephone directory for each split, and the PBX will route the calls to the correct split.

ACDs are also used for making outgoing calls. Telemarketing groups often use outgoing ACDs. Telemarketing agents make calls to sell or inform people about a product or service. When they pick up the telephone, the switch automatically connects them to one trunk in a pool of trunks set aside specifically for their use. In this way, the heavy call volume generated by the telemarketers does not interfere with other users of the PBX.

All of AT&T's PBXs offer ACD as a standard feature. There are differences, however, in their capacities. System 25 can support four separate ACDs with a total of 28 agents. System 75 can handle up to 32 splits with a maximum of 200 agents, while System 85 and Definity can manage 30 splits with a total of 1,024 agents.

Station Message Detail Recording (SMDR)

Station Message Detail Recording (SMDR) is used to keep track of calls made by users of a PBX. Every time a call is placed out of the switch or a call comes into the switch, a record is generated.

On outgoing calls, the information that is recorded is the telephone number of the station that made the call, where the call was placed to, the duration, the date, and the time of the call. SMDR information is either printed on a printer or stored in a storage unit. Calls printed on a printer are listed in chronological order; they are output as soon as the call terminates. Stored calls are later processed by a computer. Special programs compile the records into a variety of useful reports. The most common type shows all the calls made by individual users. Other reports, called exception reports, show calls that exceed a predetermined threshold. For example, an exception report may show all calls costing more the $5 or showing all calls longer than 1 hour in duration. Figure 9.9 shows a typical SMDR report.

Incoming calls can be recorded also. It is not as common to keep records of incoming calls, but there are times when it is helpful. Usually, the telephone number of the calling party is not known. That is because the telephone number is known to the originating switch and not the terminating switch. There are two exceptions to this: (1) in a Distributed Communication System (DCS) private network (see below) and (2) in an Integrated Services Digital Network (ISDN) (see Chapter 8). In both cases, the telephone number of the calling station is sent through the network to the terminating switch.

In other situations, where the calling station number is not known, SMDR records the trunk group of the incoming call. In addition, the date, time, and dura-

```
                                    ACME CO.
REPORTING:                                                              PAGE 1
1985 SEP 15                                                    TODAY'S DATE OCT 10

                        DETAIL REPORT: CHRONOLOGICAL
            KEY
$ > > > . > >    COST TOO LARGE
$ # # # . # #    COST INDETERMINABLE

                                        ACCESS      CALLED
  DATE      TIME      DUR      CODE      CODE       NUMBER       STATION    ACC'T      COST
  09/20     14:41     0:02      I          71                     6561                 $  .16
  09/20     14:41     0:01                 80     1-317-468-1234  6364                 $  .39
  09/20     14:42     0:02                 80     1-304-525-5678  6705                 $  .78
  09/20     14:42     0:01       80        86     1-312-555-1212  6650                 $  .00
  09/20     14:42     0:01                  9        629-1234     6327                 $  .18
  09/20     14:43     0:01      I          71                     6869                 $  .08
  09/20     14:43     0:03       84        83     1-605-343-5678  6807                 $ 1.17
  09/20     14:43     0:01      I          71                     6235                 $  .08
  09/20     14:43     0:02      I          71                     6451                 $  .16

                      TOTALS
                      GOOD CALL RECORDS           9
                      TOTAL DURATION           0:14
                      TOTAL COST              $3.00
```

Figure 9.9 A Sample SMDR Report.

tion of the call are recorded. This information can be used to track the number of incoming calls and to determine peak calling times. By looking at the call volume on different trunk groups, statistics can be gathered about where people call from. For instance, a business may have four different 800 numbers (incoming WATS) from four different geographical areas. By compiling reports on the number of calls from each trunk group, the company can determine where they do the most business and where they do the least.

And extension of SMDR is Centralized Station Message Detail Recording (CSMDR). CSMDR is used when a company has more than one PBX. It allows SMDR records from all switches to be collected from a single location. A storage unit, called a Local Storage Unit (LSU) or Call Detail Recording Utility (CDRU), is co-located with a PBX at each site; it stores call record information for the PBX. A polling unit, placed at a central location, periodically calls each LSU or CDRU and collects the stored records. In this way, the records from each PBX are brought to a single location where they can be processed by a computer.

Centralized System Management

Centralized System Management (CSM) is an AT&T offering that allows PBXs at different sites to be controlled from a single location. Any task that can be performed by a System Management Terminal (SMT), System Access Terminal (SAT), or personal computer with Advanced Administration Software (AAS), can be performed by CSM. Most of CSM's capabilities have also been incorporated into the Monitor I and Manager IV products used with Definity PBX's.

CSM is a software program running on a supermicro or minicomputer. It has menus and screens designed to simplify switch management. An administrator makes changes as needed to the screens which create or update records of a PBX configuration. The records are stored in a database and include all of the equipment and facilities at each PBX. When a record is changed, CSM connects to the appropriate PBX and updates it.

This procedure allows an additional benefit not available with a SMT or SAT: changes can be scheduled for a future time. With an SMT or SAT, changes occur immediately because the administrator is interacting directly with the switch. With CSM, the administrator interacts with a computer. The computer can be instructed to save updates until a specified time. At the designated time, a connection is made to a switch and the updates are downloaded. This capability can be very useful when a large number of changes must be done. For example, 200 people will move from one building to another over a particular weekend. An administrator could enter changes days in advance of the move, but not schedule them to be downloaded until after the close of business on Friday. This would allow people to use their telephones normally prior to the move and would still provide them full service at their new location the following Monday morning.

CSM has three major components: Facilities Management (FM), Terminal Change Management (TCM), and Traffic Management (TM). A fourth component, Cost Management (CM), is optional.

Facilities Management allows trunks and trunk groups to be added, moved, and changed. It also allows modification of Automatic Route Selection (ARS) tables and, where there is a private network, modification of Automatic Alternate Routing (AAR) tables. The links between modules within a System 85 can also be administered.

Terminal Change Management (TCM) lets telephone sets and data devices be added, moved, or removed. In addition, the buttons on a multibutton set can be programmed, and features such as class of service, call coverage, and hunting patterns can be administered.

Traffic Management (TM) takes advantage of traffic information available from the PBXs. Registers within the switch accumulate data about how many calls are made, how many are received, how many calls get through on the first attempt, and how many are blocked. TM software collects the data and generates reports. The reports can be used to evaluate where more trunks may be needed or where there are too many. They can also indicate the need for changes to ARS or AAR tables.

Cost Management (CM) is an enhanced form of CSMDR. It collects SMDR records from multiple switches, just like CSMDR. In addition, it has a software program to process the records and produce SMDR reports.

Messaging Services

Part of the effectiveness of a telephone system is the way in which messages are handled. When someone is unavailable for taking a call, a message is usually left. In a small office, a secretary or coworker can take messages relatively easily. In a large organization with hundreds or thousands of users, more elaborate methods are necessary.

AT&T offers several messaging options for System 75, System 85, and Definity. These include: Leave Word Calling, Audix, and Message Center/Directory. One may be used, or they may be used in combination. Combinations give callers a choice, allowing them to leave messages in the way they find most convenient. System 25 has its own offering, called the Voice Messaging System.

The services work in conjunction with telephones, display modules, and computer terminals. All of them require a telephone with a message waiting light. This is simply a light on the telephone that is lighted and unlighted by the messaging equipment. It gives users a visual indication of when messages are waiting. Display modules can be used with Leave Word Calling and Message Center/Directory. They are associated with the 7400 series Digital Telephone Sets and allow messages to be viewed from the user's telephone. The model 7405 set has a 40-character light-emitting diode display, while the models 7406 and 7407 sets have 40-character liquid crystal diode displays. The model 510D Personal Terminal is an integrated voice/data terminal (IVDT) (see Figure 9.10) with a telephone, display screen, and keyboard. The 510D, as well as other computer terminals, can be used to retrieve messages and to update information with the Message Center/Directory service.

Leave Word Calling. Leave Word Calling is a feature that lets a user leave a short, predefined message for another user. If Jim Smith called Susan Jones and

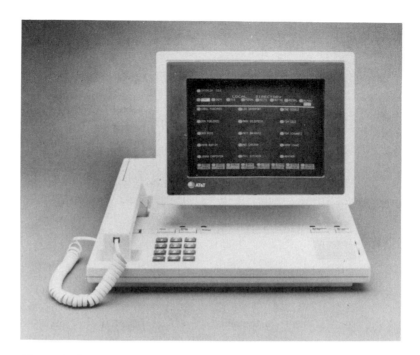

Figure 9.10: The AT&T Model 510D personal terminal, an Integrated Voice/Data Terminal (IVDT).

she was not in, he could leave a message for her by pressing a button on his telephone. The message would say something like, "Jim Smith, extension 1234, called at 2:15 p.m." When she returned, Susan would see the message light lit on her telephone. To retrieve the message, she would need a 7400 series Digital Telephone Set and a display module. System 75 offers another way to retrieve the message. It has a voice synthesizer capable of converting the message to voice form. Any telephone, then, could be used to retrieve the message.

Audix. Audix is a voice mail system. Its name is an abbreviation for *AUDio Information eXchange.* It allows callers to leave a recorded message for system users. Typically, if a user does not answer the telephone, the call routes to the Audix equipment. The caller hears a recording saying, "Susan Jones is not available, please leave a message at the tone." A tone sounds and the caller can leave a message. When the user returns, in this case Susan Jones, she would see the message light lit on her telephone. Upon calling Audix to retrieve messages, she would be told that she has one new message, and then she would be instructed how to retrieve it. When she hears the message, she is hearing the recorded voice of the caller. She can then delete the message from the machine or save it for later reference.

Audix is a specialized type of computer. It digitizes voice messages and then stores them on a hard disk. Two models of the machine are available, the larger having capacity for 4,000 users with 196 hours of storage time. There is a standard

Common Features and Equipment 253

message that callers hear when they are asked to leave a message. If users prefer, they can record their own greeting in lieu of the standard one to make call answering more personalized.

Audix also has the capability of playing an announcement without giving the caller a chance to leave a message. This feature is useful when information is to be given out, but leaving a message is inappropriate. For instance, after a business closes for the day, incoming calls could be routed to an Audix message saying, "The office is closed. Normal hours are from 8:30 a.m. to 4:30 p.m. Please call back during normal hours."

Another feature is broadcasting. This is the capability of recording a message and having it sent to more than one person. For instance, a boss could record a message saying "There is a staff meeting on Monday at 9 a.m.," and have it sent to all employees. The message light on the employees' telephones would light and they would receive the broadcast as a normal message.

Message Center/Directory. Message Center/Directory is a service used by medium- to large-size businesses. It has the advantage of having a live attendant answer calls. When a user does not answer the telephone, the call is routed to the message center. An attendant receives the call and pulls up information about the user on a computer terminal screen. The attendant can tell the caller if the user is in a meeting or out of town, based on the information displayed. If necessary, a message can be taken or the caller can be transferred to an alternative number.

Message Center/Directory makes use of a computer, called an Applications Processor, which is connected to the switch. It contains a database with the names and telephone numbers of all users. Other information, such as a supervisor's name or an alternative telephone number, are also retained. Collectively, this is used as directory assistance information. Itinerary information is also stored, keeping track of whether users are in or out of the office, where they are, and when they are due back. In this way, attendants have detailed directory and itinerary information to give about users.

Messages are also stored in the database. When a user returns to the office, the attendant can be contacted for them. Or since they are stored in a computer, users with a terminal or IVDT (such as the 510D Personal Terminal) can retrieve their own messages without going through the attendant. Another option is to have the messages printed out. The use of a terminal also gives users the means of updating their itinerary information.

Speak-To-Me Messaging is also available with Message Center/Directory. It is a voice synthesizer that converts written messages to speech. This provides an alternative way for users to retrieve spoken messages without going through the attendant. This can be useful if a user wants to pick up messages after hours when no attendants are on duty.

Unified Messaging

In cases where more than one messaging service is in use, AT&T offers Unified Messaging. It uses a "universal mailbox" to let users know where their messages are waiting. Consider, for instance, a system using all three services, Leave Word Calling, Audix, and Message Center/Directory. If a user named Jane sees the message light lit on her telephone, how does she know which service has the message? If she has a digital telephone with a display, she may check Leave Word Calling and find no messages. She would then have to check Audix and the Message Center until she locates it. To prevent users from guessing and wasting time, the universal mailbox lets users find out where all messages are from any of the services. What that means is that if Jane checks her digital display, she will see that a message is waiting at the Message Center.

There are some other convenient applications as well. Leave Word Calling messages can be converted to voice form and retrieved through Audix. Electronic mail services can be tied in to the universal mailbox. If an electronic message or document is received, users can be informed with a referral to the appropriate service. For instance, AT&T has an electronic mail service called AT&T Mail. When messages or documents are received, the AT&T Mail service can contact the user's PBX and light the message waiting light. In addition, a referral is left in the user's universal mailbox stating that a message or document has arrived in AT&T Mail.

Voice Message System. System 25 has its own messaging service called the Voice Message System (VMS). It is an integrated package designed for the small- to medium-size businesses that use System 25. Many of the features of Leave Word Calling and Audix are incorporated in it.

VMS is a software program that runs on a Unix-based AT&T personal computer. Special hardware couples the computer closely with the System 25 allowing message-waiting lamps to be lit and allowing the computer to initiate call transfers.

Five components make up VMS: voice mail, call coverage, automated attendant, announcement service, and message drop. The first three are designed for use by every company. The last two are more specialized and will be useful to some companies, but not to others.

Voice Mail is the heart of VMS. It provides the "answering machine" type services that record messages for users. Easy-to-follow prompts provide message handling functions. Users can pick up messages left for them and can leave messages for others. The standard greeting can be altered to be more individual and personal. Messages can also be composed and edited and then broadcast to a group.

Leave Word Calling is a separately available feature. Users can press a single button on their telephone and leave a short message for others to call them back.

The call coverage component is responsible for routing calls to voice mail when a user fails to answer the telephone. VMS prompts callers to leave a message or allows them to be transferred to the attendant.

Automated Attendant Service augments, but does not replace, a live attendant. When a call comes in, a recorded greeting is heard and then the caller is prompted to dial 1, 2, 0, or an extension number. Dialing 1 or 2 routes the call to a specific department, such as customer service. Dialing 0 directs the call to the attendant, and dialing an extension number puts the call through to the extension.

A customer may not want all calls to go to the automated attendant. In that case, only certain incoming trunks will be routed to it. Others will bypass it, going directly to users or to the attendant.

Announcement Service is the fourth component of VMS. It allows callers to hear a message without being given the opportunity to leave one. This can be appropriate in some cases, such as when the business has closed for the day. Incoming calls would hear an announcement that the business is closed and they should call back during normal business hours.

Message Drop Service is the last component of VMS. It lets callers leave a message, but does not allow them to transfer to another extension or to the attendant. They are limited to leaving a message and hanging up. This feature could be used by a company soliciting opinions, for instance a radio station recording listener response to one of its programs.

Networking

All of AT&T's PBXs can be part of a network. System 75, System 85, and Definity can be full nodes, while System 25 can be an endpoint only. Figure 9.11 shows how a network might be set up.

Full node switches have the ability to examine incoming calls and route them to another location, if necessary. This requires it to be able to recognize when a call is not for its location and to determine where the call should be routed. This process, called **tandeming,** requires the switch to keep tables of other locations in the network. A relatively small switch, such as the System 25, does not have enough mem-

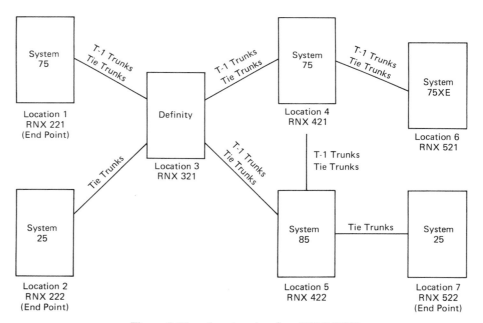

Figure 9.11: A network using AT&T PBXs.

ory for them and therefore cannot be a full node. It can, however, be an endpoint, receiving network calls that terminate at its location.

Analog tie trunks are the facility traditionally used to connect nodes. The System 75, System 85, and Definity support digital facilities as well and can use T1 trunks to link network locations.

Electronic Tandem Network (ETN). An Electronic Tandem Network, or ETN, is the most common application using PBXs in a network. The network depicted in Figure 9.11 is an ETN.

Seven-digit numbers are used to route calls through the network. These numbers are different from the telephone numbers used with the Public Switched Network. Users have two numbers: one for calls from the PSN and one for calls from the ETN.

Each node is assigned its own three-digit location code, called an RNX. For example, Location 3 in Figure 9.11 has an RNX of 321. All telephones connected to this node will have an ETN telephone number starting with 321. The last four digits of the number will be locally assigned so that every telephone has its own unique number. In this way, a user can be reached as easily through the ETN as through the PSN.

The full-node switches also have a feature called **Automatic Alternate Routing (AAR).** It allows for different routes between locations to be programmed into each node. For example, a call from Location 1 to Location 5 would normally go from Node 1 to Node 3 and then tandem to Node 5. If all trunks between Node 3 and Node 5 were busy, AAR would allow the call to go from Node 3 to Node 4 and then to Node 5. This feature is similar in function to Automatic Route Selection (ARS), except that it is applied to private network calls and not public network calls.

Main/Satellite/Tributary Networks. Main/Satellite and Main/Tributary networks allow resources, such as attendant console operators, and facilities, such as WATS or central office trunks, to be shared by more than one location. A smaller location becomes a satellite or tributary to a larger location, called the main. The main has the attendant console operators and facilities. If the smaller location has some facilities of its own— for instance, central office trunks—it is called a tributary location. If it has no facilities (except tie trunks to the main), it is called a satellite location. Figure 9.12 shows an example of a Main/Satellite, Main/Tributary configuration.

Some examples are a System 25 as a satellite or tributary to a Definity. In the same way, a System 75 might be a satellite or tributary to a System 85. A small Definity can be a satellite or tributary to a System 85 or, because of the wide size-range of Definity, a System 85 might be a satellite or tributary to a Definity. It is even possible, though unusual, for a System 85 to be a satellite or tributary to another System 85 or for a Definity to be a satellite or tributary to another Definity.

Distributed Communication System (DCS). Distributed Communication System (DCS) is an enhanced ETN. Not only does it provide network routing of

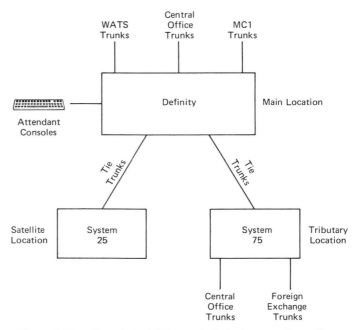

Figure 9.12: A main/satellite, main/tributary configuration.

calls, it also provides transparent use of PBX features through the network. For example, Leave Word Calling is a PBX feature that will not work through an ETN. A user at one node cannot press a Leave Word Calling button and have a user at another node receive a message to call back. In a DCS network, however, this can be done. DCS creates, in effect, one large PBX composed of multiple locations as opposed to multiple PBXs linked together.

The key to providing feature transparency is a special link among all the PBXs called a *Data Communications Interface Unit,* or DCIU. The DCIU establishes a special channel for call information to be passed. With it, every switch in the network can receive data about calls and callers from other nodes and can provide appropriate services for them.

Only System 75, System 85 and Definity have a DCIU. System 25 does not, and therefore cannot take advantage of DCS's feature transparency.

CHAPTER SUMMARY

AT&T is a major manufacturer of telephone equipment. It currently markets four PBXs. All of them have a digital switching matrix and share certain components, such as telephone sets. The Definity represents a merger of two other PBX products, the System 75, and System 85. While the System 75 and System 85 are still available, it is reasonable to expect that they will be phased out in favor of Definity.

The System 85 is a large PBX, with a line size of up to 32,000. It uses a multi-module design allowing for a wide range of sizes. It also supports a remote module that lets locations up to 2½ miles away be included.

The System 75 is a mid-range offering, ranging from 40 to 800 lines. Its features include distributed intelligence and universal carriers. Two models are available, the full-size System 75 and a smaller model, the System 75 XE, with single-carrier, stackable cabinets.

AT&T's smallest PBX is the System 25. It shares the single-carrier cabinet design with the System 75 XE. It has a line size ranging from 30 to 140 stations. Designed for smaller users, it supports telephone sets from AT&T's Merlin Electronic Key Systems.

Each of the PBXs has system administration capability. Definity has several options: Manager I for Generic 1, Manager II for Generic 2, and Manager III, Manager IV, and Monitor I as enhanced system administration products. System 85 uses a System Management Terminal, System 75 uses a System Access Terminal, and System 25 uses Advanced Administration Software running on a personal computer. System 75 and 85 can also be administered through Centralized System Management.

Many features are available on the PBXs. Among them are Automatic Call Distribution (ACD) groups, Station Message Detail Recording (SMDR), and Unified Messaging.

The PBXs can also be linked together into networks. Main/satellite networking is available on all the PBXs. Electronic Tandem Networking (ETN) and Distributed Communication System (DCS) are available on the larger PBXs.

KEY TERMS

Asynchronous Data Unit (ADU)
Automatic Alternate Routing (AAR)
Cabinet
Carrier
Circuit pack
Distributed intelligence
Port
Proc
Queue
Split
Tandeming
Universal carrier

REVIEW QUESTIONS

1. What is a universal carrier?
2. How does a 7300 series Telephone Set differ from a 7400 series Telephone Set? How does a 7400 series Telephone Set differ from a 7500 series Telephone Set?
3. What is the function of the TMS in a System 85 and Definity Generic 2?
4. What is a queue?

5. Why is a Remote Group limited to serving 23 telephones or data devices? (Hint: Remote Groups connect via T-1 facilities.)
6. Explain the difference between Station Message Detail Recording (SMDR) and Centralized Station Message Detail Recording (CSMDR).
7. How does Centralized Station Message Detail Recording (CSMDR) differ from Centralized System Management (CSM)? (Ignore the Cost Management component for the purpose of this question.)
8. Explain the difference between ARS and AAR.
9. How does a Main/Satellite Network differ from a Main/Tributary network?

TOPICS FOR DISCUSSION

1. Discuss the effect of AT&T's history as a monopoly on its position as a telephone equipment vendor. Does its past help or hinder its ability to sell in a competitive marketplace?
2. At the time AT&T was broken up in 1983, it was predicted that increased competition would lower the price of telephone service. Research the change in rates from 1983 to the present. Discuss the effect of competition on the cost and quality of telephone service.
3. AT&T sells a full range of products. Not only does it market PBXs, but it also markets electronic key systems, long distance network services, computers, telecommunications products (such as modems and multiplexers), and consumer products (such as answering machines and designer telephones.) Discuss whether this broad product line gives AT&T an advantage over its competitors. Interview an AT&T sales representative and get his or her perspective, then interview one of AT&T's competitors.
4. Is the Definity an improvement over the System 75 and System 85? Is it anything more than a re-packaging of the older systems under a new name? What does it say about AT&T's direction for the future?

Northern Telecom

CHAPTER 10

OBJECTIVES

Upon completing this chapter, you should be able to:

- describe how an IBM PC network or Macintosh AppleTalk local area network can communicate together through a linkage with a Northern Telecom private branch exchange

- describe how a LANSTAR X.25 gateway enables packet-switched data to flow from an SL-1 to a value-added network such as Telenet

- indicate how voice and data can be transmitted simultaneously on an SL-1 network

- explain how data can flow from an SL-1 network to an IBM mainframe host computer

INTRODUCTION

In this chapter we will take a close look at Northern Telecom's methods for providing voice/data integration and network management and control. We'll note how its PBXs can function as local area networks and can also provide compatibility with a variety of different standards, including the X.25 public networks, IBM mainframes, Apple computers, and DEC computers. We will also see now Northern Telecom is positioned to provide ISDN services in the immediate future.

GENERAL BACKGROUND

As early as 1980, Northern Telecom introduced integrated voice data switching. Initially this task was accomplished using an **add-on data module (ADM)** and a data line card (DLC). Today there are numerous different ways of transmitting integrated voice and data with a Northern Telecom private branch exchange (PBX). The Meridian SL-1 systems illustrated in Figure 10.1 range from 32 to 5,000 lines, while the Meridian SL-100 systems have a capacity of 30,000 lines that handle both voice and data. Figure 10.1 illustrates a Northern Telecom Meridian SL-1 with Packet Transport Equipment. One way is through use of an integrated voice/data terminal such as Northern Telecom's Displayphones, as illustrated in Figure 10.2. With a Displayphone it is possible to use standard two-pair telephone wire to transmit simultaneous voice and data to a maximum distance of 1,200 meters (4,000 feet) on 22 AWG wire.

Simultaneous Voice and Data Transmission

Simultaneous voice and data transmission is possible with a Northern Telecom PBX. Figure 10.3 illustrates a QPC 311 Data Line Card that contains two voice ports and two data ports. The user of an SL-1 Displayphone or an SL-1 telephone with co-

Figure 10.1: Northern Telecom's Meridian SL-1ST. (Photo courtesy Northern Telecom)

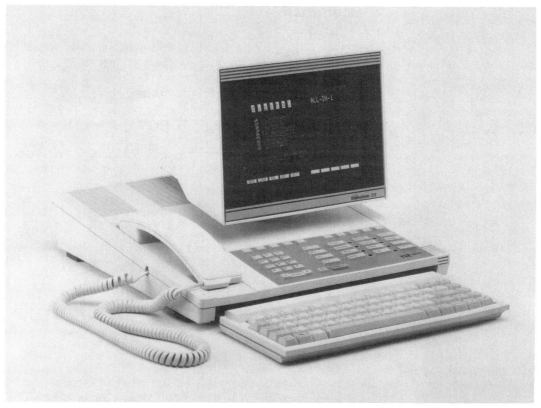

Figure 10.2: Northern Telecom's Displayphone is an integrated voice/data terminal. (Photo courtesy Northern Telecom)

located ADM can receive and send voice calls and data simultaneously. A microprocessor permits the QPC311 to separate voice and data signals and send them to the appropriate places.

A **Digital Trunk Interface (DTI)** provides an interface between an SL-1 and the North American standard T-1 interface using a 1544-Mbps circuit divided into 24 channels, each of which is available for voice or data transmission. The DTI can utilize fiber optics, microwave, satellite, or leased lines. Each channel can transmit information up to 56 kbps synchronous and 19.2 kbps asynchronous. Since the Meridian SL-1 network loop utilizes a 32-channel 2.048-Mbps bit stream, the Digital Trunk Interface is responsible for ensuring that information moves smoothly from this loop to the time slots available on the 24-channel T-1 circuit. A special clock controller is required to meet the Bell System Stratum 3 synchronization criteria. A truly redundant system will contain two such clocks. Figure 10.4 shows how information from terminals and personal computers enters the Meridian SL-1 and is then switched via a Digital Trunk Interface to a T-1 link. The data may go through a

General Background

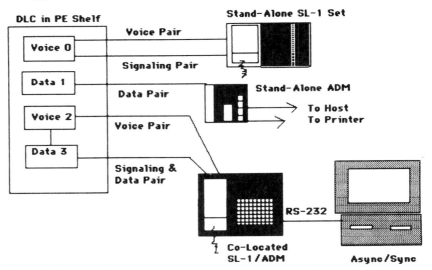

Figure 10.3: Simultaneous voice and data transmission with an asynchronous/synchronous interface.

digital Central Office (CO), another Northern Telecom PBX, or another vendor's PBX before arriving at its ultimate destination, another computer or computer terminal.

Using the SL-1 as a Data Switch

There are a number of other ways to switch data through the Northern Telecom SL-1. An add-on-data-module (ADM) can operate in either synchronous or asynchronous mode while linking together data terminals, computers, and printers. The ADM (QMT8) permits synchronous transmission from 1,200 to 56,000 bps while permitting asynchronous transmission from 50 to 19,200 bps. An advantage of switching data through the SL-1 switch is that the computers can utilize such PBX features as auto dial, speed call, and ring again. As we pointed out previously, an ADM in co-located mode can provide simultanous voice and data transmission with an SL-1 set's two-twisted-pair wiring. The ADM is connected to an SL-1 via a data line card, as seen in Figure 10.5. It is also possible to link a Synchronous Data Module (QMT12) to an Sl-1 via a Data Link Card. The Synchronous Data Module permits synchronous transmission between data terminals up to 8,000 feet apart.

Figure 10.5 also illustrates a Multi-Channel Data System (MCDS), a method of providing high-density asynchronous computer data transmission more efficiently. A maximum of 64 asynchronous ports can be configured for each MCDS, with each port capable of being configured between 110 and 19,200 bps. Each port

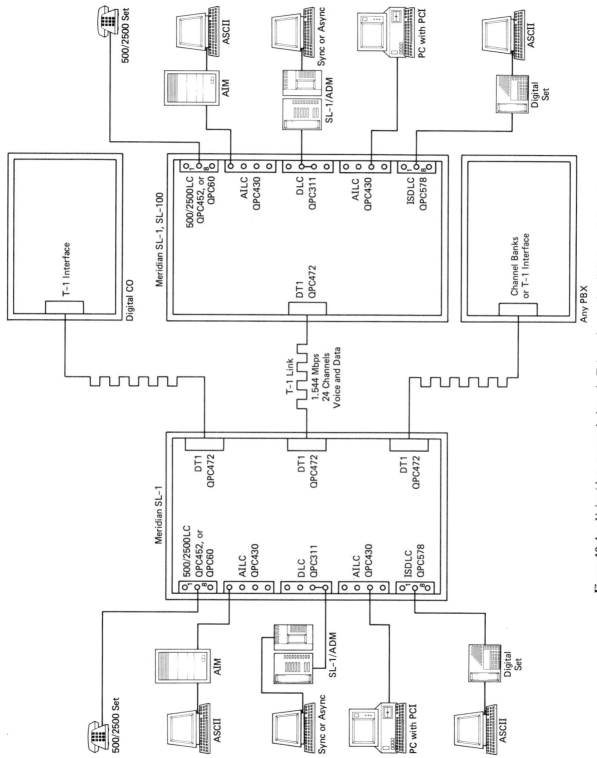

Figure 10.4: Voice/data transmission via T-1 carriers and a digital trunk interface.

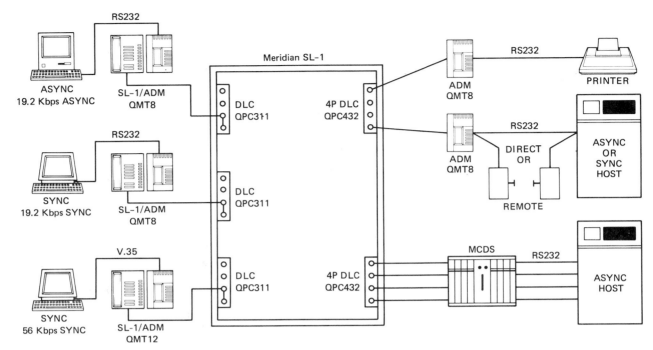

Figure 10.5: Data communications via an SL-1 switch.

can automatically adjust itself to the speed of the terminal accessing it and operate independently of the other ports.

LANSTAR

Another way of transmitting simultaneous voice and data is through Northern Telecom's **LANSTAR,** a local area network that utilizes the powerful switching capabilities of Northern Telecom's SL PBX family. LANSTAR uses standard twisted-pair telephone wiring. It requires that each IBM PC, PC-compatible, or Apple Macintosh II microcomputer on the network contain its own network interface card. As Figure 10.6 illustrates, this local area network has a star topology, since it is built around the Northern Telecom switch.

While each microcomputer must contain its own network interface card in one of its expansion slots, the Northern Telecom PBX requires a LANLINK Assembly Card in the Meridian LANSTAR cabinet. Each card is capable of supporting up to 16 personal computer connections. A single Meridian LANSTAR packet transport cabinet is capable of supporting up to 1,344 personal computer connections. A smaller cabinet is available for standalone network configurations that support up to 112 users. Figure 10.7 illustrates the hardware required.

Each personal computer that is part of LANSTAR is capable of receiving and

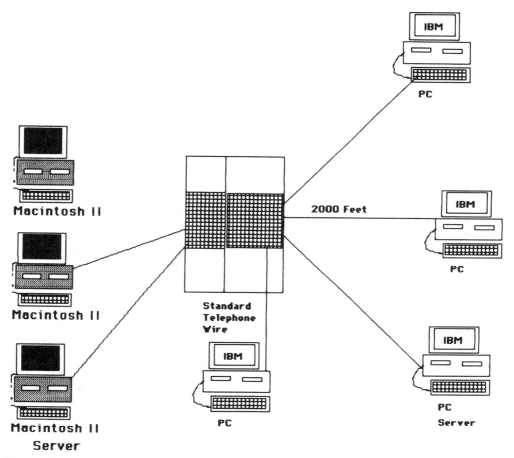

Figure 10.6: LANSTAR's network includes PCs and Macintosh computers. (Courtesy Northern Telecom)

transmitting information at 2.56 Mbps through the standard telephone twisted-pair wiring. The network uses baseband full-duplex signaling with a full 2.56 Mbps dedicated to each workstation. This is a contention network that utilizes request/grant perfect contention method. The Meridian LANSTAR packet transport provides a total of 40-Mbps bandwidth. As we mentioned, LANSTAR can function independently as a standalone local area network, but it also can be fully integrated with the Meridian SL-1 by means of a T-1 type 24-channel connection.

APPLE MACINTOSH COMPUTERS AND LANSTAR

Apple's AppleShare network file-server software runs unmodified on LANSTAR. Up to 50 users can simultaneously view and have access to the same folders. Each network workstation can transmit data at 2.56 Mbps with a dedicated connection

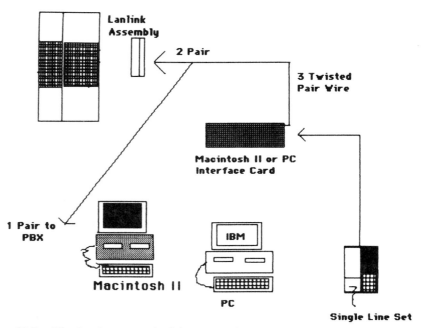

Figure 10.7: The hardware required for a LANSTAR. (Courtesy of Northern Telecom)

to the packet transport cabinet. Networked Macintosh II computers can be up to 2,000 feet from the Meridian LANSTAR packet transport. We are making the assumption, of course, that the software is specifically designed network versions. Using programs such as MultiTalk and ComServe it is possible to have remote access to LANSTAR. An AppleTalk Bridge software package is also available that runs as a background application and permits the linking together of several different AppleTalk networks, with a LANSTAR network running AppleTalk software. It is possible for IBM PCs and compatibles with AppleTalk PC adapter cards to be bridged into the LANSTAR network and access files. Figure 10.8 illustrates how a Meridian LANSTAR with LANSTAR AppleTalk can link together Macintosh computers using twisted-pair wiring.

MERIDIAN LANSTAR NETBIOS AND IBM PC NETWORKS

As we pointed out in Chapter 4, there are a number of local area networks utilizing IBM PCs and compatibles, which networks use IBM's NETBIOS as a basis for their network software. Since LANSTAR is NETBIOS-compatible when it uses a NETBIOS emulation program, it can be used in conjunction with such software as Banyan's VINES, IBM's own IBM PC Network Program, and Torus's Tapestry. This means that Meridian's LANSTAR can become a DOS-based IBM PC network.

Under this arrangement, LANSTAR permits multiple file servers on the net-

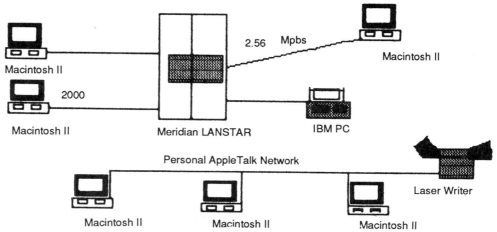
Figure 10.8: Meridian LANSTAR and LANSTAR AppleTalk.

work, each one accessible by workstations on that network. A single PC can establish up to 26 simultaneous disk-drive connections, yet the network user need not worry about the mechanics of this operation, since connections to all network resources remain transparent. Security is possible on such a network at the DOS level and also possible at the directory, subdirectory, and file levels by assigning passwords. It is also possible to assign read/write/create attributes to limit users' access to data. A LANSTAR network running the NETBIOS emulation program is capable of terminal emulation because of its Personal Computer Terminal Emulator Program (PCTE). This program permits asynchronous communications and emulates an IBM 3101 terminal, TTY, and DEC VT 52 and VT 100 terminals. While security is not sophisticated, there is password protection with up to eight characters for the password.

MERIDIAN LANSTAR VINES

Some companies will need far more network services than it is possible to obtain from a relatively low-cost DOS network software package running under Meridian's LANSTAR NETBIOS emulation program. Banyan's **VINES** program upgrades the quality of network service found under Meridian LANSTAR by offering electronic mail, communications with both asynchronous and synchronous host computers, access to wide-area networks, and integration with a number of heterogeneous networks.

With Banyan's VINES and Meridian LANSTAR it is possible to link together a variety of different company networks that have different topologies, hardware, and software. Departments utilizing IBM's Token-Ring Network and Ethernet, for example, can be tied together with the corporate headquarters network running

VINES software under Meridian LANSTAR. Figure 10.9 illustrates how several different types of networks can be linked together utilizing Meridian LANSTAR.

A network user under this arrangement views all network resources as locally attached devices. VINES includes a distributed database program called StreetTalk that keeps track of all network users and resources and dynamically updates any changes globally across all network servers. There is a browse feature that enables users to scan through the database and identify network users and resources.

In addition to electronic mail, users can also communicate using a **CHAT** feature. Messages can be broadcast from one personal computer to another or to all personal computers on the network. The CHAT feature enables several users to carry communication together simultaneously and thus hold a conference of sorts.

In a corporate environment there may be need for far more sophisticated communications capabilities than the CHAT feature. Companies may require the ability to link together several offices that each contain their own mainframe computers, local area networks, and Meridian LANSTAR networks. Figure 10.10 illustrates how two offices utilizing Meridian LANSTARs can communicate via X.25 packet switching, while at the same time they also can communicate with other offices via asynchronous 9,600-bps modems and 9,600-bps synchronous modems. Asynchronous communications include emulation of an IBM 3101, DEC VT 52, and VT 100, as well as ANSI and TTY terminals. The network terminal emulation software makes an IBM mainframe believe it is talking to an IBM 3270 controller. Workstations on the network emulate 3270 terminals, while network printers are capable of IBM printer emulation.

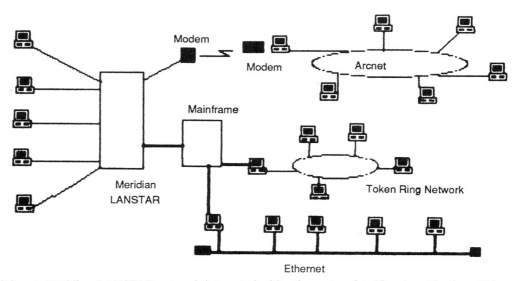

Figure 10.9: A Meridian LANSTAR network integrated with other networks. (Courtesy Northern Telecom)

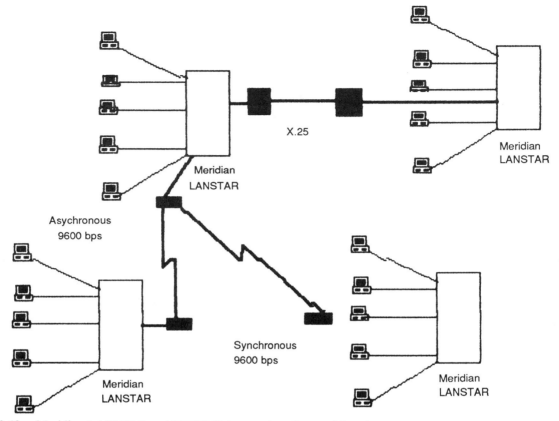

Figure 10.10: Meridian LANSTAR and VINES links together dispersed locations. (Courtesy Northern Telecom)

CONNECTING THE MERIDIAN LANSTAR TO THE IBM WORLD OF DATA COMMUNICATIONS

Asynchronous Interface Line Cards (AILCs) and Asynchronous Interface Modules (AIMs)

There are a number of different ways to link the Meridian LANSTAR with the IBM mainframe world. Remember, the non-IBM world tends to be asynchronous, while mainframe IBM computers transmit using a synchronous format. As Figure 10.11 illustrates, there are a number of different interface methods for data to enter a Meridian SL-1. Data from an Apple Macintosh, an IBM PC, or a standard RS-422 terminal can enter a Meridian SL-1 through an **Asynchronous Interface Line Card (AILC).** Each AILC (QPC430) provides four RS-422 ports. The AILC resides on a Peripheral Equipment (PE) shelf in the Meridian SL-1. As we see in Figure 10.11,

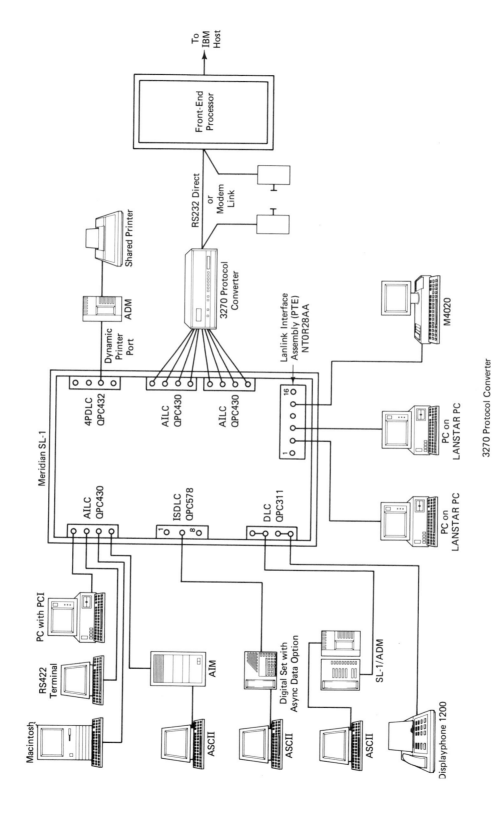

Figure 10.11: An SL-1 linked to an IBM mainframe via a 3270 protocol converter.

a standard ASCII terminal can be connected to an **Asynchronous Interface Module (AIM),** which in turn can be connected to the AILC. The AIM converts RS-232 signals to RS-422 signals before transmitting them to the AILC. It is equipped with a standard RS-232/CCITT V.24 interface and features 8-bits ASCII protocol. It may transmit at speeds ranging from 110 to 19,200 bps.

The Asynchronous/Synchronous Interface Module (ASIM)

Northern Telecom also offers an **Asynchronous/Synchronous Interface Module (ASIM)** designed to accommodate both asynchronous and synchronous modes of transmission. The ASIM (QMT11) is connected by single twisted-pair wire to an SL-1 data port on the QPC311 or QPC432 Data Line Card. The ASIM can support transmission speeds of up to 19.2 kbps in asynchronous mode and 56 kbps in synchronous mode. One advantage of such an approach is that it eliminates the need for separate synchronous and asynchronous modules.

While it is possible to dial from a keypad or keyboard on a terminal attached to an ASIM, this device also supports the connection of a 500/2500-type telephone or an SL-1 electronic telephone. Notice in Figure 10.11 that a Northern Telecom Displayphone 1200 can be directly connected to a Data Line Card on the SL-1.

Protocol Converters

Figure 10.11 also illustrates the connection between personal computers on LAN-STAR and the Meridian SL-1. The networked PCs are linked to the PBX through the LanLink Interface Assembly. Once data is within the SL-1, it must be converted into a form that the IBM mainframe world can understand. Notice in Figure 10.11 that it takes two AILCs or eight ports to link the SL-1 to a 3270 protocol converter. This link utilizes 25-twisted-pair wiring with two twisted pairs of wires required for each port. Notice that a protocol converter can be linked directly to a Front End Processor or it may be connected via a modem link. It is possible to have multidrop connections with several 3270 protocol converters sharing the same leased line. Data travels from the protocol converter to the Front End Processor at 9,600 bps. The 3270 protocol converters emulate the IBM 3271, 3274, and 3276 cluster controllers. An IBM mainframe looks at information coming from the protocol converters and assumes that it is dealing with IBM 3270 display terminals and printers. Each protocol converter permits seven ASCII terminals to hold conversations simultaneously with the mainframe; it is possible to have multiple protocol converters attached to an SL-1.

The 3270 Emulator Software

An alternative to using protocol converters is to use the SL-1's 3270 Emulator software. This software emulates an IBM 3274-1C or 3274-51C Controller and permits communication directly with a Front End Processor. The Front End Processor may be linked directly to the SL-1 with an add-on data module (RS-232 or V.35) or by use of an Asynchronous/Synchronous Interface Module in synchronous modes.

There are three different emulation modules: one to emulate a controller using bisynchronous protocol (3274 BSC CCE), one for a controller utilizing SNA/SDLC protocols (3274 SNA CCE), and a terminal emulation module imitating 3278/3277 IBM terminal emulation (3278/3277 DE). To communicate with a mainframe, an SL-1 would need one cluster controller emulation module and the terminal emulation module. Notice in Figure 10.11 that in addition to personal computers under LANSTAR that can utilize this terminal emulation, it is also possible for an integrated voice/data terminal such as the Meridian M4020 to provide such emulation. Up to 16 ASCII devices may hold simultaneous sessions with a mainframe computer under the emulation software. It is also possible to run multiple emulation packages.

The Coax Elimination and Switching System (CESS)

The **Coax Elimination and Switching System (CESS)** is a way for a company to connect its IBM 3270 terminals (3178, 3278/9, 3180) to IBM 3274/6 cluster controllers via a Meridian SL-1 PBX. In the process the company can reduce substantially the amount of coaxial cable needed, by replacing much of the cabling with ordinary twisted-pair wiring. IBM display terminals can be attached to a Meridian SL-1 Data Line Card via a Coaxial Interface Module (CIM). The CIM is connected to the SL-1 via one pair of ordinary twisted pair wiring. The CIM supports additional pairs for a Meridian Digital Telephone, SL-1 Electronic set, or a 500/2500 telephone.

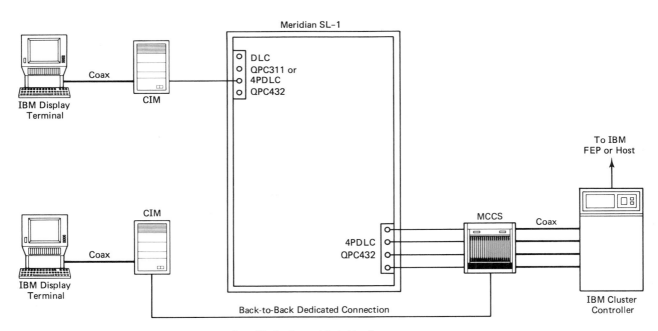

Figure 10.12: The Meridian SL-1 coax elimination and switching system. (Courtesy Northern Telecom)

As Figure 10.12 illustrates, the SL-1 is connected by standard twisted-pair wiring to a Multi-Channel Coax System (MCCS), which, in turn, is connected via coaxial cabling to an IBM Cluster Controller. Each MCCS consists of 16 Coax Interface Cards (CICs) while each CIC has two ports. It is possible, therefore, for each MCCS to provide up to 32 ports.

Notice that in addition to the switched access to the cluster controller, display terminals also can be linked back-to-back by a dedicated connection consisting of twisted-pair wiring. Both the Coaxial Interface Modules and the Multi-Channel Coax Systems can be located up to 4,000 feet from the PE cabinet of a Meridian SL-1. The MCCS can be connected to multiple cluster controllers. It is also possible to use PBX features such as keyboard dialing, auto dial, ring again, and **hotline.** The hotline feature means that for users who need call only a single destination a search for an available port is made as soon as the terminal is powered on.

CONNECTING THE SL-1 TO PUBLIC AND PRIVATE PACKET NETWORKS

Northern Telecom offers an SL-1/X.25 packet assembler/disassembler (PAD) that links a Meridian SL-1 to private and public packet-switched networks. The PAD contains two RS-232 modem ports, as seen in Figure 10.13, which each can transmit and receive data at up to 19.2 kbps. Each PAD supports eight or sixteen ports; if necessary, it is possible to have multiple PADs attached to a Meridian SL-1. Notice that the PAD is connected to the Meridian SL-1 by an Asynchronous Interface Line Card (AILC).

Meridian SL-1 users who wish to access X.25 networks enjoy all the PBX features of the SL-1, including auto dial, speed dialing, and ring again. It is also possible for remote terminals to use the PAD by using inbound modem pooling.

NETWORK MANAGEMENT: NORTHERN TELECOM'S MERIDIAN DATA NETWORKING SYSTEM (MDNS)

The **Meridian Data Networking System** is a highly system-fault-tolerant transport system using intelligent nodes for the purpose of connecting different networks. The transport system contains a T-processor and a Data Transport System (DTS). The T-Processor circuit-switches both voice and data, the latter by compressing digital voice data. The Data Transport System is used only for packet-switching data.

MDNS is a network management tool for centralizing control of information. It provides real-time surveillance, dynamic network reconfiguration, failsafe configuration updates, and non-disruptive moves and changes. Its software automatically reroutes calls if a link goes down. MDNS requires password identification for user access.

What makes MDNS particularly attractive is the number of different networks it supports. In addition to IBM's SNA, PCs and ASCII terminals, and X.25 protocols, it also supports the IEEE 802.3 and 802.5 standards as well as the OSI model's

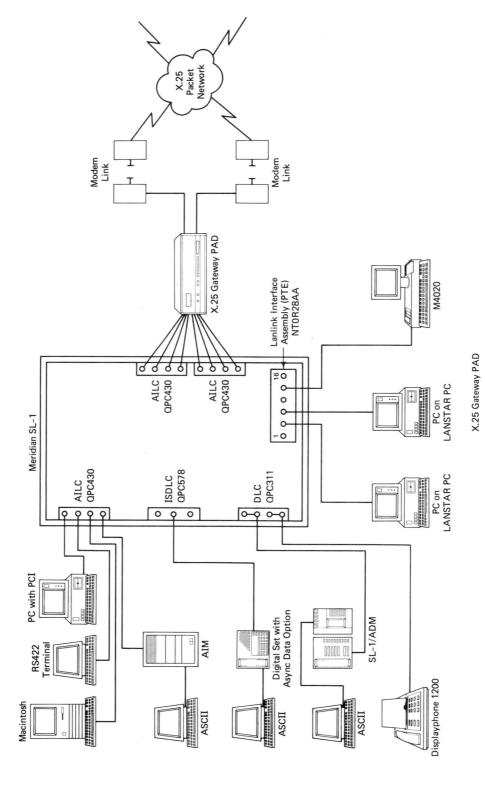

Figure 10.13: An X.25 PAD attached to a Meridian SL-1. (Courtesy of Northern Telecom)

X.400 protocol and TCP/IP. Northern Telecom also announced support for DECnet and AppleTalk; eventually MDNS will be ISDN-compatible. MDNS provides connectivity to host and PC-based applications running under Unix, MS-DOS, OS/2, DEC VMS, IBM VM, and MVS operating systems. It provides access to database applications, file transfer, electronic mail, and print services.

MERIDIAN CUSTOMER-DEFINED NETWORKING

As the Integrated Services Digital Network (ISDN) becomes a reality, there will be more and more network functions required by Northern Telecom users including complete compatibility with ISDN, the ability to interface with a variety of different vendors' products, and integrated voice and data transmission on both public and private networks. The company has developed **Meridian Customer-Defined Networking,** a plan for an open architecture with which to address these needs. Meridian Customer Defined Networking comprises three major elements: Meridian Network Services, Meridian Network Control, and Meridian SuperNode.

Meridian Network Services

Meridian Network Services is the software component of Meridian Customer-Defined Networking. It includes a number of features that enhance a network's efficiency and flexibility by permitting a Meridian SL-1 to implement its features network-wide and yet make it appear that the entire network consists of a single node. Users at separate locations can still utilize features such as Calling Line Identification, Caller's Name Display, Network Ring Again, and the message center capability of Network Call Answering Services. Future software enchancements will enable Meridian Mail to permit voice messaging to and from remote locations.

Additional tools found in Meridian Network Services include ACD Networking, Enhanced Trunking, and Network Conferencing. ACD Networking means that there can be several ACD centers that can be monitored by a central office. Changes to regional ACD configurations can be handled from this central office.

Enhanced Trunking will be a result of ISDN and will mean more efficient use of facilities. With the ability to access multiple carriers and various services more easily, costs will drop. By use of dynamic channel assignment, for example, the very same trunk can emulate Central Office/Direct Inward Dialing (CO/DID) trunks, FX, a WATS services, and private lines. **Call-by-call Service Selection** another advantage of Enhanced Trunking, means that each call's channel assignment is based on its unique requirements. Figure 10.14 illustrates this feature. Least-cost call routing is still another example of enhanced trunking ability. Calls are routed through the most economical network channels and even switched from private to public networks when a private network route is unavailable. Meridian Network Services will also include Network Conferencing. Meridian MS-1, a 64-port digital conferencing system, can be networked to provide operator-controlled voice and data sessions.

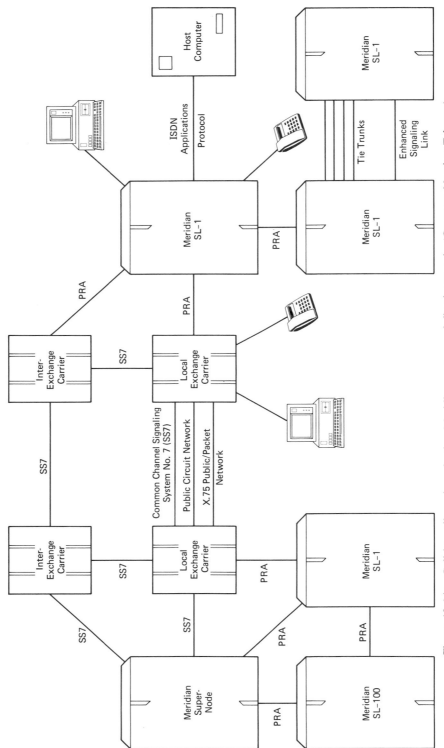

Figure 10.14: Call-by-call service selection Meridian customer-defined network. (Courtesy Norther Telecom)

Meridian Network Control

Meridian Network Control includes tools for enhanced network management, diagnostics, planning and design, and customer support services. At the moment, Meridian Network Control is designed to provide more efficient control of voice, data, text, and video images over T-1 facilities, but in the future these tools will be available for use with ISDN.

Consistent with Northern Telecom's open architecture philosophy, Meridian Network Control will support the company's **Network Operations Protocol (NOP)** which is based on the OSI model discussed in Chapter 2. Meridian Network Control is compatible with IBM's NetView. It supports T-3 as well as many 36 T-1 trunks. Equally important, this network management tool supports ISDN, since it is built around Northern Telecom's **ISDN Applications Protocol,** a protocol that the company is trying to establish as an industry-wide standard.

Figure 10.15 illustrates the Meridian SL-1 Primary Rate Interface circuit pack, one of two circuit packs and a software module necessary to add ISDN Primary Rate Access (PRA). It provides twenty-four 64-kbps channels, with one of these channels—known as the D channel—used to carry the signaling information for the other 23 channels. These channels can carry voice, data, and visual images. These channels can be configured to support standard T-1 connections or used for ISDN PRA service.

Meridian SuperNode

The third element of Meridian Customer Defined Networking is **Meridian SuperNode,** a private network node that can be added to existing corporate networks to increase processing power and network management capabilities. The SuperNode consists of a number of separate processors that "offload" the processing required for Meridian Customer Defined Networking so that the increased programming and network management features do not result in reduced network efficiency and speed. SuperNode utilizes a 32-b microprocessor and a 128-Mbps bus as well as

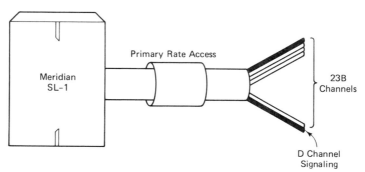

Figure 10.15: The Meridian SL-1 primary rate interface circuit pack. (Courtesy Northern Telecom)

interfaces for both Integrated Services Digital Network (ISDN) and Signaling System 7 (SS7).

Figure 10.16 illustrates the power of Meridian SuperNode in the implementation of Meridian Customer Defined Networking. Here we see both private and public networks linked together with X.75 public/packet networks and ISDN applications flowing from a host computer through a Meridian SL-1. We also see ISDN links between a Meridian SL-100 and Meridian SL-1 and between the Meridian SL-100 and Meridian SuperNode. The figure also illustrates how existing Centrex facilities can be incorporated into this custom hybrid network. The Meridian SL-1 and Meridian SL-100 PBXs provide ISDN primary and basic-rate interfaces to private network nodes and interexchange carriers as well as local exchange carriers. It is the Primary Rate Interface that provides this key digital link.

Northern Telecom is well positioned to be a major player in a market where voice/data integration is becoming a major requirement on the shopping lists of

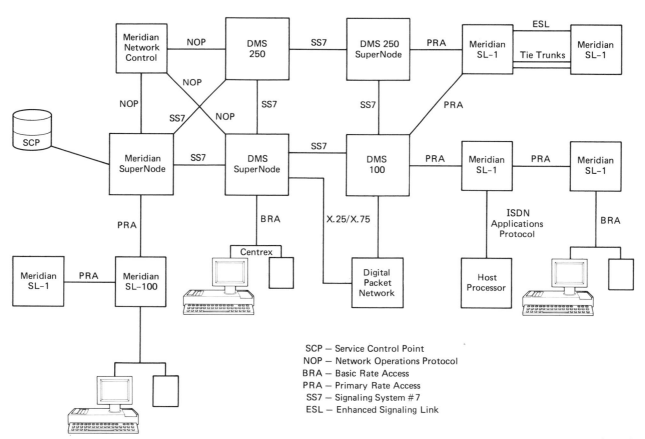

Figure 10.16: Meridian customer-defined networking incorporating Meridian SuperNode. (Courtesy Northern Telecom)

Fortune 500 corporations. Its major strength is its open architecture, which makes it possible to add features such as ISDN to its PBX equipment by simply adding additional cards and circuit packs to its cabinets. Its ability to link together seemingly incompatible vendor equipment, such as Apple's Macintosh, IBM's PC, and IBM's mainframe computers, through LANSTAR is a major advantage over some competing PBX vendors. Perhaps equally important for the immediate future, its Meridian Customer Defined Networking is a plan that ensures that the company will not be left behind when companies rush to build hybrid networks incorporating ISDN.

CHAPTER SUMMARY

As one of the leading PBX vendors, Northern Telecom offers a wide range of PBX voice/data products. Users may connect analog as well as digital peripherals and asynchronous as well as synchronous equipment to a Northern Telecom PBX. A Northern Telecom switch can be used as a local area network in conjunction with Meridian LANSTAR. It is also possible to interface a PBX to IBM mainframe equipment by use of a protocol converter and to public networks by use of an X.25 PAD.

Northern Telecom makes it possible to switch both voice and data information to a wide variety of different computer systems using Meridian Data Network System. Meridian Customer Defined Networking, Meridian Network Services, and Meridian SuperNode constitute the three basic elements of Northern Telecom's plans for providing ISDN services.

KEY TERMS

Add-on-data module (ADM)
Asynchronous Interface Line Card (AILC)
Asynchronous Interface Module (AIM)
Asynchronous/Synchronous Interface Module (ASIM)
CHAT
Coax Elimination and Switching System (CESS)
Digital Trunk Interface (DTI)
Enhanced Trunking
Hotline
ISDN Applications Protocol

LANSTAR
Meridian Call-by-Call Service Selection
Meridian Customer-Defined Networking
Meridian Data Networking System (MDNS)
Meridian Network Control
Meridian Network Services
Meridian SuperNode
Network Operations Protocol (NOP)
VINES

REVIEW QUESTIONS

1. Using Northern Telecom equipment, what are some of the different ways that a company can link its personal computers with IBM mainframe computers?
2. How is LANSTAR able to link together Apple Macintosh computers with IBM PC computers?
3. If internal corporate communications are very important, what features does VINES offer running on Northern Telecom equipment that might be beneficial?
4. What advantages would a hybrid network under Customer Defined Networking have over today's conventional voice/data networks?
5. Why does Meridian SuperNode increase network efficiency?
6. For a company with both asynchronous and synchronous products, how can they link together these units more efficiently with Northern Telecom equipment?

TOPICS FOR DISCUSSION

1. Under what conditions would it be appropriate to use a Northern Telecom PBX as a local area network? When would it be inappropriate?
2. Investigate what kinds of security protection are available with an SL-1 switch.
3. Research the type of management reports, particularly SMDR reports, available from Northern Telecom PBXs.
4. The secondary PBX market is rapidly expanding. AT&T supports resale of its equipment. Research Northern Telecom's position on this issue.
5. Investigate the costs involved in increasing the number of lines and stations incrementally for a Northern Telecom PBX.

IBM/Rolm/Siemens and other Major PBX Vendors

CHAPTER 11

OBJECTIVES

Upon completion of this chapter, you should be able to:

- discuss how IBM integrates Rolm PBX systems with its data communications products
- discuss some of the significant voice/data features of the Ericsson PBX system
- compare and contrast Harris's PBX systems with those of other major vendors

INTRODUCTION—IBM/ROLM PBX SYSTEMS

It is futile to describe specific PBX systems in much detail, since these models are constantly changing. The purpose of describing the IBM/Rolm family of PBX systems here is to provide an understanding of the general way in which the company has chosen to integrate its voice and data transmission and specific methods of achieving voice data network. These networks link together microcomputers, local area networks, minicomputers, and mainframe computers as well as provide gateways to the various wide area networks available today. Although the Rolm label will disappear as a separate entity, IBM's competitive position ensures that the company will continue to be a major player in the telecommunications marketplace. As such, this company provides a logical benchmark against which competitors will have to be measured.

Integrated Voice/Data Terminals and Station Equipment

IBM/Rolm offers a wide range of telephone station equipment including industry standard 2500-type phones as well as digital ROLMphones. A ROLMphone 400, a multiline full-function digital telephone provides a 60-character liquid crystal display. It has an integrated speakerphone and up to 29 lines. A Data Communications Module (DCM) is a circuit card containing a microprocessor, which when installed in a Rolm phone provides either asynchronous or synchronous communications. This function is accomplished by attaching an asynchronous or synchronous data terminal to the DCM through an RS-232 serial interface attached to the DCM.

ROLMlink serves as the foundation of IBM/Rolm's voice/data communications. The voice and data terminals are attached to the CBX PBX systems through their DataComm modules and twisted-pair wire. ROLMlink converts analog voice waves into digital signals and multiplexes them along with digital data over a single twisted-pair telephone wire. Customized computer chips at both the CBX end and the voice terminal end perform the control and data transmission functions.

ROLMlink operates at 256 kbps. 64 kbps is used for the digitized voice signals, while 64 kbps is used for digitized voice signals. The remaining 128 kbps is used for control signals and error detection. Using 24-gauge twisted pair, ROLMlink permits units to be separated from the CBX by as much as 3,000 feet. Extended ROLMlink and 22-gauge twisted pair extends this distance to 4,500 feet.

Rolm's Cypress terminal is a digital terminal that can emulate both a DEC VT100 or VT220 ASCII terminal and an IBM 3270 terminal. It also can emulate ADDS Viewpoint and Data General Dasher terminals. These terminals include their own microprocessor, telephone dialing pad, and function keys, as well as a serial port for printer connection. It is possible to convert an IBM PC into a voice/data terminal by attaching the Juniper II adapter. ROLMlink data switching using twisted-pair-wire technology means that the converted PC becomes a fully operational voice/data terminal with all the capabilities of the Cypress.

Rolm/IBM offers a family of PBX systems with a range of sizes beginning with the Redwood, which has a number of limitations we will discuss. The VSCBXS has a capacity of 208 lines, while the CBX II 8000 has a capacity of 600 lines. Rolm/IBM does offer very large switches including the CBX II 9000 (up to 10,000 lines) and the CBX II 9000AE (up to 20,000 lines).

T-1/D-3 Gateways

The Rolm T-1/D-3 feature enables companies to send both voice and data over leased lines or private lines using either a T-1 rate of 1.544 Mbps or the standard D-3 24 channels. A group of 10 circuit cards installed in a CBX cabinet become the functional equivalent of a D-3 or D-4 (mode 3) channel bank. A Rolm CBX II with a T-1/D-3 Interface eliminates the need for analog-to-digital (A/D) and digital-to-analog (D/A) conversion requirements, since it is handled internally by the switch.

The T-3/D-3 Interface is able to process digital information directly from the CBX II TDM bus without the necessity of A/D and D/A conversions. Figure 11.1 illustrates this arrangement. The 24 channels can carry voice information only or a combination of 22 voice channels and two data channels.

For point-to-point communications, a data channel can be connected directly to a T-1/D-3 data card pictured in Figure 11.2 to achieve speeds of 9.6 kbps for

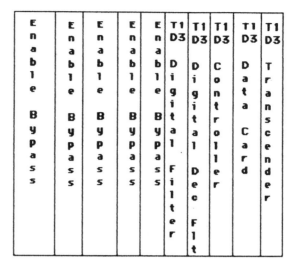

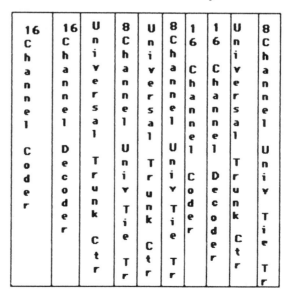

Figure 11.1: The Rolm T-1/D-3 circuit cards.

Introduction—IBM/Rolm PBX Systems   285

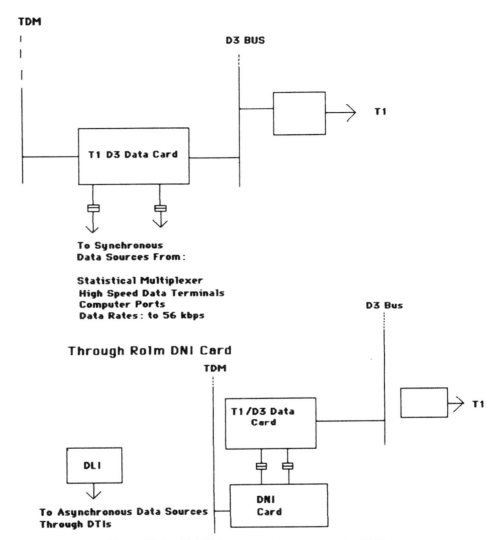

Figure 11.2: T-1/D-3 transmission on a Rolm CBX.

asynchronous data and 56 kbps for synchronous data coming from a computer or multiplexer.

The **Data Network Interface (DNI)** card permits up to 63 simultaneous asynchronous data connections with speeds up to 19.2 kbps. Since a Rolm T-1/D-3 gateway supports up to two data network interface cards, we are really looking at a maximum of 126 simultaneous data connections along with 22 voice channels. The voice and data signals are integrated into one voice/data digital stream of information. Figure 11.2 illustrates how a data channel can be connected directly to a T-1/D-3 Data Card and how it can be connected through a Rolm DNI Card.

Since our primary concern in this book is integrated voice/data networks, it is significant that this Rolm T-1/D-3 Gateway does provide effective voice/data integration. Figure 11.3 illustrates how two data devices such as computers located at both ends of a T1 facility with CBX IIs equipped with T-1/D-3 interfaces can communicate. Note that interface converters are needed for other than RS-449/422 interfaces because of the balanced (to ground) signal characteristics of the RS-422 transmission.

Switched Networks Using a T-1/D-3 Gateway and Intelligent Network Processors

Intelligent network processors (INPs) dynamically allocate T-1 bandwith. These devices can be used in conjunction with Rolm CBX II switches and T-1 Gateways to

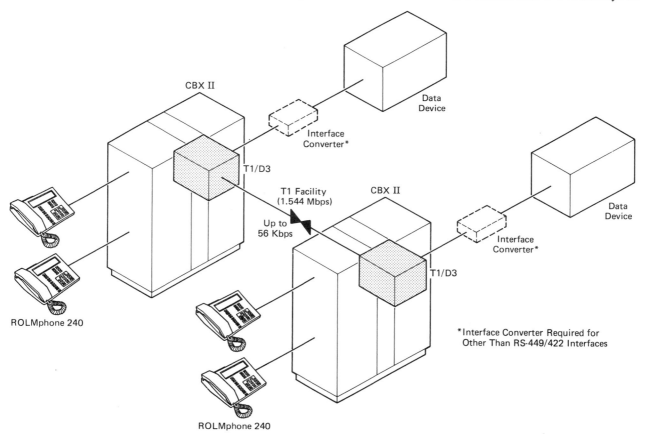

Figure 11.3: Integrated voice/data transmission with a CBX II over T-1 lines.

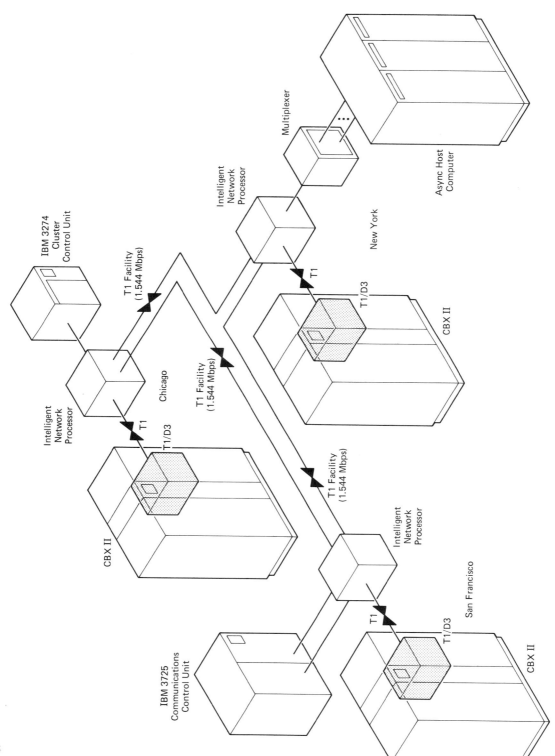

Figure 11.4: Switched networking options.

provide alternative routing in the event of a facility failure as well as both synchronous and asynchronous communications. Figure 11.4 illustrates how intelligent network processors can dynamically allocate certain T-1 bandwiths so that, in the event of a failure along one link between the Chicago and San Francisco offices, another link can be used. Notice how it is possible to transmit and receive both synchronous data from the IBM 3274 cluster control units and 3275 communications control unit and asynchronous data from an asynchronous host computer.

ROLM NETWORK MANAGEMENT

The Automatic Facilities Test System (AFACTS)

The Rolm **Automatic Facilities Test System (AFACTS)** provides automatic internal testing of Rolm CBXs out to analog and digital network facilities. AFACTS provides automatic internal testing of such network facilities as tie trunks and central office trunks. It is capable of testing such transmission and signaling characteristics as continuity, frequency response, and noise level. Exception reports provide information on the type of failure, that is, the reference limit exceeded with the actual deviation from the expected level. They also provide trunk status information, including whether the trunk is in danger of failing, failed, or taken out of service. A summary report provides the status of all test-designated trunks in the system. Since the trunks are tested only when they are idle, AFACTS does not interrupt normal service. AFACTS can also test individual voice channels within a T-1/D-3 trunk group.

AFACTS is also capable of automatically removing any trunk from service if it does not meet performance specifications. AFACTS restores this trunk to service when it does meet specifications. The user can specify an upper limit on the number of trunks taken out of service, so that AFACTS will simply indicate them as failed but not remove them from service. Figure 11.5 indicates a typical AFACTS network configuration, while Figure 11.6 shows a typical AFACTS trunk test report.

Data Communications with the CBX PBXs

The CBX voice/data controller is designed to handle both voice and data transmission. Personal computers can transmit and receive files through the switch without the need to link them together into a local area network. It is also possible to maximize the efficiency of existing modems by pooling the modems and providing access through the PBX. This system permits the network administrator to establish data group names for computer resources and modems so that these resources can be accessed by name. The CBX supports both full- and half-duplex synchronous transmission as well as full-duplex asynchronous operations.

Figure 11.7 illustrates the data connectivity available through a CBX system. Rolm phones equipped with synchronous/asynchronous data communications modules (DCMs) can be connected to the CBX voice/data controller through a ROLM-link Interface card (RLI). Data communications software residing in the CBX works

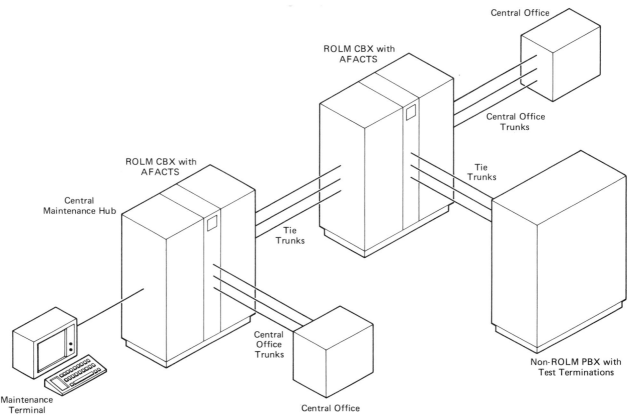
Figure 11.5: An AFACTS network configuration.

in conjunction with an Enhanced Communications Processor (ECP) card to offload the data call-setup function from the processor. A Rack-Mount Data Communication Module (RMDCM) provides a synchronous or asynchronous connection from the CBX to data terminal equipment such as a host computer.

It is also possible to install an Asynchronous Rack-Mount Data Communications Module (ARMDCM) in the CBX family. The ARMDCM provides two asynchronous ports on a single card.

ROLM gateways provide another method of interfacing asynchronous terminals connected to a CBX and IBM's SNA environment or BSC environment. Each ROLM BSC Gateway can support up to eight types of asynchronous devices, while each SNA Gateway can support up to six types of asynchronous devices. Both gateways can support one or two host lines to a 37XX front-end processor. A maximum of 15 asynchronous devices can be connected to a single gateway, and multiple gateways can be connected to a CBX switch.

A system administrator can reconfigure ROLMphones equipped with data

	Trunks Assigned	Busy	Tested	Analog Warn	Analog Fail	Signaling Fail	TOS*	BBS**
Current Pass								
No Test	0							
Test #1	0	0	0	0	0	0	0	0
Test #2	25	1	24	0	0	1	0	0
Test #3	0	0	0	0	0	0	0	0
Test #4	15	1	14	2	0	2	2	0
First Pass								
No Test	0							
Test #1	0	0	0	0	0	0	0	0
Test #2	25	0	25	0	0	0	0	0
Test #3	0	0	0	0	0	0	0	0
Test #4	15	0	15	1	0	0	0	0

Maximum taken out of service limit = 2
Current taken out of service count = 2

*Number of trunks Taken Out of Service.
**Number of trunks Brought Back into Service

Note:
Test #1: Seizure
Test #2: Seizure, 1004 Hz Level
Test #3: Seizure, 1004 Hz Level, Frequency Response
Test #4: Seizure, 1004 Hz Level, Frequency Response, Noise Level

Figure 11.6: An AFACTS trunk test statistics report.

communications modules (DCM) to either synchronous or asynchronous transmission using the Keypad Call Setup (KCS) feature. Figure 11.8 illustrates modem pooling using both synchronous and asynchronous modems. It also illustrates how IBM personal computers as well as the Cypress voice/data terminal and ROLMphones can communicate with asynchronous and synchronous computers. Transmission can take place with speeds of up to 19.2 kbps for asynchronous transmission and 64 kbps for synchronous transmission.

Integrated Wide-Area Networks Using the CBX Family

It is possible to design several network "hubs" that are linked together by dedicated T-1 lines. Network Equipment Technologies' IDNX acts as a digital bandwidth manager in this arrangement to maximize the available bandwidth and to ensure backup routing in the event of a facility failure. Outlying CBX switches can access the network backbone via dedicated lines or public-switched networks.

Figure 11.8 illustrates how three network hubs can be linked together. Notice how these hubs include network management functions performed by IBM's NetView/PC software, phone-mail access to all three hubs, access to a local area net-

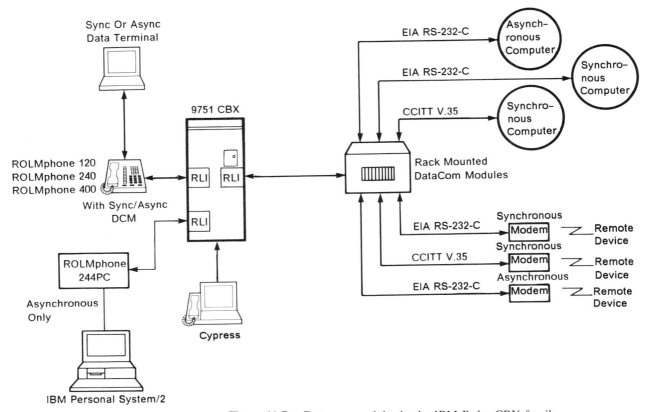

Figure 11.7: Data connectivity in the IBM Rolm CBX family.

work, and access to both analog and digital PBXs. There is a gateway to X.25 public networks as well as access to an IBM mainframe computer featuring IBM's NetView network management software.

IBM Communication Network Management Architecture

We have just observed how it is possible to link together an IBM mainframe computer, microcomputers, an IBM Rolm CBX switch, and non-IBM PBX systems. What makes it possible to accomplish these tasks is IBM's commitment to an open, modular communications architecture that accommodates both SNA and non-SNA components. Under this architecture, IBM defines a **focal point** as a network component that facilitates the consolidation of network management data from all components in a network. IBM's NetView software, running on its mainframe computers, often serves this function. IBM defines **entry points** as distributed locations for control of SNA resources. IBM 3174 cluster controllers are examples of entry points. It is here that network data (related to distributed SNA addressable resources) becomes available to a focal point. Finally, IBM defines **Service points**

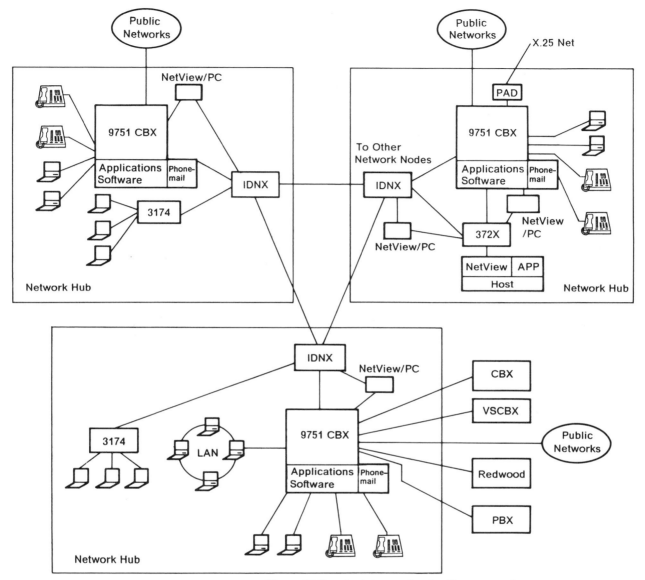

Figure 11.8: An integrated voice/data network.

as locations for the management of non-SNA resources. Service points have interfaces to link together SNA and non-SNA environments. IBM's NetView/PC network management application software serves as an example of a service point.

The different components of IBM's communication network can all contribute information and provide some level of network management. The IBM/Rolm CBX switch provides the AFACTS system (discussed earlier), which can handle trunk management and produce management reports (pictured earlier in this chapter). The CBX also provides traffic statistics and call detail recording reports.

An IBM PS/2 microcomputer running NetView/PC can collect data for Rolm's call detail reports, maintain the CBX attendant directory, and provide Rolm Alert Monitor functions. It can list the alerts, including information about each alert, provide the probable causes of each alert, and even recommend action. It also can provide the option of generating a problem record for the specific alert data in question. In addition, NetView/PC provides the ability to create, delete, or modify network problem records. It even can serve as an alarm clock providing reminders about network scheduled services such as scheduled configuration changes. Since NetView/PC supports LU 6.2 and 3780/RJE file transfer protocols, it is able to transfer information to a host IBM computer for further processing.

The IBM mainframe computer running NetView provides overall information management, while running customer application programs. It routinely receives all alerts concerning network failures, provides this data to the network operator, and provides status information on all network components.

Figure 11.9 illustrates IBM's communication network management architecture. Notice how the IBM mainframe host computer with its various application programs and operating systems serves as the focal point for this communications network. It is running the NetView network management program under Virtual Terminal Access Method (VTAM) and the Network Control Program (NCP). The Service points here serve as interfaces to non-IBM PBX systems as well as non-IBM data communications devices that are not running under SNA.

ERICSSON VOICE/DATA INTEGRATION

Ericsson offers a modular MD 110 system that can serve as few as 160 stations (MD 110/20) or as many as 10,000 stations (MD 110/90). The two major building blocks of the Ericsson modular design are the **Line Interface Module (LIM)** and the **Group Switch Module (GSM).** Each Line Interface Module contains its own microprocessor-based control system and is capable of functioning as an autonomous PBX or as a part of a much larger integrated system when several LIMs are linked together by Group Switch Modules.

A Line Interface Module supports up to 172 voice and 172 data ports. Analog and/or digital lines and trunks can be connected to any LIM in the system. Each LIM can have up to four PCM lines (120 channels) connecting it to a Group Switch. The PCM lines provide up to 32 serial channels carrying voice or data at 2.048 Mbps, with each of 30 channels capable of carrying a full 64 kbps of voice/data signals. Two channels are reserved for synchronization and signaling.

A Group Switch is required to link together three or more LIMs. It is possible to have one LIM processor access resources such as trunks and service circuits under the control of another LIM processor. Since each LIM is independent, the failure of one LIM does not cripple the entire system. Each LIM contains regional software

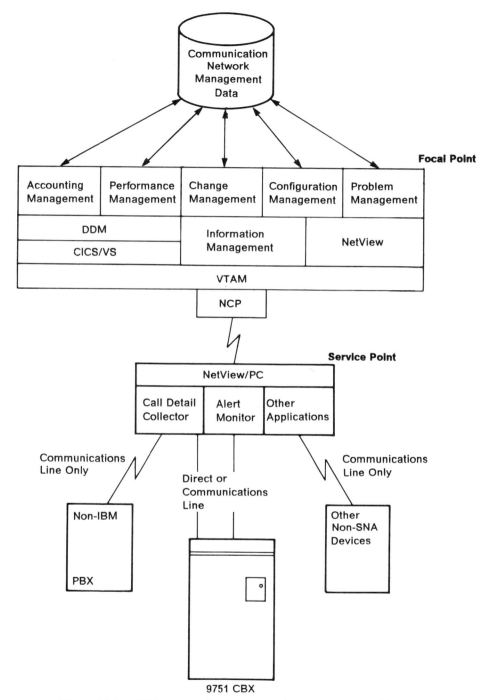

Figure 11.9: IBM communication network management architecture.

for intra-LIM communications, while inter-LIM communications are handled by central software that can be accessed by a LIM originating a call to another LIM. This central software is duplicated in multiple LIMs to provide greater reliability through redundancy. Figure 11.10 illustrates how redundant software increases LIM reliability.

Merging Voice and Data

Voice and data communications are transmitted simultaneously using the same digital line card along a single twisted pair of wires on the Ericsson MD system. This system is capable of handling both asynchronous and synchronous transmission.

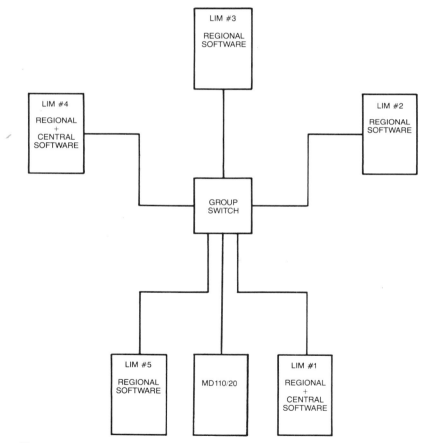

Figure 11.10: Ericsson reliability is increased through software redundancy.

Terminal adapter units (TAUs) support both full-duplex and half-duplex transmission as well as supporting asynchronous operations up to 19.2 kbps and synchronous operations up to 64 kbps.

There are a number of different types of terminal adapter units that serve as building blocks for linking together voice and data equipment. A terminal adapter unit for telephones (TAU-T) attaches to the rear of an Ericsson digital telephone and permits asynchronous and synchronous data transmission up to 9.6 kbps through an RS-232C interface.

A terminal adapter unit for standalone operations (TAU-S) is designed for data rather than for voice operations. It connects an MD 110 to printers, computer ports, and terminals, and supports transmission speeds up to 19.2 kbps for asynchronous applications and 48 kbps for synchronous applications using an RS-232C interface.

A terminal adapter unit for digital multiplexed interface (DMI) operations (TAU-D) supports the DMI standard. It can be connected to other TAU-Ds or data can be multiplexed over an MD 110 DS-1 Digital Trunk Card.

A terminal adapter unit for personal computers (TAU-PC) permits IBM PCs and compatibles to link directly with MD 110 over the telephone's two-wire extension line. The TAU-PC is installed in one of the PC's expansion slots and serves as a serial communications port supporting standard communications software. The TAU-PC supports asynchronous and synchronous transmission up to 19.2 kbps.

Ericsson also offers a terminal adapter unit for local area networking (TAU-LAN), which provides transmission at 64 kbps. This card uses the MD 110 two-wire voice cabling and is consistent with IEEE 802.5 token-ring standards. Using 3COM LAN software, it is possible to create from 1 to 64 logical LANS with one MD 110 system with up to 20 workstations in each LAN.

Finally, there is also a TAU available for IBM terminals. Marketed as the CAC-3278/3274/3287, these units provide coax elimination over single twisted-pair wire for connection to IBM host mainframe computers, while a TAU-5251 provides access to IBM System 34/36/38 minicomputers.

Figure 11.11 illustrates how these various terminal adapter units can be used as part of the MD 110 system to support multiple data applications. Notice how data terminals, shared printers and local area networks are able to be linked together by this one digital PBX system. Modem pooling is also possible with a modem access unit (MAU).

The MD 110 system is referred to by many telecommunications managers as a fourth-generation PBX, since it integrates voice and data communications on the same single twisted-pair wire and provides direct digital interfaces with no analog channel banks required. Because of these features as well as its modular approach, a number of major universities have adopted it. One example might show how far we have come in voice data integration.

Figure 11.12 illustrates how the University of California's San Diego campus (UCSD) has incorporated the various MD 110 building blocks we have discussed to provide a full range of voice and data functions. Using Ericsson's digital telephone instruments and data modules when appropriate, UCSD has found that it has saved approximately 25% of the $18,000 it spent monthly on moves and changes.

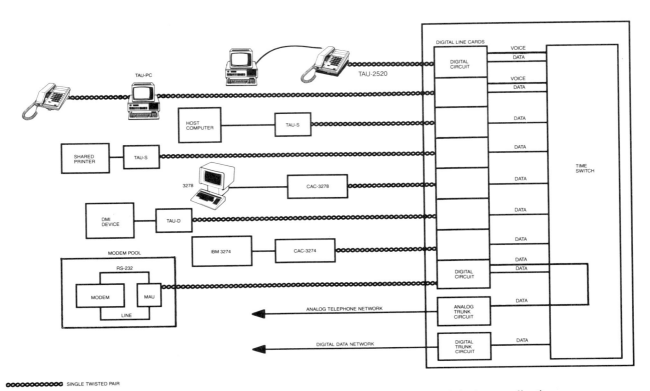

Figure 11.11: The MD 110 supports multiple data applications.

THE HARRIS FOURTH-GENERATION PBX

Harris offers its 20-20 PBX system, which features either alternate voice/data or simultaneous voice/data transmission. We will see that Harris also offers IBM 3270 terminal emulation, modem pooling, and X.25 network access, as well as PC file sharing and printer sharing. While models are always subject to change, Harris does offer a family of 20-20 products ranging from a 20-20S (250 ports) to 20-20D (8,000 ports).

Harris Workstations

The Harris Optic Teleset is a family of digital telephones that provides both voice and data transmission. An optical data interface plugs right into the phone and connects to a **digital line unit (DLU)** via standard twisted-pair wire. The Optic Teleset interfaces with asynchronous data terminals at speeds of up to 19.2 kbps. Microcomputers and ASCII terminals can exchange information with each other and with IBM host computers as well as with X.25 networks through a Harris 20-20 switch and the Optic Telesets. A "PinkSlip" messaging feature permits messages to be

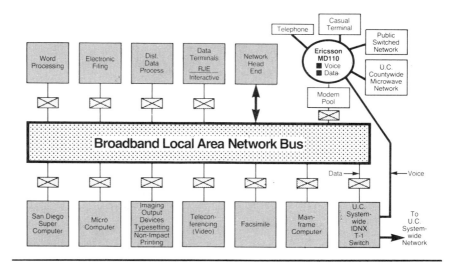

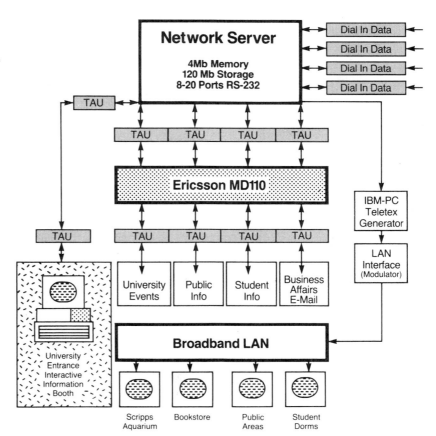

Figure 11.12: UCSD's local area network and dial information system.

stored electronically and then forwarded from the Harris Attendant Workstation to the Optic Teleset. Figure 11.13 illustrates on Optic Teleset with PinkSlip messaging.

Harris Data Communications

Terminals, microcomputers, and modems can be attached to the 20-20 switch through the RS-232C port on the Optic Telesets. They also can be attached to the 20-20 through a Data Communications Adapter (DCA) interface, which attaches inside the 20-20 to a Digital Line Card (DLC). Speeds of up to 19.2 kbps in asynchronous mode and 64 kbps in synchronous mode are possible over twisted-pair wire.

The 20-20 is capable of converting ASCII data to 3270 BSC or 3270 SNA/SDLC for communications with an IBM mainframe computer. It is also possible to attach a PAD to the 20-20 for X.25 communications. The 20-20 architecture also contains a DMI interface that is compatible with standard T-1/D-3/D-4 transmission. The basic 20-20 architecture also includes ISDN and LAN interfaces as well as the capability of handling modem pooling.

Figure 11.13: An optic teleset.

Harris' SOPHO-NET and Data Connectivity through Automatic Protocol Conversion

Harris offers **SOPHO-NET,** a packet-switched, open-system, wide-area network to link together disparate networks. It automatically translates protocols, so that existing equipment can communicate more efficiently. What makes this system so intriguing to large companies is that it has the built-in capability of handling many of the major protocols, including BSC and SDLC, DEC, Wang, and Hewlett-Packard as well as Harris/Lanier terminals and computers. It is capable of integrating a number of message handling systems and archiving systems—including Philips' MEGADOC and Lanier's Library Services—with local area networks, including HarrisNet, as well as LANs from other vendors. Finally, it is able to integrate SNA, ISDN, and X.25 protocols as well as telex and facsimile group IV transmissions.

SOPHO-NET can transmit text, data, and images over the same line. Important for the future, it is compatible with the OSI model discussed in Chapter 6. Figure 11.14 illustrates how SOPHO-NET can link together such disparate vendor products as a DEC minicomputer, IBM mainframe, a LAN, and a PBX.

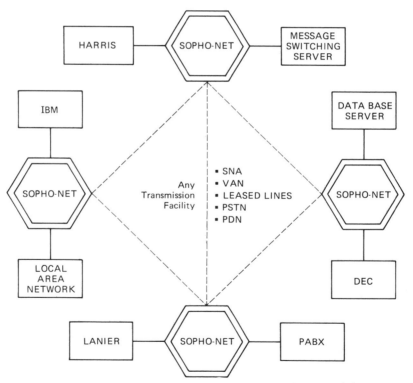

Figure 11.14: SOPHO-NET increases network connectivity.

The Harris Fourth-Generation PBX

CHAPTER SUMMARY

IBM/Rolm's PBX systems are designed to handle both voice and data. They can be linked to IBM mainframe computers through NetView/PC software. AFACTS is a system designed to provide detailed reports on trunk activity. Ericsson's PBX systems are designed to be modular. Voice and data ports can be added as needed. These switches can be used to link together local area networks as well as voice communications. Finally, Harris/Lanier's SOPHO-NET is a system for protocol conversion that permits their switches to connect such disparate networks as DEC and IBM together. It handles both voice and data.

KEY TERMS

AFACTS
Data communication module (DCM)
Data Network Interface (DNI) card
Digital line unit
Entry point
Expansion module group
Focal point

Group Switch Module (GSM)
Interface module group
Intelligent network processor (INP)
Line Interface Module (LIM)
Service point
SOPHO-NET

REVIEW QUESTIONS

1. Explain how Rolm's PBX systems can be linked to X.25 networks.
2. Discuss how Rolm's PBX systems can be linked to IBM mainframe computers using BSC or SDLC protocols.
3. What is the role of intelligent network processors in a Rolm PBX system?
4. What is the function of AFACTS in a Rolm PBX system?
5. Discuss your understanding of IBM's open communication network management architecture.
6. Ericsson's PBX systems are built around the concept of modularity. How does a network manager expand his or her system?
7. How would one link local area networks to an Ericsson PBX system?
8. What is significant about Harris's SOPHO-NET?

TOPICS FOR DISCUSSION

1. Given the speed limitations of a PBX, under what circumstances would a Rolm/IBM CBX prove to be a cost-effective alternative to a local area network?

2. The IDN units utilized in designing Rolm/IBM T-1 networks are particularly valuable for transaction-oriented data such as information flowing from a bank branch to a main office. What major benefits do these IDN units appear to serve?

3. Does SOPHO-NET offer any benefits over simply attaching a series of protocol converters?

Designing a Voice/Data Network

CHAPTER 12

OBJECTIVES

Upon completion of this chapter, you should be able to:

- gather relevant facts and data necessary for designing a network
- organize facts and data into logical groupings
- determine the components needed to satisfy user needs
- implement components into a voice/data network

INTRODUCTION

Businesses have long recognized the need to communicate and share information. Telephones allow communication without the parties having to be physically present. To recognize how important this is, imagine a working world in which telephones did not exist. Co-workers that worked in different offices in different parts of town would have to arrange a meeting simply to discuss a problem. How would they set up the meeting? Messages would have to be sent to one another by mail or by courier. It could take days to set up a simple meeting. How inconvenient and inefficient! Most businesses would avoid this problem by keeping employees in one office, but this is not a good answer either because businesses would be hampered from expanding to new locations of opportunity.

Telephones allow an efficient and economical way to communicate. Workers

Figure 12.1: A world without telephones.

pick up the telephone to talk to co-workers dozens of times a day. Important information is shared, and the flow of work is not interrupted. This integrated means of communicating and sharing information allows a business to be more competitive than it otherwise could be.

The past decade has seen a tremendous increase in the use of computers in business. Personal computers are used to write letters and create reports. Mini- and mainframe computers keep track of products and supplies. Businesses today are recognizing a need to share information electronically the same way they share information verbally.

THE CONSULTANT

Most businesses do not have people with the expertise to decide what and how to change their offices to allow for this new type of information sharing. Often, consultants will be hired to assist with decisions. In this chapter, the steps a consultant takes to advise clients will be traced.

An assumption is made that should be noted. It is assumed that the customer's business is medium to large in size. Small customers will not usually need a consultant, because their choices are relatively inexpensive and straightforward. Voice systems will normally be electronic key equipment, and data systems will normally be personal computers.

For the rest of this chapter, consider yourself a consultant. Endeavor to think as a consultant and to ask the questions a consultant might ask. This requires identifying a goal and working out a way to get there.

What and How

What changes should be made and *how* should they be implemented? Your goal is to answer these questions for a client. To reach your goal, several steps must be taken. The steps are as follows:

I. Data gathering
    A. Purpose of the business
    B. Operational flow
    C. Existing equipment
    D. Future needs
II. Organizing
    A. Identifying tasks
    B. Grouping users by task
    C. Assessing the groups
III. Determining the components needed
    A. Data components
    B. Voice components
    C. Network components
    D. Integration
IV. Implementing the components

DATA GATHERING

Data gathering helps prepare a roadmap. It identifies where you are, where you want to go, and how you want to get there. The roadmap is your plan and can be used as a guide. Four steps are involved in data gathering:

1. identifying the purpose of the business
2. understanding the business's operational flow
3. listing existing equipment
4. identifying future needs.

Some terms must be defined before continuing. **User** refers to a person who makes use of a system. Different users have different needs. For instance, employees who use the telephone at work are users of the telephone system. User A, a telemarketing sales representative, makes only outgoing calls and never receives calls. User B, a customer service representative, only answers calls and never initiates them. Users A and B have different needs. User A frequently uses "automatic redial" to redial busy numbers, while user B never uses this feature. User B, on the other hand, relies on distinctive ringing to distinguish whose telephone is ringing in the customer service department.

The term **task** refers to a specific job to be done. It may be simple, such as opening the door for customers as they enter a business, or it may be complex, such as tracking the changes in the price of a stock as it is traded over the course of a day in the stock market.

The term **system** refers to the equipment used to help tasks get done. A system is a tool no different from a hammer, a car, or a typewriter. A hammer helps build houses, a car helps you get from place to place, and a typewriter helps you write.

This book focuses on two types of systems: voice systems and data systems. Voice systems, telephone systems, and voice networks are tools for communicating verbally. Data systems, computers, and data networks are tools for communicating electronically.

Users and tasks are the building blocks around which a system is built. Understanding the users and the tasks allows for the development of a system.

Purpose

Before all else, a consultant must understand the nature of the business involved. What is its purpose? Why does it exist? What function does it perform?

A business must provide a useful function. Whether the business is a bank, a radio station, or a hamburger stand, it must provide something that people want, or it will cease to exist. A bank provides a secure place for money, a radio station provides news and entertainment, and a hamburger stand provides food in a hurry. Understanding the purpose of a client's business is the first step in helping.

Operational Flow

Tasks. Once you know what a business does, you must understand how it does it. Every business will perform a set of tasks. A bank makes loans and takes deposits. A radio station selects programming and sells advertising. A hamburger stand takes food, cooks it, and serves it.

Set of Rules. Each task is performed by following its own set of rules. The rules are the steps that must be followed to accomplish the task. A bank will have rules its employees follow when making loans, taking deposits or giving withdrawals to customers. A radio station will have rules for airing music or advertising products. A hamburger stand will have rules describing how food will be prepared.

Operational Flow. These rules are the operational flow for the task. Understanding the rules and why the business chose the rules leads to understanding the business. Figure 12.2 is an example of an operational flow of a bank giving a withdrawal to a customer.

A business will have an operational flow for each task. First, understand each

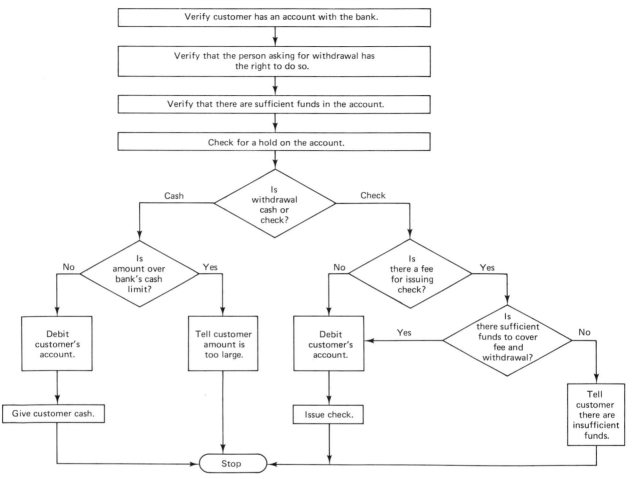

Figure 12.2: Bank withdrawal operational flow.

operational flow, then think about how each operational flow interacts with the others.

Information Flow. One of the tasks every business must deal with is getting needed information from place to place. How does a bank, for instance, let its branches know when a customer makes a deposit or withdrawal? This information must be distributed quickly and accurately or the bank could face the sorry situation of having the same funds withdrawn from an account at different branches.

Let us define this distribution of information within a business as information flow. Traditionally, information flow has meant paper flow. That is, paper documents such as invoices, memos, and reports have carried information from place to place. Couriers, mail, and overnight delivery services are routinely used to transport the information.

Electronic Information Flow. When a business begins to use computers, the opportunity to transfer information electronically becomes available. Electronic transfers are immediate. This is highly desirable to businesses such as a bank. There are other advantages to electronic information transfer. For example, a radio station can receive news stories created at another location and then transmitted to it. Large hamburger chains have regional warehouses to supply the stores in the area. Each store will send in orders for hamburgers, buns, french fries, and so on. With computers, requests can be directly keyed in instead of being written down and sent in.

Electronic information flow can occur when compatible computers are placed at locations wishing to share information. *Compatible,* in this context, means computers that support the same communications protocols. A network is needed to transfer information among computers.

Of the many operational flows within a business, electronic information flow is the one that involves networking. As such, we will focus on electronic information flow from this point on.

Existing Equipment

It is a rare customer who has no existing equipment and will purchase everything brand new. Far more typical is a customer who has a variety of equipment acquired over years of doing business. A consultant must consider the financial impact of proposals made to the customer. Often, there is great expense involved, and it is a consultant's responsibility to keep the cost as reasonable as possible.

Can equipment be reused? One of the best ways to do this is to reuse the equipment a customer already has. A good consultant will consider every piece of equipment and question whether it can be reused or whether it should be disposed of.

The decision to reuse equipment does not always mean it is the best technical solution. A customer may have data modems that operate at 1,200 bps. The best technical solution would be to upgrade to 9,600 bps or faster modems. Though they may be at least eight times faster, high-speed modems are expensive. Does the customer need the increased speed? There must be a solid reason to recommend throwing out the slower modems in favor of the faster ones. (One such reason would be the time saved by employees, if the employees spent several hours per day sending and receiving information through the modems.)

Customers often will have invested thousands of dollars in equipment. If it cannot be reused, consider other ways of protecting a customer's investment. Ask, "Is the equipment purchased or leased?" If purchased, "Can the equipment be sold to offset the cost of buying new equipment?" If leased, "Can the lease be terminated without financial penalty?" and "Can the equipment be upgraded to more suitable equipment?"

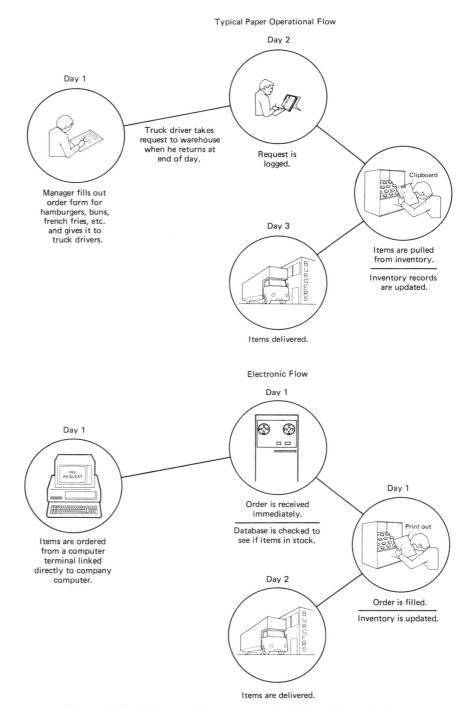

Figure 12.3: Diagram of paper operational flow/electronic flow.

Is the data environment synchronous or asynchronous? A company's existing data environment will either be synchronous or asynchronous. When designing a solution, the type of environment is important because equipment may work in one environment but not the other.

Future Needs

Customers call in consultants because they want change in their business. They have a concept of what the result should be, but they do not know how to get from here to there. You must probe with questions to determine and assess the customer's goals. What do they want to do that they are not now doing? What are they now doing that they want to do differently? Are the goals realistic? What is the cost? How long will it take to implement? What can you suggest that the customer has not thought of?

ORGANIZING

Identifying and Grouping Tasks

One of your most important functions as a consultant is to recognize tasks that can be made more efficient and economical by changing from a paper flow to an electronic flow. Note that this does not mean changing every paper-oriented task. Tasks requiring signed or notarized documents cannot be transferred electronically. Also, some tasks may prove uneconomical to change. For example, a title insurance company generally keeps records of all documents affecting real estate in a given county. The records may date back many decades. Converting to an electronic flow would require keying all the records into a computer—a massive effort! The worker-hours spent on such a conversion may far outweigh the worker-hours saved by a computerized system.

After the data gathering process is complete, users should be organized into logical groups. Look at the tasks each user performs. How does this person use voice systems? How does this person use data systems? Consider this for all the users. Patterns will emerge where users use a system in the same way.

These users can be grouped together. Some users may be grouped with respect to use of voice systems, but not data systems. Others may be grouped with respect to data systems, but not voice systems. Still others may be grouped with respect to both voice and data systems, while some users cannot be grouped at all.

Often users that can be grouped have the same job, such as all secretaries or all salespeople. However, users with different jobs may have similar work patterns. For instance, secretaries may access a company computer for a customer's address and telephone number. Sales clerks may access the same computer for status of the customer's sales orders. Even though the information needed is different, both users use the computer in the same way.

Assessing the Groups

What do users need? Once groups are defined, lists should be made of the group's requirements for voice and data. The voice list should answer such questions as:

- "How much of a user's time is spent on the telephone?"
- "Are calls incoming or outgoing?"
  If incoming:
  "Where do they come from? The local area? Nationwide?"
  "What percentage of calls are from other employees?"
  If outgoing:
  "Where do they go to? The local area? Nationwide?"
  "What percentage of calls are to other employees?"
- "What voice system features can user's use to be more productive?"

The data list should answer such questions as:

- "How much of a user's time is spent interactively with a computer?"
- "Does the user log on and log off after each interactive session or does he or she stay logged on all day?"
- "Does the user make computer requests that put a significant load on the processor? If so, how often?"
- "Does the user need access to more than one computer?"
- "Is the computer (or computers) local or remote to the user?"
- "Does the user require access to a printer?"
- "Does the user need to communicate electronically with other users? Are they in the same office or remotely located?"
- "Does the user need access to other users' computer files?"

What do users not need? While it is important to provide what users need, it is just as important to restrict them from accessing what they do not need. For instance, a company's computer may have a customer database with names, addresses, telephone numbers, credit limit, and current account balance. It may also contain payroll information with employee names, salaries, home addresses, and telephone numbers. The payroll information is confidential and should not be seen by most employees. Before employees are allowed access to the customer database, some form of security must be put in place to keep them from the confidential information.

Modifying the Operational Flow

When organizing the data about a customer, realize that operational flows can be dynamic. That means they can be changed. It makes sense for a business to make a change if its overall efficiency would improve. Suppose, for example, a lumber company's operational flow calls for sales people to call the main office for a credit check whenever a customer buys more than $100 worth of lumber. If the sales people get a computer terminal as part of an upgrade plan, it would make sense for them to check credit through the computer terminal rather than by making a telephone call.

DETERMINE THE COMPONENTS

After the user's requirements are known, components must be chosen to fulfill them. Three questions should be asked at this point. Do the users have data communication requirements? Do the users have voice communication requirements? Is a network involved? The answers to these questions determine the components used in a final system design. Four steps must be taken:

1. determining data components
2. determining voice components
3. determining network components
4. determining voice/data integration

DATA COMPONENTS

There are several options to consider in the design of a data system. Users may need personal computers. They may need access to a minicomputer or to a mainframe computer. There may be more than one minicomputer or mainframe computer, and users may need access to them. A local area network may be needed. Is data networking needed?

It is most helpful at this point to determine if networking will be a part of the data system. The customer's layout and users' need to share files or resources will determine if networking is needed. What is the layout?

Layout involves where users and computers are located. Are the users in one spot or are they spread over two or more locations? Is there one computer or more than one? Where are they located?

If all users are at one location and share a minicomputer or mainframe computer, then a network is not needed. However, if users need access to more than one computer, a local area network can be the best solution. If users or computers are at multiple locations, other types of data networking may be needed. These include leased private lines, public data networks, and private networks. Private networks will be discussed in a later section.

Do users need to share files or resources?

Resources refers to physical devices, such as printers and modems, that perform a specific function. If users have their own files and their own resources, then there is little need for a network. However, if users must share files or resources, then a network is usually the best solution.

Table 12.1 shows the options described above and lists when a data network is useful. If a network is not to be used, then further questions regarding data networking or voice/data integration can be skipped.

What volume of data traffic is involved?

If data networking is to be used, then the volume of data traffic must be considered. Previously, users were grouped according to the type of tasks they performed. The groups' data usage patterns can now be looked at to determine the volume of data traffic that a system must accommodate. The volume should be categorized as heavy, medium, or light.

The three categories are based on average usage each day. For instance, User A spends 2 hours on the computer each morning and then none the rest of the day. User B spends 15 minutes each hour on the computer throughout the day. User B is categorized as a heavier user than User A. Generally, categorization can be done by group. Categorization by individual users is only necessary if the user's pattern

TABLE 12.1
Data Networking Options

Options
Single Location = Users at one location
Mult Location = Users at more than one location
No Host = No mini or mainframe computer; Personal computers only
Single Host = One computer shared by users
Mult Host = More than one computer
Sharing = Users share files or resoruces
No Sharing = Users do not share files or resources

Local Area Network	Other Data Networking	No Network
Single location	Mult Location	Single Location
No host	Single Host	Single Host
Sharing	Sharing	Sharing
Single Location	Mult Location	Single Location
Mult Host	Single Host	Single Host
Sharing	No Sharing	No Sharing
	Mult Location	Single Location
	Mult Host	Mult Host
	Sharing	No Sharing
	Mult Location*	
	Mult Host	
	No Sharing	

*This assumes that some locations must access computers at other locations. If all users access the computer at their location no network is needed.

of usage is different from the group's. Groups should be categorized as heavy data users, medium data users, or light data users.

Heavy data users are groups that interact with a computer 6 to 8 hours a day. Data entry operators are an example. They sit at a terminal all day entering data into a computer.

Medium data users are groups that use a computer 2 to 6 hours a day. Sales clerks are an example. They may check customer information in a computer several times a day. If they spend 10 minutes each time, their cumulative use for the day would make them medium users.

Light data users are groups that use a computer less than 2 hours a day. Managers typically fall into this category. They need specific information to run a company. With properly designed software, they can get it in a matter of minutes. They may access the computer several times a day and still be classified as light users because of the short duration of each session.

The components that can be used to satisfy heavy, medium, and light data usage groups are listed in Figure 12.4. The components can be used individually or in combination with one another. Figures 12.5 and 12.6 show the possible combinations of components. Figure 12.5 shows possible combinations when there is no host computer, and Figure 12.6 shows possible combinations when there is a host computer.

Personal Computers

Personal computers put computing power at the user's desktop. That power is available only to the user and cannot be shared with others. Depending upon the user's needs, this can be a benefit or a limitation. Users have complete control over the contents of their files and do not have to compete with others for processor time. On the other hand, files and resources cannot be easily shared unless the computer is connected to a local area network. Groups that do not need networking are the most likely candidates for personal computers. Groups that have heavy computing needs but light to medium networking needs are candidates for personal computers connected to a local area network. Users with heavy networking needs will limit the number of users a LAN can support. This makes a LAN inadequate for many heavy user applications.

Local Area Networks

Local area networks will be the strongest contenders in a light to medium data traffic environment. Most popular LANs are designed to support medium traffic loads without noticeable delays in user response time. Many support personal computers, others support terminals and host computers as well. When users need access to more than one host computer, LANs are the easiest way to provide connections.

Dedicated Connections

Dedicated connections to a host computer provide heavy usage groups the best computer access when sharing is needed. Dedicated connections may also be given to medium and light usage groups when sharing is needed, but there are too few users

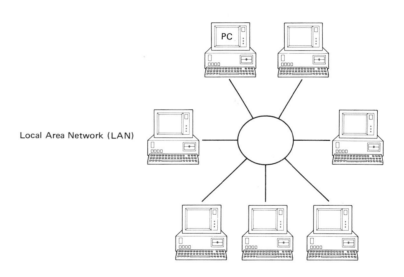

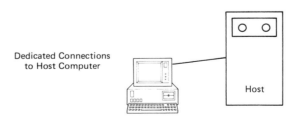

Figure 12.4: Data system components.

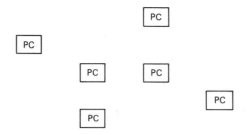

Stand Alone Personal Computers

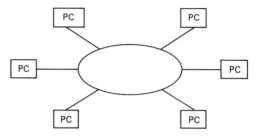

Personal Computers on a Local Area Network

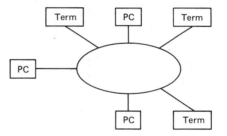

Terminals and Personal Computers share a LAN

Figure 12.5: Data component combinations—no host computer.

to justify the cost of a local area network. Normally users are given access to the host via a terminal. Some personal computer users may want access also. This can be done by running a terminal emulation software program on the personal computer. Terminal emulation causes the personal computer to act like a dumb terminal that can communicate with the host computer. When terminal emulation is used, the personal computer's own processing capabilities cannot be used. Some software programs allow users to hot key between programs in the personal computer and sessions with the host.

System designs for most companies will involve a combination of components. A company's existing equipment can have a significant impact. If a company has

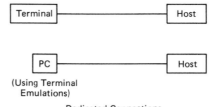

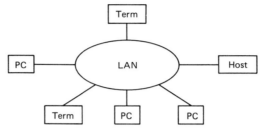

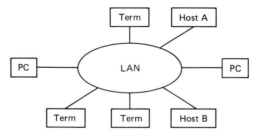

Figure 12.6: Data component combinations—with host computer.

personal computers, a local area network may be the lowest cost and most flexible solution. If, however, the company has a minicomputer or mainframe computer, dedicated connections may be a better solution.

VOICE COMPONENTS

Determining the voice components to use depends upon several factors. What type of dialtone is needed? What features are needed? Is the design and capacity of the system suitable? What is the cost?

Dialtone

There are three basic categories of dialtone: intercom, local-access, and toll. Intercom dialtone connects a user to others within the same telephone system. Local-access dialtone connects users to the Public Switched Network in the local calling area. Toll dialtone connects users to the Public Switched Network outside the local calling area. Table 12.2 offers further description of each category.

Local-access dialtone is provided by the local exchange company (LEC). Calls are placed over the Public Switched Network and are charged at direct distance dial (DDD) rates. There is usually a calling area for which there are no charges beyond a flat monthly fee. Some large, metropolitan areas have zone unit message (ZUM) rates, which amount to toll charges within the local calling area. LECs often offer discounts when ZUM usage exceeds a threshold. LECs also offer foreign exchange (FX) service, allowing calls to and from an area beyond the local calling area. Again, the fee is less than normal DDD rates.

Toll dialtone is provided by common carriers, such as AT&T, MCI, or Sprint. These carriers offer wide-area telephone service (WATS) and 800 service (inbound WATS). WATS offers reduced rates over DDD when call usage exceeds a threshold. "Inbound WATS," or 800 Service, allows the called party and not the calling party to pay for the call at reduced rates. The calling area for these services may be part or all of the United States.

TABLE 12.2
Dialtone

Intercom Dialtone	Local Access Dialtone	Toll Dialtone
Intercom dialtone allows calls from station to station within a telephone system. The dialtone comes from the user's telephone system and is independent of the public-switched network. Intercom calls may occur between users at the same location or, if a private network is in place, between users at different locations.	Local access dialtone allows calls to the local calling area. By definition there are no charges for calls beyond a set monthly fee. Dialtone comes from the Local Exchange Company (LEC).	Toll dialtone allows calls outside the local calling area. There is normally a basic monthly fee plus a charge for each call made. The additional charge is normally based on the duration of the call and destination. Dialtone can come from any long distance vendor, such as AT&T, MCI or Sprint. It may also come from the local telephone operating company if the service is Zone Unit Message (ZUM) or Foreign Exchange (FX).

Features

PBXs and Centrex services offer literally hundreds of features. Most users will find a few worthwhile and will not use the rest. Popular features, such as call transfer, call hold, and call forwarding, may well be given to all users. Other features, such as automatic call distribution, voice mail, and modem pooling, may be a great benefit to some and useless to others. The rich variety of features allows for tailoring the system to users' specific needs.

Selecting which features are appropriate is best achieved by working with the system's vendor. The vendor can explain how features are used and how features affect one another. They can also provide valuable reference documents. Different vendors will offer different solutions, highlighting the strengths of their system. By listening to all the vendors, a rounded viewpoint of options can be obtained. Once a list of features has been compiled, the systems offered by all vendors should be examined to ascertain that the desired features are offered.

Design/Capacity

No two systems are designed alike. Design features to note are the system type, switching matrix, modulation technique, architecture, and system capacity.

Type. The system type is either analog or digital. Older equipment tends to be analog, while newer equipment tends to be digital. Also, medium to large systems tend to be digital; small digital systems are rare. Analog systems are generally fine

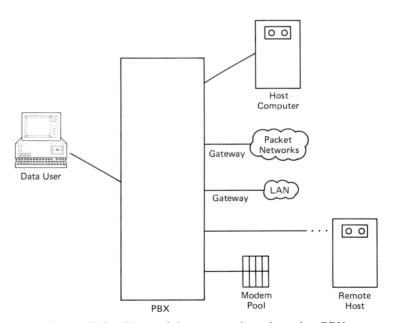

Figure 12.7: Types of data connections through a PBX.

for handling voice traffic, while digital systems tend to be more flexible in handling combined voice and data traffic. The modulation technique used by a system defines whether it is classified as analog or digital.

Modulation Technique. Modulation technique refers to how calls are handled as they pass through a system. Systems using a space-division switching matrix do not modify a signal at all. They receive an analog signal and it remains analog through the matrix. Systems using a time-division switching matrix "slice" the call into small "packages." Each package passes through the matrix like a boxcar in a train. Pulse amplitude modulation (PAM) and pulse width modulation (PWM) are analog techniques, while pulse code modulation (PCM) and delta modulation (DM) are digital techniques. Pulse code modulation is becoming the most commonly used technique.

Switching Matrix. Though most PBXs on the market today are digital and have a time-division switching matrix, there are still a number of older generation PBXs still around. This is particularly true in PBXs of smaller size. They may use a frequency-division technique instead of time division.

The significant feature of a switching matrix is the number of talkpaths it allows. This becomes particularly important when data traffic is to be carried through the matrix, because of the increased likelihood of blockage. So long as it is kept within reason, data through a PBX is acceptable.

Using a switch with non-blocking design can solve the problem entirely. To be non-blocking, there must be a talkpath for every user on the system. This can be extremely expensive, but it ensures that calls will always be completed. A middleground approach is the virtually non-blocking matrix. There are a limited number of talkpaths, therefore blocking can occur. But there are a sufficient number that blockage is less than one call per hundred. Unless a system has a non-blocking matrix, the volume of data traffic through the system should be limited to no more than 20% of the total traffic.

Architecture. Systems are designed with either a centralized or distributed architecture. A centralized architecture places system intelligence and control at the center of the system. The Northern Telecom SL-1 PBX has a centralized architecture. In this system, all calls must be handled by the main processor.

A distributed architecture distributes intelligence and control to lower levels of the system. The AT&T System 85 has a distributed architecture, in which users are grouped into modules. Calls within a module are switched by the module itself, while calls to other modules or outside the system are handled by the main processor.

There are advantages to both centralized and distributed architectures. A centralized architecture tends to be less expensive and easier to maintain than a distributed architecture. A distributed architecture can be more reliable because the distributed intelligence allows calls to be made even if the central processor fails.

Capacity. Systems are designed to support a range of numbers of users. For instance, the Toshiba Perception II is a small PBX supporting from 8 to 120 users. The IBM/ROLM CBX II 8000 is a medium-size PBX supporting from 16 to 800 users. The Northern Telecom Meridian SL-100 is a large PBX supporting from 1 to 30,000 users.

When deciding the capacity needed in a system, you must know how long the customer wants to keep the system. If it is to be kept for 5 years, then look not only at the number of users the system has today, but look at the number of users expected 5 years from now. The system should accommodate the expected growth and still have 15% to 20% spare capacity.

For example, the ABC Distributing Co. currently has 150 employees and wants a new system that they plan to keep 5 years. Upon questioning, you find that the company plans to merge with the XYZ Distributing Co. within a year. XYZ has 95 people in its main office and has several remote offices with a total of 350 people. Within 3 years of the merger, all employees at remote offices will be brought to ABC's location. How large a system do they need? To find out, look at the number of employees using the system in 5 years:

```
  150 ABC employees
   95 XYZ employees at the main office
+ 350 XYZ employees at the remote office
  595 Employees total
```

Then calculate the spare capacity needed:

```
   595   Employees
×   20%  Spare capacity
   119   Capacity for additional employees

   595 Actual Employees
 + 119 Employees, Spare Capacity
   714 Employees, Total Capacity
```

The system must have a capacity of at least 714 users. Less than that and the system could be undersized.

Cost. Cost of the system is always an important factor. Prices are provided by the vendor of the system. Comparing the cost per user of different systems is a common way of determining the relative value of each system. The cost per user is found by dividing the total cost of the system by the total number of users.

Once decisions are made about the type of dialtone and features that users need and the design and cost has been examined, you have the elements with which to make a decision about a voice system. The next question is, "Will data transmission be a part of the system?" If so, then additional factors must be examined.

NETWORK COMPONENTS

Previously, we examined whether networking should be part of a data system. If a network was to be used, it meant a local area network. Local area networks work well for linking users at a single location, but how can users at different locations be linked together? The answer in through a private network. Originally, private networks linked locations for voice communications only. In recent years, private networks have been used increasingly for data communications as well.

Private Network

When is a private network appropriate? When is it not? There must be at least two switches for a private network to exist. The switches may be located hundreds of miles apart, within the same city, or even within the same building. Most private networks are used by medium to large companies with locations dispersed geographically.

Cost is the major factor in determining whether or not a private network is appropriate. Are there many calls among locations? If so, then a private network may save the company money. If not, then a private network may be more expensive than the public network.

Nodes

A PBX or Centrex can fulfill the requirement of being a node. There are a multitude of PBX vendors in the telecommunications marketplace in addition to the variety of Centrex services available from the local exchange company. What is the best system to use? A combination of technical and non-technical factors should be considered when making a choice.

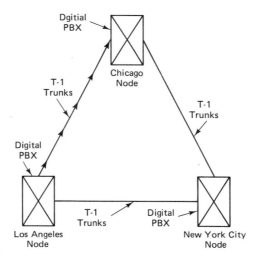

Figure 12.8: Private network.

Technical factors. These include the equipment's design for use as a node, the type of trunking it can support, its ability to transport data with voice, and compatibility among all nodes.

Not all Centrex systems and PBXs are designed to be used as nodes in a private network. The software must support tandeming, and the hardware must support tie trunks.

When voice and data will be integrated in the network, a digital architecture simplifies the transmission of voice and data through the PBX. Most PBX vendors offer a digital PBX; however, the method of transmitting voice and data varies with each one.

Tie trunks used to connect nodes may be analog or digital. Digital trunks have an advantage when both voice and data will be transported because they allow data transmissions without converting from digital to analog. T-1 facilities are the most prevalent digital trunks available for PBXs.

Compatibility addresses the ability of each node to communicate with the others. This can be a problem if different equipment will be used as nodes at different locations. A PBX vendor may recommend different size switches at different locations. All switches must be compatible and all must support the same trunk interface.

Customers may have existing equipment that is different at different locations. Beware of mixing different PBX equipment in the same private network. While it is true that most vendors offer similar capabilities, the method of implementation will vary greatly. For example, PBXs pass routing information with a call when it is first set up in a network. Some AT&T networks pass the information along a separate channel apart from the actual call. Northern Telecom passes the information in the same channel with the call. The equipment from these two vendors could not operate together in the same network.

Non-technical factors. These include cost, reliability, a company's reputation, and an evaluation of the locations where nodes will be located.

Cost will be provided by the vendor or telephone company. Reliability statistics are available from the vendor or telephone company and are also available from technical reference books, such as *Datapro* (Delran, NJ, Datapro Research Corporation, 1987). It is wise to check the claims of a vendor against an impartial source.

Reputation can be evaluated by talking to other companies that use the system. Technical reference books and trade journals can also provide information.

Location is important for two reasons. If Centrex is being considered, service is limited to the LEC's operating area. A PBX vendor, on the other hand, can place its equipment anywhere needed. Its ability to service and support all locations, however, may not be satisfactory. When equipment fails, most businesses can accept a few hours out of service, but no more. Quick response is required from a vendor. If service personnel are not locally available, repair may take days, not hours.

On-Net/Off-Net Calls

Calls among users on the network are called *on-net calls*. Calls from a user to non-network telephone numbers are called *off-net calls*. The network consists of nodes

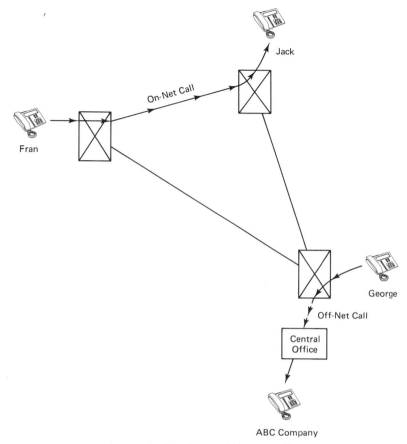
Figure 12.9: On-net/Off-net calling.

and the tie trunks that connect them. On-net calls stay in the network, while off-net calls do not. This is significant primarily when the network is digital.

On-net calls remain digital end-to-end. This benefits data communications by increasing reliability and reducing transmission errors. Off-net calls leave the network and pass through the Public Switched Network. The Public Switched Network is analog, therefore, off-net calls must be converted from digital to analog before being completed. Thus they lose the benefits of digital end-to-end transmission.

INTEGRATING VOICE AND DATA

Decisions have been made regarding whether a voice system, data system, and private network will be part of the final system. The components for each system have been chosen. The next step is to decide whether data should be integrated with voice.

Integration involves filling one system's requirements with the components of another. Figure 12.10 shows that data requirements can be fulfilled by a voice system or a private network. For purposes of this discussion, components of a voice system or private network will collectively be referred to as *voice components.*

Notice that integrating voice into a data system will not be discussed. That is because the option does not make sense. Today's data systems do not support voice communications, therefore integrating voice into a data system is not possible. Integrating voice into a private network is redundant. If there is a private network, it will have been chosen as an extension of the voice system. As such, it is already integrated.

Up to this point, data requirements have been fulfilled with computers, local area networks, leased data lines, and public data networks. When can these components be replaced by Centrex systems, PBXs, or tie trunks? If they can be replaced, should they be?

There are virtually no instances when a computer can be replaced by voice components. Other data components link computers together and are better candidates for replacement.

A local area network links data users at a single location. A Centrex or PBX can fill this function and thereby replace the LAN. Figure 12.11 shows an example of replacing an Ethernet LAN with a PBX. Replacement makes sense if the cost of adding users to the Centrex or PBX is less than the cost of installing a LAN. This most often occurs when there are few users, or users are scattered at different locations, or users need access to computers at different locations.

Centrex and PBX systems work best when data traffic volume is light. The volume should be no more than 20% of the combined voice and data traffic volume total. Beyond this threshold, blockage through the switching matrix becomes unacceptable or the cost of the equipment becomes prohibitive.

Leased data lines and public data networks link data users at different locations. Tie trunks linking nodes in a private network may support data transmission. If they do, the private network can replace the leased lines and public networks.

The tie trunks are a fixed monthly cost. The customer pays the same monthly fee whether just voice communications or voice and data communications are sent over the trunks. It makes good sense to replace leased lines and public data networks with private network connections wherever possible.

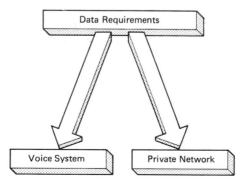

Figure 12.10: Satisfying data requirements by voice system or private network.

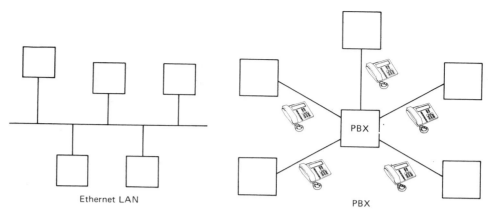

Figure 12.11: Replacing an Ethernet LAN with a PBX.

IMPLEMENTATION

A complete system has now been designed. It may be just a voice system, just a data system, or a voice and data system. With the design complete, the system is ready to be implemented. This means taking all the components and putting them into place.

Planning, Changing, Sequencing, Timing

Implementation is a disruptive procedure by its very nature. Changes must be made to existing operations and changes are disruptive. A plan should be worked out with the customer to identify the changes, set the sequence they will be carried out, and determine the timing of each change. Doing this allows the customer to understand the implementation's effects on its operations and to minimize disruption.

Each system component represents a change. Listing each one and then examining what is needed for implementation will properly identify them. For each component, ask, (1) "What is needed to install it," and (2) "Does old equipment need to be removed?"

For example, say that a customer's design calls for installation of a new PBX, increasing the number of ports on the company's mini-computer, and installation of a local area network. The company presently has Centrex telephone service that will be removed. Some users have no data equipment and will be getting a terminal directly connected to the minicomputer. Other users have personal computers that will be replaced by terminals directly connected to the minicomputer. Still others have personal computers that will be connected to the local area network. The changes the customer will have to make are shown in Table 12.3.

Setting the sequence means looking at each of the components and scheduling them in the most logical order. Common sense and knowledge from experience dictate the order used. Some components will not work unless other components are installed first. For instance, the additional ports on the customer's minicomputer

TABLE 12.3

Example of Implementation Changes

List of Changes	What is Needed for Installation?	Remove Equipment?
Install new wiring	Twisted pair for PBX, coaxial cable for computer ports, twisted pair or coaxial cable for LAN	Reuse twisted pair where feasible
Install PBX common equipment	Convert a store room to house the PBX common equipment	Disconnect the Centrex®
Install new telephone sets	Telephone sets and cords	Remove old telephone sets
Install additional ports on the minicomputer	Computer port hardware	None
Install new terminals	Computer terminals	Remove personal computers from some locations
Install the local area network	Transceivers, interface boards, LAN software	None

should be installed and activated before terminals are given to the users. Table 12.3 is ordered according to the sequence that items should be implemented.

Determine the Timing

Once the sequence is set, a separate question of timing must be addressed. Timing reflects when an installation is to occur. Customer needs and preferences and the availability of equipment usually dictate the timing.

It is helpful to think of it in terms of first the overall implementation and then individual component implementations.

With regard to the overall implementation, a customer has two choices: a changeout or a phased implementation.

A changeout involves installing and removing equipment all in one step. New equipment is put in place and tested, but not activated until the day it will be put into service. If it is replacing old equipment, both old and new set side-by-side until the day of the changeout. When all is ready, the new equipment is activated. Old equipment, if any, is deactivated and removed. Changeouts are most often used when only one or two changes will be made. From the users' point of view, there is no change in routine until the day of the changeout. At that time users start using the new equipment. A changeout gives a clean break between old procedures and new.

Phased implementation means installing or replacing equipment in stages. When a customer's list of changes includes many items, a phased implementation usually makes the most sense. With a phased implementation, the first item on the list is scheduled, then the second, the third, and so on. Intervals are scheduled between each item so that the entire implementation may take months, or even years to complete. From the users' point of view, phased implementation allows the opportunity to get used to one type of new equipment at a time.

Timing the implementation of individual components requires considering a different set of issues. Some implementations interfere with a customer's normal business routine. The customer in the example must have new wiring installed. Installing wiring can be very disruptive to an office environment. This is because it requires pulling cable through walls, drilling holes, and installing jacks. It can be noisy and distracting to employees. To alleviate this kind of disturbance, an implementation plan may call for wiring to be done in the evenings or on weekends when employees are not around.

If the customer has a large group of users moving into its location, it will want the users to have telephone service when they move in. The implementation plan should have the PBX installation completed prior to the move.

Another consideration is old equipment. New telephones for the PBX will most likely be installed before the common equipment becomes activated. The telephones connected to the Centrex cannot be removed at the time the new telephones are installed or the employee will have no telephone service at all. In this case, old telephones should stay in place until after the PBX becomes active.

As a contrast, consider a user who will get a terminal connected to the minicomputer and will be giving up a personal computer. Since the minicomputer is already active, the user can use the terminal as soon as it is installed. The personal computer should be removed at the same time because it is no longer needed. Think of the problem that would be created by not removing the personal computer. Both the terminal and personal computer require a fair amount of desk space. The user will have to find deskspace for both devices.

Training

Every item in the list of changes will change a business's operational flow when implemented. Users must be told in advance of the changes, and they must be trained on new equipment and procedures. Training classes and information seminars are most helpful in disseminating information. They should be coordinated with the implementation plan. It is a general rule that the better the training, the smoother the implementation.

CASE STUDY: THE ELEGANT OAK FURNITURE CO.

Elegant Oak Furniture is a fictitious company that has hired you as a consultant. The company's executive officers want advice on a major project. They want to modernize the telephone systems used in their offices around the country. They also would like to make it easier and more economical for the offices to communicate with one another. During recent years the use of computers and data processing has steadily increased. They would like the ability to transfer data throughout the company as an integrated part of any new system that you recommend.

Elegant Oak is a manufacturer of solid oak desks and chairs. The company has been in business for 57 years and has an established reputation for producing high-quality furniture. All products are made at its factory in Rochester, New York. The company has warehouses at the factory and in Atlanta and Denver. Sales offices are located in Boston, Atlanta, and Denver. The main corporate office is co-located with the sales office in Boston.

DATA GATHERING

Operations

Elegant Oak's operations are clearly defined. The factory creates the furniture from raw materials. An inventory must be kept of wood supplies, screws, glue, stain, varnish, and so on. The woodworking equipment must be serviced on a regular basis. Calling patterns at the factory are split between calls to company warehouses and calls to suppliers. A few calls are made to corporate headquarters. Once a week

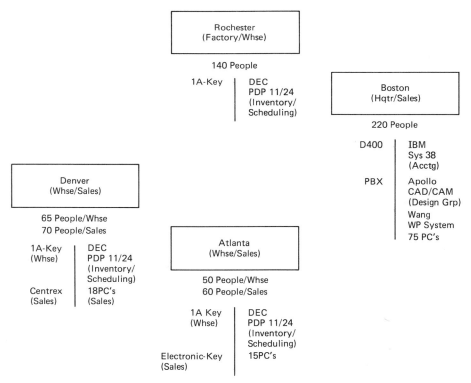

Figure 12.12: Inventory operational flow/database.

an inventory report is sent to headquarters. It is a computer printout created by batch processing and then sent via an overnight delivery service.

The warehouses maintain a stock of desks, chairs, and replacement parts. Inventory levels are monitored to maintain adequate supplies at each location. Shipping and delivery schedules are made daily to optimize the flow of stock into and out of the warehouses.

The sales offices are staffed by sales representatives and sales support clerks. Each sales representative is responsible for sales in a geographical area. The support clerks assist in the sales effort by taking customer calls, typing letters, and writing orders when a sale is made. The Atlanta and Denver offices send completed orders by overnight delivery service to corporate headquarters in Boston. The Boston sales office hand-carries orders to the order entry department each day. After a sale is made, the sales representatives track the order until it is completed. The support clerks coordinate delivery schedules with customers and handle expedites when a "rush" job is needed.

Top management operates from corporate headquarters. Also at headquarters are the accounting department, the sales-order department, the buying department, and the design group. The accounting department is responsible for accounts payable, accounts receivable, and payroll. The sales order entry department inputs orders from the sales offices into the headquarters computer and tracks orders until completion. The buying group is responsible for purchasing the wood and other materials used to manufacture the furniture. The design group designs the furniture and creates the product line that will be sold each year.

Existing Equipment

Two hundred twenty people work at Elegant Oak Furniture's corporate headquarters in Boston. A Western Electric Dimension model 400 telephone system is used for voice communications. The Dimension 400 is about 15 years old. It uses analog technology and does a fine job of handling telephone traffic, but is not designed to handle more than light data traffic.

The accounting department has an IBM System/38 minicomputer performing three distinct tasks: sales order entry, accounting, and payroll. Sales order entry requires keying sales from all offices into the computer. Customer bills are generated, and commissions to sales representatives are made. The accounting function involves accounts receivable and accounts payable. Under accounts receivable, records of customer payments are kept. Under accounts payable, creditors are paid and a log of payments is made. The computer also handles payroll, keeping track of wages and salaries and issuing paychecks on a weekly or monthly basis. Hourly workers such as secretaries and factory workers are paid weekly, while salaried management employees are paid monthly.

The design group has 15 Apple Macintosh personal computers, one for each of its designers. The Macintosh computers are used to create graphic images of new furniture designs. Recently, these computers were all linked together with the AppleTalk local area network. This allows designers an easy means of sharing information and of sharing the office graphics plotter.

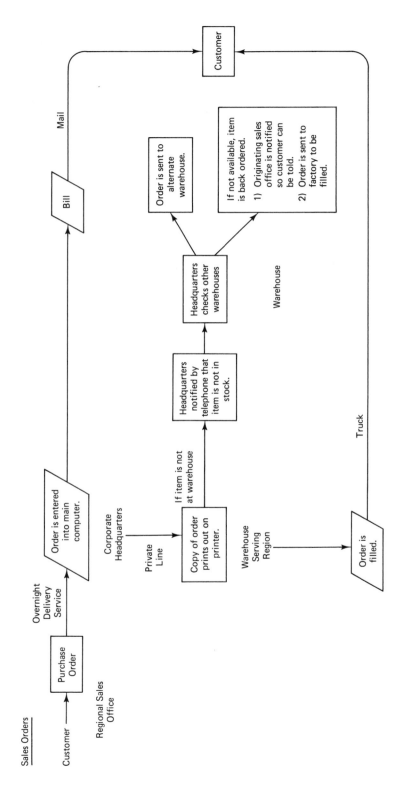

Figure 12.12: Order processing and fulfillment.

The group is also considering the purchase of an Apollo CAD/CAM workstation. This specialized computer performs the same function as the Macintosh, but has color graphics, greater speed, and the ability to display images three-dimensionally.

Twenty IBM personal computers are in use by managers at headquarters. These computers perform various tasks, primarily spreadsheet applications projecting sales and tracking earnings. A Wang word-processing system is used for corporate correspondence. Executives dictate letters to a Lanier Dictaphone dictation machine. Word processing clerks transcribe the dictation tapes to a finished document.

The sales department has an additional 20 IBM personal computers used to create sales aids such as spreadsheets and viewgraphs. It also makes use of headquarters' Wang word processer and Lanier Dictaphone.

The Atlanta location has a sales office and a warehouse. The sales office has 60 people responsible for sales activity in the southern United States. It has a Tie Communications electronic key telephone system and 20 IBM personal computers. The personal computers are used to create sales aids and to type letters and memos. The warehouse has 50 people and has a mechanical 1A-Key telephone system leased from the local telephone company. A Digital Equipment Corp. model PDP 11/24 computer is used for inventory tracking.

Rochester is the factory location, where the furniture is made. Co-located with it is a warehouse. These installations employ a total of 140 people. They have mechanical 1A-Key telephone equipment provided by the local telephone company. They also have a DEC PDP 11/24 minicomputer that handles scheduling for the warehouse delivery truck drivers and inventory for the factory and the warehouse.

Denver is the Western regional office. A sales office and a warehouse are located there. The warehouse employs 65 people. Their telephone system is mechanical 1A-Key leased from the telephone company. A DEC PDP 11/24 computer is located there also and handles delivery scheduling and warehouse inventory. The sales office has 70 people. They use a Centrex telephone system leased from the telephone company. They also have 18 IBM personal computers.

Future Needs

Elegant Oak Furniture has a long-term corporate objective of linking and integrating its locations and operations. Top management looks at this project as a major opportunity to fulfill this objective. The following "wish list" has been provided:

1. Link the telephone systems together at all locations to allow calling Denver from Atlanta as easy as calling from one desk to the next
2. Lower long distance charges by installing WATS lines that can be used by anyone in the company
3. Centralize customer service so that a single group handles routine requests and the mailing of sales brochures for the entire company
4. Allow sales offices to enter orders remotely into the computer at corporate headquarters

5. Allow remote offices access to information in the headquarters computer
6. Let users of personal computers electronically transfer information—for example, sending a letter composed on a personal computer to word processing for typing and mailing
7. Share inventory data among the systems at each warehouse

GROUPING

Many employees in Elegant Oak Furniture's offices across the country do similar jobs (secretaries, sales representatives, truck drivers, as well as others). These employees can be grouped as shown in Table 12.4. The voice and data needs of each group are included.

The final network design must account for the needs of each group. It should facilitate and augment a worker's ability to perform his or her job.

TABLE 12.4
Grouped Jobs and Needs

Group	Needs	

Top Management
-President/CEO
-Chief Financial Officer
-Vice-presidents for:
    Product Development
    Marketing
    Production
    Data Processing
	Voice:
	Heavy voice usage
	Most calls are made to other managers at corporate headquarters
	Calls also made to managers at remote sites
	Calls outside the company to locations worldwide
		Data:
		Moderate data usage
		Facsimile/Telex transmission
		Inventory data and production schedule data needed from computers at remote sites
		Accounting, sales activity, and operating expense data needed from central computer
		Word processing

Accounting -Accounts Receivable Clerk	Moderate voice usage Most calls made to customers asking for payment of delinquent bills Other calls split between remote sites and calls within corporate headquarters	Moderate data usage Reconcile payments received with bills in the central site computer
-Accounts Payable Clerk	Light voice usage Most calls split between departments at remote sites and departments at corporate headquarters	Heavy data usage Input data into cental site computer for generation of payment to creditors
-Payroll Clerk	Light voice usage Most calls split between departments at remote sites and departments at corporate headquarters	Heavy data usage Entry of timesheet data into computer for generation of paychecks Updating records of salaries and deductions

TABLE 12.4
Grouped Jobs and Needs

Group	Needs	
-Order Entry Clerk	Light voice usage Primarily answer inquiries from remote offices about order status	Heavy data usage Enter new sales orders into headquarters computer. Monitor order status until order is filled and customer can be billed
Raw Material Buyers	Moderate voice usage Buyers use the telephone constantly when they are in the office. They travel often, making their telephone usage sporadic Most telephone calls are to suppliers Other calls are made to factory checking stock levels Occasionally calls are made to the design group to coordinate purchasing	Light data usage Weekly printouts from the factory's computer are used to monitor stock levels of raw materials
Designers	Light voice usage Most calls to suppliers discussing materials available for use	Moderate data usage MacIntosh computers used in furniture design
Sales Representatives	Heavy voice usage Most calls split between outgoing calls to customers and incoming calls from customers Some calls to warehouses to check availability of furniture Some calls to sales support clerks	Moderate data usage Personal computers used to type letters, create spreadsheets and make viewgraphs
Clerical -Secretaries	Light to moderate voice usage Primarily answer incoming calls	Light data usage Use personal computer to type letters
-Sales Support Clerk	Moderate voice usage Primarily answer incoming calls from customers and sales representatives Some outgoing calls to corporate headquarters to check order status and to warehouses to check availability of stock	Light data usage Use personal computer to type letters and create sales aids
Furniture Craftspersons	Light voice usage Business telephone calls rarely made	Light data usage Receive computerized printout of production schedules on a weekly basis

Grouping

Production Schedulers	Moderate voice usage Most telephone calls made to corporate headquarters to determine production requirements and to advise buyers of raw material inventory levels	Heavy data usage Create production schedules on a weekly basis Review inventory reports from the warehouses
Warehouse Workers	Light voice usage Outgoing telephone calls rarely made. Most calls are incoming from other offices checking the availability of stock	Light data usage Refer to computerized inventory printouts to locate stock
Transportation Schedulers	Moderate Voice Usage Most telephone calls are split between incoming and outgoing calls to customers in the delivery area	Moderate data usage Schedule delivery dates and routes daily
Truck Drivers	Light voice usage Most telephone calls made from the field to the warehouse	Light data usage Receive a computerized printout of daily delivery schedule

SUMMARY

A fictitious company has been presented in this chapter. It's locations and methods of operation have been described, as have it's equipment inventory. Also described are the objectives it intends to achieve by re-designing it's voice and data systems.

You now have enough information to design a system for Elegant Oak Furniture. Review the steps to follow that were discussed in the first part of the chapter. Then refer to the case study as often as necessary. Your final design should incorporate both voice and data elements. Good luck!

KEY TERMS

Change out
Phased implementation
System

Task
User

Network Management and Security

CHAPTER 13

OBJECTIVES

Upon the completion of this chapter, you should be able to:

- distinguish between centralized network management and decentralized network management
- describe the facets of network management
- list voice network and data network management tools
- explain how security devices work in a network

INTRODUCTION

Network management is a catchall term covering several different activities. First there is normal network maintenance. This includes routine administration and diagnostic procedures. Then there is network optimizing, tuning the network to get maximum performance from it. Finally, there is security, maintaining the integrity of a network by limiting access to those who are authorized.

Partly due to the breakup of the Bell System, there has been an explosion of interest in network management in recent years. In the post-divestiture world, customers have more responsibility for the operation of their networks. This is welcome news to many, and networks are increasingly seen as a strategic corporate asset. However, with increased control comes the burden of determining how best to manage things.

This chapter examines the issues involved in network management decisions. We will also look at tools available to maintain and optimize a smooth-running operation.

NETWORK MANAGEMENT CHOICES

Two fundamental issues determine the way a network will be managed: Will the network be built around a single vendor or multiple vendors? Should the network have centralized or decentralized management? A network manager's first task is to examine the options and to determine the best approach for the company.

Single or Multiple Vendors

Prior to 1983, most companies leased networks from the Bell operating companies (BOCs). This included the facilities and the equipment tying them together. A company used the system, but the BOC was responsible for maintaining it. With the trend toward deregulation of the industry culminating in the breakup of the Bell System, this scenario has changed. Many vendors now compete for a telecommunications manager's business.

As in most highly competitive fields, there is a dazzling array of products to choose from. Some are quite specialized, while others attempt to encompass the entire network. The choice of whether to select services and products from a variety of vendors or to choose a single vendor for all requirements is critically important.

Multiple Vendors. There are pros and cons to either decision. Choosing multiple vendors has the advantage of allowing a manager to select precisely the right product for a given need. It also leads to getting products at the best price because of competition among vendors.

The problem with using multiple vendors stems from the lack of standards (or variety of standards) in many areas. For example, a company with a Corvus local area network using a CSMA protocol cannot easily add an IBM Token-Ring Network and have the users on both networks share information or resources. Network management equipment presents an even greater difficulty. There are no network management standards at all. Each vendor has its own proprietary offering. For example, products from Racal-Milgo for managing modems will not work with AT&T modems, and vice versa.

Some of the incompatibility problems should get resolved over time. Many areas that involve networking are relatively new, and standards are just now evolving. As vendors respond to pressure from customers, they are likely to find it in their interest to rally around a particular standard rather than pursue a proprietary approach.

Another problem with using multiple vendors is **"fingerpointing."** This amounts to each vendor denying responsibility when a problem arises. An example of what can happen is depicted in Figure 13.1. It shows how a user with a Compaq microcomputer communicates through a Northern Telecom PBX to the Telenet

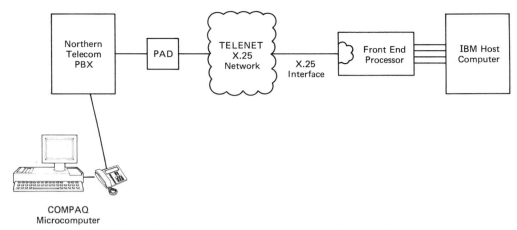

Figure 13.1: A circuit using multiple vendors' equipment.

packet switched network. Also connected to Telenet is an IBM Mainframe computer, which is linked to the Compaq. If the user cannot access the IBM host, who does the telecommunications manager call to clear the problem—Compaq, Northern Telecom, Telenet, IBM? All too frequently, the vendors will claim there is no problem with their equipment and point a finger to the others. It can be very frustrating and time-consuming to get the vendors to cooperate in efforts to determine the true cause. Time is the one thing a telecommunications manager does *not* have when the network malfunctions. The manager's goal is to restore service as quickly as possible.

Single Vendor. To avoid the problem of fingerpointing, some managers elect to get all products and services from a single vendor. Then, when there is a problem, there is only one vendor to call. A single vendor also helps ameliorate problems of incompatibility among systems. There is no guarantee of this, though, since vendors, especially large ones, tend to offer products incompatible with others in their own sales line.

No vendor encompasses the entire breadth of telecommunications options, but some come fairly close. AT&T and IBM probably come the closest. Each has an extensive line of services and equipment. AT&T is generally stronger in networking products, while IBM offers more in terms of processing ability and user applications. The strengths of one tend to be the weaknesses of the other.

Several vendors have comprehensive offerings in a particular area. For instance, Codex is strong in data communications, while Northern Telecom is strong in voice communications.

The decision to use a single vendor will depend, to a great extent, on the requirements of the company. If a company has no data communications requirements, then a company like Northern Telecom is as appropriate as AT&T, IBM, Siemens, or any other voice vendor. If, however, a company has mixed voice and data communication needs, the field of vendors narrows. A company's decision to go with a single vendor or multiple vendors can and should be influenced by these factors.

Other factors are the flexibility and growth potential of the equipment and the support the vendor will provide. Are the vendor's products capable of fulfilling user's needs and can they be expanded and enhanced as the company grows? What is the vendor's reputation for support? If the main reason for using a single vendor is for quick response to problems, then the vendor must be able to provide adequate service or the purpose of selecting them vanishes.

Centralized versus Decentralized Management

In spite of vendor support, network managers will find a substantial amount of time spent on identifying and correcting network problems. To deal with it, technicians familiar with the network and its equipment will be needed.

The second major factor affecting network management involves determining how to deploy management people and resources. Two options are available: having a centralized management center with all personnel at a single location or having decentralized centers with personnel dispersed among them.

A centralized management center has the advantage of dealing with the entire operation from a single location. All resources can be concentrated there, meaning that when a problem arises anywhere in the network, the best people and equipment will be available to work on it.

The negative side of centralized management is that it may be less efficient than having several sites. There are many routine tasks to be done that do not require the expertise of topnotch troubleshooters. These tasks can be done by local personnel at the equipment sites. Centralized management may cause unnecessary inconvenience to users, too. For example, if the management center is in Philadelphia, a user in San Diego may have trouble getting assistance in the late afternoon because of time zone differences. If there were a management center on the West Coast, there would be no problem.

One of the big advantages of remote sites is having personnel on-site when a problem occurs. An on-site technician can fix the problem or be the eyes and ears of more experienced technicians located elsewhere.

The negative side of remote management centers is cost. It is likely that more management personnel will be needed, because if there is a site, there must be people to run it. A single, central site allows for economies-of-scale. One technician can service several sites by telephone or with remote diagnostics. However, if there are several centers, there must be at least one technician at each.

To make a decision as to whether network management should be centralized or decentralized, the factors discussed above should be considered in light of the size and type of network involved and with regard to the level of expertise of management employees. Small networks can be easily handled from a single site. As networks increase in size and complexity, the advantages of decentralization hold more sway.

NETWORK MANAGEMENT FUNCTIONS

The responsibilities of network management fall into four categories:

1. network administration
2. network maintenance
3. network optimization
4. network security

Each covers an important aspect of overall network function.

Network Administration

Network administration handles day-to-day activities such as adding new users, removing old ones, and modifying the features available to them. In addition, it covers tasks like bringing new equipment into service and upgrading or expanding the capacity of existing equipment.

Administration of a PBX can include setting up new trunk groups, changing a user's class of service, or removing dialtone from offices that users have vacated. For example, if T-1 facilities are newly added to a network, they must be activated in each PBX that will terminate them. A user may have a class of service restricting the ability to make long-distance calls. The administrator can change the class of service associated with that telephone to one with long-distance dialing capabilities. In the same way, an administrator can deactivate telephones altogether and turn off dialtone to offices that users have vacated.

Network Maintenance

Network maintenance is the aspect most commonly associated with the name *network management*. It deals with facility and equipment problems such as failing circuits or malfunctioning terminals.

Two steps are involved in maintenance: isolating problems and correcting them. Isolating them is usually done with the help of diagnostic equipment or with the assistance of a vendor. Correcting the problem can take many forms. It may require bringing a vendor on-site with replacement parts, if the problem is in a key piece of equipment like a PBX or a computer. If the problem is a modem or a multiplexer, a **hot standby** may be used. A hot standby is spare equipment that can immediately be put into service when the original component fails. The failed component can then be sent for repair, while the network operates normally. If the failure is in a circuit such as a leased private line, correction may include the use of **dial backup** equipment. It works this way: dial backup equipment is located at both ends of a four-wire private line facility. When the circuit fails, the equipment is used to establish two switched network connections, and all traffic is routed over the switched network until the private line facility is back in operation. Newer dial backup equipment will automatically swing over to switched network lines without requiring human assistance. This minimizes the amount of time that users will find the connection unavailable.

You can see that a common procedure for maintenance is to have spare equipment or redundant capability on tap for immediate use. This approach bypasses the need to actually wait for a repair. A replacement is operating while the original equipment or facility is being repaired.

Another important part of network maintenance is preventive diagnostics. This involves performing regular checks on network components to identify problems at an early stage. Often this can "nip a problem in the bud," before it becomes a major failure affecting users.

Network Optimization

The function of network optimization is to use network resources in the most efficient manner. This involves both monitoring network activity and shifting traffic to achieve a balanced distribution. Equipment for gathering statistics is very important to this type of work. Equally important is equipment to manage the flow of information.

Network Security

Security is concerned with unauthorized access to the network and its information and how to control it. There are two ways to approach security. One is from the physical aspect, the other from the technical aspect.

Physical Aspects of Security. The physical aspect involves physical barriers to access. For instance, access to a voice network is through telephones connected to a node switch. If telephones are secured in the evenings and on weekends, then protection is provided against illegitimate use. The normal way to secure telephones is to lock the offices they are in after working hours.

Since the telephones are assigned to employees, unauthorized use during working hours is not a major problem. The telephone is dedicated to use by one employee. If someone else wants to use it, the employee can effectively screen access to it.

A similar situation exists with data networks. Access must be through a computer terminal, personal computer, or intelligent workstation. The offices they are in can be locked up during non-business hours, and employees can monitor their use during the day. Devices may not be dedicated to employees the way telephones are. Instead they may be in a terminal room or in a common area. In this case, employees will be expected to know who their coworkers are and to be aware if a stranger is attempting to use a terminal.

This type of security provides the first level of protection. Virtually all businesses do these things without consciously thinking of them as security measures. They are simple and effective, but far from complete. For instance, a telephone in a public lobby or in an elevator cannot be locked up. It is also difficult to lock up a computer terminal located on the floor of a factory. Another problem is that many networks provide remote access to telephone common equipment or to computers. This capability thwarts any physical barriers. To cope with these situations, other methods of security are needed.

A different aspect of physical security involves the facilities. Metallic-based cabling emanates an electromagnetic field when electricity passes through it. Sensitive detection equipment can be used to pick up these waves and monitor the information passing through. Microwave and satellite present the same problem. Transmission waves literally pass through the air, making it relatively easy to tap into the signals. Shielding, such as the type used in coaxial cable, provides some measure of security, but it is not complete. Fiber-optic cable offers the best security because the lightwaves carrying information generate no electromagnetic field at all.

Technical Aspects of Security. Physical measures provide a first level of security, but, as we have seen, there are shortcomings. Further steps can be taken through the use of technical methods.

Technical methods involve the use of software or equipment to protect the contents of or access to information within a network. One of the most commonly used forms is the **log in** and **password.** These require a user to supply two pieces of information before gaining access to a data network. They may be thought of as two keys needed to unlock the terminal. The first is the log in. It is a word that the system recognizes as the name of a user. It is usually the person's first name or a combination of first and last name. A list of valid users are stored in a table and the log in entered at a terminal is compared to the list. If a match is found, the user is prompted for the second key, the password. It is a string of characters known only to the user and constitutes a private code between the user and the network. With both pieces of information properly given to the system, a user is allowed access.

Authorization codes perform a similar function in voice networks. They are numbers assigned to users. The network can be set so that before telephone calls can be made, the user will be prompted to enter an authorization code. If a user does not have one, the call will not be placed. Generally, the numbers are four to seven digits in length. They may be related to a person's birthdate or social security number to make them easier to remember. Some examples would be 6602345, 2892, and 204060.

Network access control devices are commonly used in data networks to provide protection on dial-up lines. The devices connect to the dial-up modem and intercept calls as they come in. The caller is prompted for a log in and password. If incorrect information is given, the caller is disconnected. If the information is correct, the device reads a telephone number from a table and calls the number. The number should correspond to the location of the caller, and at this point, the caller gets access to the network.

Another way to provide security is with **encryption.** It involves coding or scrambling a voice or data transmission so that it is unintelligible to an interceptor.

At the receiving end, the transmission is unscrambled. Pay cable television channels are a well-known example of encryption. The signals are encrypted so that only wavy lines and blurred images can be seen without the use of decryption equipment.

NETWORK MANAGEMENT PRODUCTS

There are many network management products available in today's market. Some are equipment only, while others are software programs or a combination of hardware and software. Some are designed only for voice networks, others strictly for data networks. They run the gamut from small, relatively inexpensive devices, such as circuit testers, to complex, integrated system management products. Regardless of their size or complexity, they are all simply tools to help in the difficult task of network management. Descriptions of representative products appear below.

Voice Network Systems

Management tools for voice networks provide for **facilities management, station management, traffic management,** and **call management.** These capabilities are inherent in each node switch and can be administered locally. However, this requires someone at each site to perform the tasks. Managing from a central site requires equipment there that can keep track of all locations and access them with changes to the node's configuration.

Centralized System Management. AT&T's voice network management tool is called Centralized System Management (CSM). It provides an apt example of products of this type and will be used to illustrate similar offerings from other vendors. Differences among vendors are in the method of implementation. For instance, IBM implements its voice network management as part of NetView, primarily a data networking product at this time (see description later in this chapter). Northern Telecom and others offer either their own packages or resell products written by other software developers.

CSM is a software program that runs on AT&T 3B2 computers (a super-microcomputer). Up to 30,000 stations can be managed on a single system (this figure is reduced to 10,000 stations if the cost-management option is included).

Internally, it contains a database with records for all stations and facilities. It also contains software allowing it to dial up a node switch and make changes or retrieve traffic information without human intervention. Though it is designed primarily for use with System 85, it can also interface with System 75, but in a much looser way. Essentially, the operator bypasses CSM software and interacts directly with the System 75 (fortunately the screen layouts and command structure are very similar).

There are three components in the main package: Facilities Management (FM), Terminal Change Management (TCM) and Traffic Management (TM). An optional Cost Management (CM) component is also available.

Facilities management allows for monitoring and reconfiguring network trunks and trunk-related features. Network capabilities that are available include:

- adding, moving, or rearranging trunks and trunk groups
- administering trunk features such as queueing
- assigning names to trunk groups, such as voice, data, modem pool, or host computer access
- defining on-net and off-net call routing patterns.

Trunk group testing and diagnostics can be run also, such as:

- trunk testing with the **automatic circuit assurance (ACA)** feature
- trunk testing with the **automatic transmission measurement system (ATMS)** feature
- taking trunks out of service or putting them back in service.

Reports of test results can be printed on demand. Individual node switch features can be administered, too. For example,

- setting up trunks for automatic call distribution groups
- administering centralized attendant service
- setting facility restriction levels
- creating authorization codes.

CSM's database keeps track of the current network configuration and provides reports of current trunk assignments.

Terminal change management is AT&T's term for station management. It relates to the management of equipment on the line side of node switches. These include telephones, modems, computers, and terminals. CSM's capabilities allow

- adding or removing telephones and other terminal equipment
- changing features and attributes assigned to equipment
- administering node switch system features such as class of service, call forwarding, call coverage, speed dialing, and station ringing assignments
- establishing automatic call distribution groups.

In addition, CSM's database can keep track of spare capacity in node switch cabinets and spare equipment such as extra telephones or modems.

Traffic management looks at network and node usage statistics and compiles reports of traffic patterns. It can be very helpful in optimizing a network. Analysis of the reports can identify bottlenecks where traffic is being blocked. It can also show areas of low usage, providing a complete picture of network traffic patterns. From these reports, decisions can be made about changing traffic routing tables to more evenly balance traffic among all facilities.

These three applications—facilities management, terminal change management (station management), and traffic management—come together to provide an integrated tool for monitoring a network and modifying it as needed. The database within CSM can serve as an added tool providing a place to store useful information that can then be manipulated into meaningful reports.

Cost Management, AT&T's call management program, is an optional application that can be added to CSM. Its function is twofold. First, to collect and store station message detail recording (SMDR) records, and second, to calculate the cost of the calls.

SMDR records are generated by the node switches each time a user places a call. Information is kept about who made the call, where it was made to, the time of the call, the duration of the call, and the facility that was used. The records are stored at the node and are periodically retrieved by a polling unit connected to the CSM computer. The records are stored in the computer, and typically, a report is generated on a monthly basis. In the report the cost of each call has been added and the calls have been sorted by date and time, by calling number, or by some other operator-defined field. Figure 13.2 shows a typical report sorted by date and time.

```
                                   ACME CO.
REPORTING:                                                          PAGE 1
1985 SEP 15                                              TODAY'S DATE OCT 10
                        DETAIL REPORT: CHRONOLOGICAL
        KEY
$>>>.>>  COST TOO LARGE
$# # # .# #  COST INDETERMINABLE
                                    ACCESS   CALLED
DATE    TIME    DUR    CODE         CODE     NUMBER         STATION   ACC'T   COST
09/20   14:41   0:02   I            71                      6561              $ .16
09/20   14:41   0:01                80       1-317-468-1234 6364              $ .39
09/20   14:42   0:02                80       1-304-525-5678 6705              $ .78
09/20   14:42   0:01   80           86       1-312-555-1212 6650              $ .00
09/20   14:42   0:01                9        629-1234       6327              $ .18
09/20   14:43   0:01   I            71                      6869              $ .08
09/20   14:43   0:03   84           83       1-605-343-5678 6807              $1.17
09/20   14:43   0:01   I            71                      6235              $ .08
09/20   14:43   0:02   I            71                      6451              $ .16

               TOTALS
               GOOD CALL RECORDS       9
               TOTAL DURATION       0:14
               TOTAL COST          $3.00
```

Figure 13.2: A sample SMDR report sorted by date and time.

On the report is the name of the company (in a network, it would be the name of the node), the reporting period, the current date and then information about each call. The date and time are shown, then the duration is shown in the format

hours : minutes tenths (of minutes)

so a call with a duration of 0:02 lasts 2 tenths of a minute, or 12 seconds. The code field indicates special conditions about a call. For instance, the "I" shown in some of the records indicates an incoming call. The access code field identifies the number of the trunk group that carried the call. Where two numbers are shown, the call was attempted over the first trunk group, but did not get through. It was then routed over the second trunk group and was successful. The called number is shown and the extension number of the station within the node. The account code, if there is one, is shown and, finally, the cost of the call. A summary totaling the number of calls, duration, and cost is shown at the bottom of the report.

CSM is designed for ease of use. It is menu-driven, with help screens available for each function. The user screens are displayed in a plain-English format to simplify interaction between the user and the program.

A hierarchical command structure is used to move from task to task. For example, the command to add a new station is

tcm : admin : extension : add

The first command in the hierarchy is "tcm." It says that the function to be performed is part of the Terminal Change Management application (see description below). The second command, "admin," refers to an administrative task involving a change to a System 85 node switch. "Extension" identifies that the change will be to an extension (as opposed to a feature), and "add" determines that a new extension will be added.

Once an operator learns the command structure, it is straightforward to jump from one task to the next. For instance, if an operator wants to remove a station, the command is the same as the one shown above, except that the word "remove" replaces "add."

The screens have a uniform layout with status information at the top and the screen display below. Figure 13.3 shows an example of a screen.

At the top is the CSM Release number (the version of CSM software that is being used). In the middle is the model and version of the switch being worked on. At the right is the name of the node (called the *target*). Below it is current command. The output from the command fills the rest of the screen.

The output shown is from a command to display a service request. A service request is the way an operator makes changes to a switch configuration. The information shown includes the service request number, an arbitrary number assigned by CSM to help track the request. The date and time refer to when the request will be changed in the switch. The transaction numbers list each step that CSM is to perform. The target identifies the switch that will be updated. In this case, the switch

CSM Release	Sys85:R2V4	target
PATH:tcm:admin:service-request:display		

THE SCHEDULED TIME IS IN THE TARGET'S TIME ZONE
SERVICE REQUEST NO.: jer0602a0   DATE: 12/12/99   TIME: 6:55

TRANSACTION NO.	TARGET	OBJECT	VERB	STATUS
1.0	9992244	12-button	add	n
1.1	9992244	terminal	add	p
1.2	9992244	button	add	p
1.3	9992244	button	add	p
1.4	9992244	button	add	p
1.5	9992244	button	add	p
1.6	9992244	abbreviated-dial	add	p

Figure 13.3: An AT&T centralized system management service request screen.

is identified by the number 9992244. The object and verb identify the equipment within the switch that will be affected and the action that will be taken. The status indicates whether the transaction has been worked and whether it was successful. The "n" shown indicates that transaction 1.0 has not yet been scheduled. "P" status means that the transaction is pending (scheduled, but not completed).

This example points out an important feature of CSM. Transactions can be scheduled for a future time. An operator has the option of having a task executed immediately or held for a period of time. This capability is extremely useful for scheduling routine facility diagnostic checks.

Data Network Systems

On the data side of the network, there are also integrated systems. Vendors of these systems have different views about how best to fulfill network management requirements. Modem vendors such as Codex, Racal-Milgo, and AT&T emphasize incorporating diagnostics in the equipment itself. Test equipment vendors such as Atlantic Research and Digilog stress products that are independent of a single vendor's hardware. They usually market systems that "wrap around" network equipment, that is, the systems monitor both the input side and the output side of network equipment. Computer companies, such as IBM, favor a third approach: focusing network management on the processor and having the intelligence and diagnostics reside in software.

There are trade-offs with each approach. Incorporating diagnostics into the equipment locks a company in to a single vendor. It is a particular problem if there is a mixture of equipment on hand before a management system is purchased. A system using wrap around equipment requires the purchase of a diagnostic device for every network device. When diagnostics are built into network equipment, this is unnecessary. Focusing network management on the processor requires a great deal of computing power. With IBM's NetView software, a mainframe computer is needed. Most other packages run on a personal computer or supermicro. The differences in approach are broad. They stem from the different backgrounds of the vendors coming to this market. Potential buyers should assess their situation well to determine the one best suited to them.

Three systems have been selected as representative of data management systems. Each represents a different approach. Paradyne, a modem vendor, incorporates diagnostics in their equipment. Atlantic Research, a test equipment manufacturer, uses the wraparound technique. Finally, IBM, with its emphasis on the processor, offers its NetView software.

Paradyne Analysis Systems. The Paradyne Analysis series of equipment is a collection of three microprocessor-based systems. The primary distinction among them is size. The model 550 can handle up to 10 lines, the model 5500 can handle from 10 to 150 lines, and the model 5530 can handle up to 900 lines. With each line able to support up to 256 devices, even the smallest model can handle a sizable system.

The two larger systems provide printed network status reports. Top-of-the-line model 5530 includes a minicomputer with a relational database providing inventory and management reports and trouble ticket generation in addition to status reports. It also offers a query language allowing users to develop their own reports.

The central computer monitors the network by constantly polling devices. To perform self-checks and respond to a poll, devices must be outfitted with an optional diagnostics board, the Diagnostic Microcomputer Control (DMC) board. The central computer communicates with devices over a secondary channel separated from the main data communications channel. In this way, normal data flow is not interrupted by network monitoring.

When a problem is found and diagnostic tests must be run, transmissions over the main data channel will be disturbed. The line must be taken out of service, or alternatively, diagnostics can be run during off hours when the circuit is idle. The Analysis systems provide for delayed initiation of tests. They could be set up during the day and run overnight, for instance.

Procedures for correcting problems include selecting a slower transmission speed, switching to a hot spare, or switching to dial-backup facilities. The central site can activate all of these procedures on remotely located equipment. The ability to switch to a lower speed is inherent in the device. A Remote Hot-Standby Switch can be used to switch to a hot spare. If the problem is with the facility, a Multidrop Auto Call Unit (MACU) can be used in conjunction with a Dual Call Auto Answer (DCAA) to changeover to dial-backup facilities.

Atlantic Research Network Test System. Atlantic Research is well known as a vendor of data analyzers, patching equipment, and various other kinds of trouble analysis and test equipment. Its Network Test System (NTS) offering comes in four models that are actually modules from other Atlantic Research products. They have been packaged according to the most commonly needed elements. While this may be an unusual approach, it has some real benefits. The primary benefit is that modules in one system can be reused in other systems. Should a customer upgrade or change systems, much of the investment in the original system will carry over to the new one. Another benefit is that there is great flexibility in the systems. If a particular module is needed, it is very likely that the system will accommodate it.

All of the models share certain features. In addition to the modular design described above, all NTS models work with any type of network equipment and perform their tasks without interfering with normal network transmissions.

The smallest model is the NTS 1000 Network Restoration and Test System. It is a manually operated system providing patching equipment, switching equipment, and test equipment. The patching and switching equipment are used to reconfigure a network manually. The test equipment can be used to perform diagnostic tests.

Model 2000, the Distributed Network Restoration and Test System, comes with a controller, patching or switching equipment, a computer terminal, fallback switches, and the RCS-100. RCS-100 is a set of remotely controlled switches. They can be activated through the computer terminal or by a tone-dial telephone. This adds remote restore capability to the features of the model 1000. It has a maximum capacity of 768 circuits.

The model 3000 Network Restoration, Test and Management System uses much the same hardware as the model 2000. There are additional software capabilities available. In addition to system maps and help screens, it has programmable circuit alarms, trouble ticket reports, and management reports. The model 3000 can accommodate up to 1,000 circuits.

The model 4000 Distributed Network Restoration, Test and Management System is designed for medium to large networks. It offers such features as distributed network control, remote testing, remote restoration of service, and control of unattended network nodes. Up to 16 remote sites can be controlled from a central site. With a capacity to handle 1,000 circuits at each site, the model 4000 can support up to 16,000 circuits.

IBM NetView and NetView/PC. IBM's network management product is actually two software products. NetView is a compilation of existing IBM software that has been pulled together under the NetView umbrella. They are programs written to be used under SNA on a mainframe host. They include the Network Communications Control Facility (NCCF), the Network Problem Determination Application (NPDA), and the Network Logical Data Manager (NLDM). The products they support are the 3720 Communications Controller, the 3725 NCP Token-Ring Interconnection, the 5860 series of modems, and the 3728 electronic matrix switch. The consolidation of products under NetView has meant the addition of user-friendly menus, help screens, and tutorials. Functions on function keys and the use of color have been standardized across the products as well.

NetView/PC is a personal computer-based software program that runs on an IBM PC/XT or IBM PC/AT. It is basically a communications package that provides a gateway for the transfer network management data from various types of equipment to NetView software running on a mainframe.

An important aspect of IBM's network management strategy is to broaden its support to non-IBM products. NetView/PC appears to be the centerpiece of that thrust. Currently, NetView/PC extends IBM's support to non-SNA products like the IBM Token-Ring Network and voice networks. IBM would like to continue the expansion to include non-IBM products as well.

NETWORK SECURITY EQUIPMENT

There are two types of network security equipment for use with data networks, network access control devices, and encryption devices. A representative example of each type is described below.

LeeMah Datacom TraqNet

LeeMah originated the callback security device market in the early 1980s with the introduction of its Secure Access Unit (SAU). They are now the leading vendor in this market, with a well-established reputation and product line.

Their current product is the TraqNet, an enhanced version of previous offerings. In addition to callback security, it provides an audit trail and random password generation.

An audit trail is a listing of all attempts to access the network through the device. Information listed includes the date, connect time, disconnect time, reason for disconnect, port and user identification name.

A random password generator provides an extra degree of security by creating a random password used for one log-in sequence only. The password can be added to or can replace normal security passwords. When a user calls up, the password is generated for that log-in sequence. When the device calls the user back, the random password must be entered to complete the log-in sequence.

The TraqNet comes in three models: the 2008, the 2032, and the 2128. The model 2008 can protect up to 8 ports; the model 2032 can protect up to 32 ports; and the model 2128 can protect up to 128 ports.

Com/Tech Systems Secre/Data

Com/Tech Systems provides encryption devices under the name Secre/Data. Three model lines are available: the 102 series, the U102 series, and the 1102 series. They support asynchronous communication up to 19,200 bps and synchronous communications up to 1 Mbps. Each bit in the datastream is encrypted separately, therefore the Secre/Data device is not sensitive to protocols and any protocol can be used successfully with it.

Com/Tech employs proprietary encryption algorithms in its equipment. The implementation used on the 102 and U102 series provides 200 million code settings. The 1102 series uses a more elaborate scheme and provides 40,000 trillion code settings.

The 102 series provides the standard model. It supports a wide range of interfaces including EIA RS-232C, CCITT V.35, 20-milliamp current loop and 60-milliamp current loop. The current loop interfaces are commonly used by the U.S. government.

The U102 series is a smaller version of the 102 series, designed for stand-alone operation or for mounting in a data rack. It comes only with an EIA RS-232C interface.

The 1102 series differs from the other models in that it is intended for applications in which data will remain encrypted for long periods of time. Toward this end, it offers 200 million times more codes than the other units.

CHAPTER SUMMARY

Network management emcompasses the methods of monitoring a network's condition and methods for identifying and correcting problems when they arise. Prior to the breakup of the Bell System, many companies leased networks from a BOC. With the breakup came a shift. Many companies stopped leasing and elected to manage their own network.

Once the path of network management has been chosen, the issue of centralized versus decentralized network management must be addressed. Network size and location as well as the level of expertise of employees are factors that should be weighed when making a decision.

Network management functions are grouped into four categories: network administration, network maintenance, network optimization, and network security. Monitoring and control equipment is available for each function.

KEY TERMS

Authorization code
Automatic Circuit Assurance (ACA) feature
Automatic Transmission Measurement System (ATMS) feature
Call Management
Dial Backup
Encryption
Facilities management

Fingerpointing
Hot standby
Log in
Network access control devices
Password
Station management
Traffic management

REVIEW QUESTIONS

1. What are the functions of network administration?
2. What are the functions of network security?

3. What is the purpose of network optimization?
4. What is network maintenance?
5. How do encryption devices differ from network access control devices?
6. How do facilities management, station management, call management, and traffic management help perform network administration, network maintenance, network optimization, and network security tasks?

TOPICS FOR DISCUSSION

1. To some companies, managing their network means having the right to bypass the local exchange company's central office and connect directly to an interexchange carrier. Discuss the legalities and economic issues of bypassing a LEC.
2. Computer viruses are a phenomenon affecting networked computers. What are viruses and how do they work their damage? What steps can be taken to guard against them?
3. White collar crimes (for example, embezzling or altering data records) often go unpunished because either the person cannot be proved guilty or the crime is deemed too small to warrant action. As the owner of a business that relies on computers, what is your reaction?

Voice and Data Periodicals for the Telecommunications Manager

APPENDIX A

Voice Periodicals

Business Communications Review (monthly)
BCR Enterprises
950 York Road
Hinsdale, Illinois 60521-2939
Comments: This is a management-oriented look at the current state of telecommunications. The articles tend to be in-depth and very current.

Communications News (monthly)
124 South First Street
Geneva, Illinois 60134
312-232-1400
Comments: This newspaper-format periodical is excellent for keeping telcom managers abreast of the latest trends and also for providing coverage of all major telcom conventions. The articles are generally not very technical.

Communications Week (weekly)
600 Community Drive
Manhasset, New York 11030
Comments: This weekly newspaper covers the financial, marketing, and legal aspects of telcom management. It also provides information on new product announcements.

Inbound Outbound (monthly)
12 West 21st Street
New York, New York 10010-6997
800-LIBRARY
Comments: This periodical is very specialized. It is written for the professional who makes a living using the telephone. Since it is telemarketing-oriented, it provides the latest information on such products as ACDs, which are critical in this industry.

Network World (weekly)
375 Cochituate Road
Framingham, Massachusetts 01701-9171
508-820-2543
Comments: This is probably the single most important voice-oriented periodical for the telecommunications manager. It provides up-to-the-minute information on the entire field of telecommunications as well as weekly surveys that compare and contrast

major products such as PBXs or local area networks.

Procomm Enterprise Magazine (monthly)
Box 886
San Anselmo, California 94960
415-459-4669
Comments: This journal provides short articles written from a management perspective that seek to solve a specific problem. An article such as how to save money using Centrex services would be typical of this periodical.

T&M (monthly)
124 South Fat Street
Geneva, Illinois 60134
312-232-1400
Comments: This is the most technical of voice periodicals on this list. Articles frequently are written from a telecommunications engineering perspective.

Telecommunications (monthly)
685 Canton Street
Norwood, Massachusetts 02062
617-769-9750
Comments: This periodical provides short articles that range from general commentary on the legal environment for telecommunications to brief looks at new technology. An article on how hybrid wide-area networks can save money is typical of the level of articles found here.

Teleconnect (monthly)
12 West 21st Street
New York 10010-6997
800-LIBRARY
Comments: This is a telecommunications magazine written for people who sell telecommunications equipment. As such, it emphasizes major features of new products at a very non-technical and entertaining level. A typical article might provide a survey of leading middle-range PBX vendors with a list of major features for each vendor.

Telephony (weekly)
Intertec Publishing
55 East Jackson Blvd.
Chicago, Illinois 60604
Comments: This weekly periodical provides news briefs on new developments in the field. Articles are rarely more than one page in length.

TPT (monthly)
119 Russell Street
Littleton, Massachusetts 01460
508-486-9501
Comments: This periodical emphasizes voice/data networking from a managerial perspective. A typical article might compare and contrast cost benefits of selecting a T-1 multiplexer versus connecting several PBXs directly together.

Data Communications Periodicals

Computerworld (weekly)
375 Cochituate Road
Framingham, Massachusetts 01701-9171
Comments: This weekly periodical provides information on legal developments, personnel changes, and marketing moves in the computer industry.

Connect (monthly)
3Com Corporation
3165 Kifer Road
P. O. Box 58145
Santa Clara, California 95052-8145
Comments: This periodical is published by 3Com and emphasizes 3Com products. It provides excellent information on local area networks and gateways between LANS and mainframes.

Data Communications (monthly)
McGraw Hill
1221 Avenue of the Americas
New York, New York 10020
212-512-2000
Comments: This is the single most important data communications periodical for the telecommunications manager, since the

line between data and voice is rapidly disappearing. This magazine publishes in-depth articles on topics such as the OSI model, SNA, and T-1 connectivity. Many articles are on PBXs or LANS.

Information WEEK (weekly)
CMP Publications
600 Community Drive
Manhasset, New York 11030
516-5000
Comments: This weekly publication features short, Time-like magazine articles on trends in the computer field. Lately more and more articles have emphasized telecommunications, particularly wide area networks.

InfoWorld (weekly)
1060 March Road Suite C200
Menlo Park, California 94052
415-328-4602
Comments: This is one of the "bibles" of the computer field. Lately it has added a special section on networking. It provides excellent information on new product announcements. It also provides in-depth reviews of both software and hardware with each issue usually emphasizing a particular topic such as word processing or facsimile machines.

LAN (monthly)
12 West 21 Street
New York, New York 10010
212-691-1191
Comments: An excellent source of information on such LAN related topics as cabling, protocol, and network software.

LAN Technology (monthly)
P. O. Box 3713
Escondido, California 92025-9843
Comments: This journal publishes technical articles on local area networks. A typical article might provide an in-depth look at how a LAN manager works.

LAN Times (monthly)
122 East 1700 South
Provo, Utah 84601
801-375-0735
Comments: Published by Novell, this periodical emphasizes Novell software, but it provides excellent articles on all areas of local area networking, including micro-mainframe communications. Each issue runs over 100 pages and is packed with information.

MacWEEK (weekly)
P. O. Box 5826
Cherry Hill, New Jersey 08034
Comments: This weekly periodical is an excellent source on the latest information on Macintosh hardware and software. It has started to cover Macintosh networking in more depth, and generally at least two articles on this subject appear each week.

MacWorld (Monthly)
501 Second Street
San Francisco, California 94107
415-243-0500
Comments: A good magazine for keping up with major Macintosh trends. Excellent reviews of major software and hardware products are found in each issue.

PC Magazine (biweekly)
P. O. Box 54093
Boulder, Colorado 80322
Comments: This magazine is one of the best sources of information on new microcomputer hardware and software. Special issues on such topics as printers or modems may provide more than 100 pages on these subjects alone.

PC Week (weekly)
800 Boylston, Street
Boston, Massachusetts 02199
Comments: This weekly paper provides the most in-depth coverage of the latest hardware and software developments with a PC orientation. Its Connectivity section covers voice and data integration and includes a good deal of telecommunication information each week.

PC World (monthly)
501 Second Street
San Francisco, California 94107
415-243-0500
Comments: This monthly magazine provides well-written articles on the whole spectrum of microcomputing. It tends to emphasize trends rather than simply report on new products. A typical article might discuss the future of micro-mainframe communications or look at current uses for electronic information services.

The Telecommunications Manager's Bookshelf

APPENDIX B

BLACK, UYLESS. *Computer Networks: Protocols, Standards, and Interfaces.* Englewood Cliffs, New Jersey: Prentice Hall, 1987.

Comments: This book provides the telecommunications manager with excellent coverage of X.25 networks, upper-level protocols, the PBX as a LAN, and switching and routing in networks. There is also a fine chapter on public networks and carrier offerings.

GURRIE, MICHAEL L. and PATRICK J. O'CONNOR. *Voice/Data Telecommunications Systems.* Englewood Cliffs, New Jersey: Prentice Hall, 1986.

Comments: This technically oriented book provides excellent treatment of switching, signaling, station equipment, and PBX features.

MARTIN, JAMES and JOE LEBEN. *Data Communications Technology.* Englewood Cliffs, New Jersey: Prentice Hall, 1988.

Comments: This book provides an excellent treatment of the OSI model as well as packet-switched networks and ISDN.

———and KATHLEEN KAVANAGH. *SNA: IBM's Networking Solution.* Englewood Cliffs, New Jersey: Prentice Hall, 1987.

Comments: This is the most readable treatment available on this very complex subject. It is particularly useful for the telecommunications manager because it contains chapters on advanced peer-to-peer communications and LU6.2 as well as material on gateways to other networks.

MCCONNELL, JOHN. *Internetworking Computer Systems.* Englewood Cliffs, New Jersey: Prentice Hall, 1988.

Comments: Probably the best single book on this subject at this time. It contains substantial chapters on internetworking issues, including routing and security, as well as in-depth treatment of the OSI model. There are also excellent chapters on wide-area networks and the latest OSI Application layer protocols.

SCHATT, STAN. *Microcomputers in Business and Society.* Columbus, Ohio: Merrill Publishing Co., 1989.

Comments: Over 25% of this text is devoted to the key telecommunications subject of connectivity. This book has extensive chapters on micro-mainframe communications as well as local area networks and the use of productivity software. It contains tutorials on WordPerfect, VP Planner, and dBASE III Plus.

———. *Understanding Local Area Networks.* Indianapolis, Indiana: Howard Sams, 1987.

Comments: This book provides a general survey of local area networks as well as specific information on the major LANS such as IBM's Token-Ring, Novell, and 3Com.

TANENBAUM, ANDREW S. *Computer Networks.* Second edition. Englewood Cliffs, New Jersey: Prentice Hall, 1988.

Comments: This is the most comprehensive and most technical treatment available on local area networks. Tanenbaum approaches LANS using the OSI model and moving from the Physical layer to the Applications layer. He includes excellent chapters on ISDN, FDDI, satellite networks, and internetworking.

TOMASI, WAYNE and VINCENT F. ALISOUSKAS. *Voice/Data with Fiber Optic Applications.* Englewood Cliffs, New Jersey: Prentice Hall, 1988.

Comments: This book is very strong in the areas of both voice and data transmission. It contains excellent chapters on BSC, asynchronous protocol, packet switching, and fiber-optic communications. It is a technically oriented book that requires some mathematical background.

On-Line Information Services and Databases

APPENDIX C

BRS Search, BRS After Dark
BRIS Information Technologies
1200 Route 7
Latham, New York 12110
800-468-0908

CompuServe
5000 Arlington Center Blvd.
Columbus, Ohio 43220
800-848-8199

Delphi
General Videotex
3 Blackstone Street
Cambridge, Massachusetts 02139
800-544-4005

Dialog
Diaglog Information Services
3460 Hillview Avenue
Palo Alto, California 94304
800-334-2564

Dow Jones News/Retrieval
Dow Jones and Company
P. O. Box 300
Princeton, New Jersey 08543
800-522-3567

EasyNet
Telebase Systems
763 W. Lancaster Avenue
Bryn Mawr, Pennsylvania 19010
800-327-9638

GEnie
General Electric Information Services Corp.
401 North Washington Street
Rockville, Maryland 20850
800-638-9636

Lexis
Mead Data Central Inc.
P. O. Box 933
Dayton, Ohio 45401
800-543-6862

NewsNet
NewsNet Inc.
945 Haverford Road
Bryn Mawr, Pennsylvania 19010
800-345-1301

Nexis
Mead Data Central Inc.
P. O. Box 933
Dayton, Ohio 45401
800-543-6862

The Source
Source Telecomputing
1616 Anderson Road
McLean, Virginia 22102
800-336-3366

WestLaw
West Publishing Co.
50 West Kellogg Blvd.
P. O. Box 64526
St. Paul Minnesota 55164
800-328-0109

Manufacturers of Local Area Networks

APPENDIX D

Apple Computer Inc.
20525 Mariani Avenue
Cupertino, California 95014
408-996-1010

AST Research
2121 Alton Avenue
Irvine, California 92714
714-863-1333

AT&T
295 North Maple Avenue
Room 2326F2
Basking Ridge, New Jersey 07920
201-221-2888

Fox Research
7016 Corporate Way
Dayton, Ohio 45459
800-358-1010

Gateway Communications
2941 Alton Avenue
Irvine, California 92714
714-553-1555

IBM
900 King Street
Rye Brook, New York 10573
800-IBM-2468

Novell Inc.
122 East, 1700 South
Provo, Utah 84601
801-379-5900

Proteon Inc.
2 Technology Drive
Westboro, Massachusetts 01581-2800
617-898-2800

3Com Corp.
3165 Kifer Road
Santa Clara, California 95052
408-562-6400

Tops
950 Marina Village Pkwy.
Alameda, California 94501
415-769-8700

Glossary

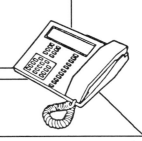

800 Service *See* Inbound WATS.

Accumulator The computer register in which the results of an arithmetic or logical operation are stored.

Add-on-Data Module (ADM) A Northern Telecom interface for handling digital information on one of its PBXs.

Advanced Point-to-Point Communications (APPC) A method by which all computers and computer programs can communicate directly with each other as peers using special programming commands.

Aliasing A condition in which a bad message looks like a good message.

Alphabet The characters that have corresponding bit patterns in a specific data coding scheme.

American Standard Code for Information Interchange (ASCII) A common standard for coding data used extensively with microcomputers.

Amplitude The height or size of an analog wave. It correlates to the loudness, volume or intensity of an analog signal.

Analog Signals that vary continuously over a range. Music and human speech are analog in form. (*See* digital as a contrast.)

AppleTalk An Apple serial bus local area network utilizing a CSMA/CA access method with a speed of 230 kilobits/second.

Application Layer That layer of the OSI model concerned with providing services to application processes including protocols for virtual terminals, electronic mail, and file transfer and management.

Arithmetic Logic Unit (ALU) That portion of a computer where the arithmetic and logical operations are performed.

Asynchronous A family of data communications protocols that send one byte of data at a time. Typical of low to medium speed data communications environments.

Asynchronous Balanced Mode (ABM) Under the HDLC protocol, the Asynchronous Balanced Mode defines the use of combined stations on a point-to-point link where transmission and link level error recovery procedures are shared equally.

Asynchronous Data Unit A device used with AT&T PBXs and local area networks to extend the distance data signals can travel. Voltage levels are also regulated to keep signals from interfering with other traffic. This allows data calls to be carried in the same cable with voice calls.

Asynchronous Interface Line Card (AILC) An interface for a Northern Telecom PBX providing up to four RS-422 ports.

Asynchronous Interface Module (AIM) This Northern Telecom interface converts RS-232 signals to RS-422 signals before transmitting them to an asynchronous interface line card.

Asynchronous/Synchronous Interface Module (ASIM) This Northern Telecom PBX interface accommodates both asynchronous and synchronous transmission modes.

Asynchronous Response Mode (ARM) Under the HDLC protocol, the Asynchronous Response Mode allows a secondary station to initiate transmission without permission from a primary station. ARM can only be used on point-to-point links. (Contrast with Normal Response Mode.)

Asynchronous Terminals Terminals designed to receive and send data using asynchronous transmission.

Asynchronous Transmission Transmission consisting of a start bit followed by a byte of information and followed by a stop bit.

Attenuation The gradual weakening of a signal over distance. As a signal moves farther from its source, energy dissipates and the signal weakens. Attenuation in a wire is usually due to resistance presented by the wire's atoms. Attenuation of a microwave signal is usually due to the tendency for a wave to disperse.

Authorization Code A PBX feature that can be used in two ways: 1) to override restrictions on a telephone so that a denied call can be placed and 2) to provide an additional level of security by requiring entry of a code before calls will be completed. The code itself is a string of numbers assigned to individual users. It is entered into the telephone along with the number to be dialed.

Automatic Circuit Assurance A network facilities test that checks for trunks with abnormally short or abnormally long holding times (holding time is the amount of time a trunk is being used on a call).

Automatic Facilities Test System (AFACTS) A system for providing automatic internal testing of IBM/Rolm/Siemens PBXs.

Automatic Transmission Measurement System A network facilities test that checks for impairment of transmission loss, voice, and echo.

B Channel *See* Bearer Channel.

Backbone A group of bridges connected together to form a very fast switching station to connect several local area networks together.

Bandwidth A measure of the carrying capacity of a facility. The greater the bandwidth, the more information that can be carried. It may be thought of as a pipe. A narrow bandwidth corresponds to a small-diameter pipe, only a limited amount of information can pass through at any one time. A wide bandwidth is like a large diameter pipe, a great deal of information can flow through it.

Baseband Coaxial Cable   Coaxial cable designed to handle high-speed data transmission through one channel.

Basic Rate Interface   The interface for a single access point into the Integrated Services Digital Network (ISDN). The interface is defined as two bearer channels and one data channel (2B + D).

Batch Job   A body of information created by batch processing. Also, the "package" of raw data to be processed.

Batch Processing   Processing raw data that has been packaged into "bundles" or "batches." An efficient way to process large volumes of raw data. Batches are processed as a unit.

Batch Transmission   The transfer of a batch job between two computers or between a computer and a remote printer.

Bearer Channel   A 64 Kbps clear channel used for the conveyance of user information under ISDN.

Bel   A measure of the relative strength of a signal. It is the logarithm of the ratio between a signal's input power and output power. The formula for computing it is:

$$\text{bels} = \log \left(\frac{\text{Input Signal Power}}{\text{Output Signal Power}} \right)$$

Binary   Referring to a base two numbering system.

Binary Synchronous Protocol (BSC)   A synchronous, half-duplex, character-oriented protocol used on many IBM computers.

Bit-Oriented Protocol   A protocol utilizing bits in specific fields to indicate data link control functions.

Bit-Robbing   A transmission technique associated with T-1 facilities. One bit out of each byte transmitted (the least significant bit) is altered to carry signaling information rather than user data. Altering bits does not have a noticeable effect on voice or image transmissions, but does distort data calls. Therefore, this technique should not be used with data. Contrast with clear channel.

Bit Stuffing   The process of inserting an 0 bit following any stream of five or more 1 bits in a HDLC frame. Bit stuffing is used to prevent control or user data from being misinterpreted as an end-of-frame flag.

Bits   A common term referring to binary digits.

Block Check Character (BCC)   A number calculated using a complex algorithm that provides a means of error checking for the information included in that transmission.

Blockage   The inability of a system to complete a call due to lack of talkpaths or facilities.

BRI   *See* Basic Rate Interface.

Bridge   The physical link required to connect two different local area networks.

Broadband Coaxial Cable   Coaxial cable designed to handle high-speed voice and data transmission flowing through a channel simultaneously at various different frequencies.

Brouter   A hybrid that combines the best features of a bridge and a router.

Bus A high-speed physical pathway used for the transfer of data. They are commonly found within computers to move data from one area of the computer to another. Some PBX's also use buses.

Byte Eight bits of information.

Cabinet Cabinets house and protect the components of a PBX. Standard cabinets are 6 feet high by 30 inches wide by 2 feet deep. They can accommodate 4 to 5 carriers. Smaller cabinets are available that hold fewer carriers.

CAD Computer-Aided Design. *See* Computer-Aided Design.

CAE Computer-Aided Engineering. *See* Computer-Aided Design.

Call Management The component of voice network management that deals with collection of station message detail recording (SMDR) records and calculating the cost of them.

CAM Computer-Aided Manufacturing. *See* Computer-Aided Design.

Carrier The rack in a PBX that holds circuit packs. There are electrical connections along the back of the carrier (the backplane) that supply power to the packs and allow them to be interconnected. Carriers hold 10 to 20 circuit packs.

Carrier Sense Multi-Access (CSMA) A contention scheme designed for local area networks in which workstations sense whether another workstation is using the network.

Carrier Sense Multi-Access with Collision Detection (CSMA/CD) A contention scheme for local area networks in which workstations sense not only if other workstations are using the network but also if there has been a data collision.

Cathode Ray Tube The fundamental building block of computer terminals; the term CRT is often used as another term for a computer terminal.

Central Office The network end of a local loop. End user facilities feed into the central office and are connected to local exchange company (LEC) switching equipment. The LEC equipment, in turn, connects to the rest of the network. Commonly thought of as the building or location where this equipment resides.

Central Processing Unit (CPU) That portion of a computer in which instructions are interpreted and executed.

Centralized Central control resting in one location.

Changeout An implementation procedure where the removal of old equipment and installation of new equipment is done simultaneously. (Compare with phased implementation.)

Character-Oriented Protocol A protocol in which ASCII or EBCDIC character sets are used for data link control functions.

Circuit Packs Electronic circuit boards used in PBXs to provide a particular function, such as connect telephones to the common equipment or connect trunks coming in from the Public Switched Network. Circuit packs contain ports, from 2 to 16, depending on the type of pack.

Clear Channel A channel that has no restrictions as far as the format or type of information that may pass through it. Clear channel signaling is associated with ISDN and T-1 facilities. For example, the primary rate interface (see primary rate interface) has 24 channels divided into 1 channel for signaling information and 23 channels for user information. The signaling channel carries transmission information for the other 23. The other channels are then completely free to carry user information (they are clear channels). Contrast with bit-robbing.

Cluster Controller A computer that is used generally to control the functions of several terminals and offload this workload from a mainframe computer to which it is connected.

Coax Elimination and Switching System (CESS) A device for replacing much of the coaxial cabling required to link 3,270 terminals with an IBM cluster controller with twisted pair wire by using a Northern Telecom PBX.

Codec   A microchip used to convert analog voice signals to binary digital code and to reconvert binary digital code back to an analog signal. PCM is the most common digital code used. Codecs are primarily used with digital PBXs.

Combined Station   Under the HDLC protocol, a combined station can act as both a primary and a secondary station. They can operate only in point-to-point mode. The X.25 Recommendation for packet switched networks uses combined stations. (*See* Primary Station and Secondary Station.)

Common Carrier   A company that provides inter-lata services, such as AT&T, MCI or Sprint.

Common Equipment   Located on a user's premises, it is the heart of a telephone system. In a key system, it contains the hardware that allows several station lines to appear on one telephone and allows one station line to appear on several telephones. In a PBX system, it contains the switching matrix along with the line interface and trunk interface equipment.

Communications Protocol   *See* protocol.

Computer-Aided Design   A specialized computer with advanced graphics capabilities used to display graphic images. Usually with the ability to display 3-dimensional representations and to rotate images to allow viewing from different perspectives.

Contention   A situation in which either a computer and terminal, or two workstations on a network, compete for the right to use the same line or transmission channel.

Contention Network   A network in which several workstations contend or compete to use the network.

Coordinate (Switch)   A switching technique that places incoming talkpaths and outgoing talkpaths on metal bars perpendicular to one another. When an incoming bar and an outgoing bar are made to touch, a connection is made.

CPE   *See* Customer Premises Equipment.

Crest   The highest point of a wave; its maximum positive value.

Customer Premises Equipment (CPE)   Equipment located on a customer site. The term comes from the days when all telephone service and equipment was owned by the telephone company. It refers to equipment placed at the telephone customer's location, such as key telephone or PBX equipment.

Cyclic Redundancy Check (CRC)   A complex form of error checking commonly used with synchronous transmission.

D Channel   *See* Data Channel.

Data   Information.

Data Channel   A 16 Kbps (Basic Rate Interface) or 64 Kbps (Primary Rate Interface) channel used for signaling and control information under ISDN.

Data Code   The code that translates information into a form that computers can understand.

Data Link Layer   The layer of the OSI model responsible for segmenting bits into frames, error detection and correction and data flow.

Data Network Interface Card (DNI)   A Rolm circuit card permitting up to 63 simultaneous asynchronous data connections.

Data systems   Computers and data networks.

Datagram A standard message not requiring any acknowledgement.

dB Loss *See* Decibel Loss.

DDD *See* Direct Distance Dialing

Decibel A measure of the relative strength of a signal. It is 1/10 the magnitude of a bel (see bel) and is calculated from the ratio of an input signal to an output signal. The formula for computing decibels is:

$$\text{decibels} = \log \left(\frac{\text{Input Signal Power}}{\text{Output Signal Power}}\right) \times 10$$

Decibel Loss The amount of a signal that is lost between its input and its output. Loss is usually due to attenuation.

Despooling A method used on disk servers to send information back and forth between volumes and to printers.

Dial Backup A fallback mechanism for private data line facilities allowing transmission to continue in spite of a problem with the private line. To keep a transmission path available, the dial backup equipment calls through the PSN over two separate lines and establishes connections with dial backup equipment at the other end of the circuit. The two dialup connections provide a full duplex communication path that can be used until the private line facility is repaired.

Dialtone The electrical connection from a telephone or data device to a switch.

Digital Signals characterized by discrete states. Bytes are digital in form.

Digital Communications Protocol AT&T's proprietary protocol that allows its PBX's to simultaneously transmit voice and data between an end user and the common equipment. Two 64 kilobit/second information channels and one 8 kilobit/second signaling channel is used.

Digital Line Unit A component of a Harris PBX system into which data interfaces are plugged.

Digital Subscriber Line The loop from Customer Premises Equipment (CPE) to the local exchange. *See* Basic Rate Interface.

Digital Trunk Interface (DTI) An interface between an SL-1 PBX and the North American standard T-1 interface.

Direct Access Storage Device (DASD) A random access secondary storage device used in conjunction with mainframe or minicomputers.

Direct Distance Dialing Dialing via the public switched telephone network.

Disk Server A hard disk organized into volumes for the purposes of distributing and storing files on a local area network.

Distributed Divided up in contrast to centralized.

Distibuted Intelligence The placement of microprocessors throughout the common equipment in a PBX or Central Office switch. Distributed intelligence allows different call processing functions to be handled at different levels rather than all functions being handled by a single, central processor.

Distributed Network Architecture A network architecture which uses a layered approach.

Domain Network resources activated by a specific SSCP reside in that SSCP's domain.

DSL *See* Digital Subscriber Line

DTMF *See* Dual Tone Multi Frequency

Dual Tone Multi Frequency A method of telephone dialing using unique tones to represent the digits 0 through 9 and the characters * and #. Two frequencies are blended to create each unique tone. This technique gives the method its name. Digital in form. (*See* analog as a contrast.)

Dumb Terminal A terminal capable only of the most basic elements of viewing and sending information.

EDSL *See* Extended Digital Subscriber Line.

Electronic Key Telephone Systems Small-sized telephone system providing basic switching functions. Provides some features, such as call transfer, conferencing and automatic dialing. Uses analog technology.

Emulate Another term for "imitate."

Encryption A security device that scrambles voice patterns or data streams so that they are unintelligible. Encrypted communications must be decrypted at the receiving end.

Enhanced Trunking The more efficient use of trunks including dynamic channel assignment.

Entry Point Distributed locations for control of SNA resources on an IBM network.

Ergometric Conducive to physical and psychological comfort.

Even Parity An error checking method in which the number of one-bits transmitted are always an even number.

Exchange A node within a network where switching occurs. Normally associated with the switching nodes of the Public Switched Telephone Network.

Execution Time The time required for a computer instruction to be executed and then stored in memory.

Extended Binary-Coded Decimal Interchange Code (EBCDIC) An eight-bit coding method commonly used on IBM's mainframe and minicomputers.

Extended Digital Subscriber Line The loop from Customer Premises Equipment (CPE) to the local exchange. *See* Primary Rate Interface.

Facilities Management The component of voice network management that deals with adding, changing, testing and removing trunks and trunk groups, and with establishing call routing patterns.

Facility Media for transporting information electronically. The media may be copper cable, co-axial cable, fiber optic cable, microwave, or satellite. Facilities near the ground are called terrestrial facilities. These include cables placed underground, on poles or underwater. Microwave is also considered terrestrial. Satellite facilities are considered non-terrestrial. Transmissions over non-terrestrial facilities travel a much longer distance than terrestrial transmissions. This is because non-terrestrial transmissions must travel thousands of miles into space to a satellite and then thousands of miles back to earth. Such a route is called a "hop." Some non-terrestrial transmissions can have several hops. This has little effect on voice communications, but can have a serious effect on data communications. A good rule of thumb is to limit data communications to no more than one hop.

Facsimile A method of transmitting a copy of a document over telephone lines. Facsimile machines work in pairs, one transmits and the other receives. The transmitting machine "reads" light and dark areas of the original document and transmits a signal to the receiving machine. The receiving machine places an appropriate light or dark spot on the copy it is making. In this way, the complete document is transmitted. Facsimile is often referred to as fax.

Fiber Distributed Data Interface (FIDDI) An ANSI standard for a counter-rotating token ring network capable of transmitting data at over 100 megabits/second across fiber cabling.

File Allocation Table A table containing key information on files including their locations on disk.

File Server A computer with a hard disk used to process network commands and store network files.

File Transfer Protocol A protocol for the transfer of files between networks utilizing TCP/IP protocol.

Fingerpointing A problem prone to multivendor networks. When facilities or equipment malfunction, each vendor claims its component is operating properly and the problem must be with other components (points a finger toward other vendors).

Focal Point An IBM network component that facilitates the consolidation of network management data from all components in a network.

Foreign Exchange Service A service provided by local exchange carriers allowing calls to and from an area beyond the local calling area.

Frame The structure defined under ISDN for passing information between the Network Termination and Terminal Equipment. Also, the structure defined by Level 2 of the X.25 Recommendation for packet-switched networks. Frames contain packets and control information necessary for routing data through a network.

Frequency The number of cycles made by a wave in a given period of time.

Frequency Division Multiplexing The use of frequency modulation to create multiple channels within a facility.

Front End Communications Processors A processing unit attached to a mainframe to off-load many communications functions from that mainframe.

Full-duplex A transmission mode in which information flows simultaneously in both directions.

FX *See* Foreign Exchange Service.

Gateway A network workstation with appropriate hardware and software that serves as a gateway or link to other networks with different protocols.

GOSIP The government OSI profile or set of protocols required so that vendors will provide the govenment with products that are compatible.

Grade of Service The number of calls blocked through a PBX or in a network. P.01 grade of service means 1 call in 100 calls is blocked.

Group Switch Module An Ericsson PBX component used to link together three or more line interface modules.

Half-duplex A transmission mode in which information flows alternatively in both directions.

Hertz The unit of measure of a wave's frequency. One hertz equals one cycle per second.

High Level Data Link Control (HDLC) A bit-oriented data link protocol that serves as a superset of SDLC under the OSI model. Both point-to-point and multipoint operations are included.

Hop *See* Satellite Hop.

Horizontal Redundancy Check An error checking method that checks for errors after an entire block of information has been sent.

Host *See* host computer.

Host Computer A computer that supports multiple users. Users use terminals to access a host computer. Host computers are virtually always mini or mainframe computers.

Hot Key The ability of a personal computer to toggle in and out of a software program by pressing a key on the keyboard.

Hot Standby Spare equipment kept on hand for emergency use. Should a component of a system or network fail, the hot standby equipment can immediately be put into service to maintain normal operations.

Hotline A Northern Telecom feature for users who only need to dial a single destination in which a search for an available port is made as soon as the terminal is powered on.

Hub Architecture A network topology that features passive and active hubs to which network workstations are connected.

IEEE 802.3 A standard for a CSMA/CD bus local area network that is very close but not identical to the Ethernet standard developed by DEC, Intel, and Xerox.

IEEE 802.4 A standard for a token bus local area network. This standard is often associated with a MAP network.

IEEE 802.5 A standard for a token ring local area network.

IEEE 802.6 A standard for a metropolitan area network.

Inbound WATS A service offered by common carriers allowing for the called party to pay for a telephone call, thus making the call free to the calling party. This service is patterned after WATS with respect to calling areas and costs.

Institute of Electrical and Electronics Engineers (IEEE) An organization that has developed a number of communications standards. In this book we will focus on its 802 Committee's local area network standards.

Instruction Register The computer register holding the instruction to be executed.

Instruction Time The time necessary for the control unit to fetch an instruction and for this instruction to be decoded and for any appropriate data to be brought into the ALU for processing.

Integrated Services Digital Network (ISDN) A digital, public network providing an array of voice, data and image services. Users dynamically control their own access to the services.

Integrated Voice/Image/Data Terminal (IVID) End user equipment that incorporates voice communication, image communication and data communication capabilities. The terminal has a handset or speaker and microphone for voice communications. It has a keyboard and monitor for data communications. It also has a built-in graphics capability to display video images and computer-generated pictures. Although there are terminals today that provide some of these capabilities for a particular vendor's equipment, IVID terminals will be most useful under ISDN because uniform standards will allow them to work with all vendor's equipment.

Intelligent Network Processor (INP) A device for dynamically allocating T-1 bandwidth.

Interactive Terminal Interface (ITI) The informal name given to CCITT Recommendations X.3, X.28, and X.29. These three collectively define the standards for low speed, asynchronous terminals and devices to communicate with packet networks.

ISDN *See* Integrated Services Digital Network.

ISDN Applications Protocol A protocol developed by Northern Telecom for supporting ISDN.

ISO Model Also known as "The OSI Model," a set-layered protocol for data communications developed by the International Standards Organization.

IVID See Integrated Voice/Image/Data Terminal.

Jam Signal A network signal generated to indicate that a collision has occurred on a local area network.

Kermit An error-checking protocol developed at Columbia that does not require eight bit byte blocks like XMODEM. This protocol can be used on EBCDIC as well as ASCII computers.

Key Equipment Allows the interconnection of telephones with station lines. Several station lines can appear on one telephone and one station line can appear on several telephones. Differs from a PBX in that key equipment does not switch calls. Station lines are fixed in place and cannot be transferred by users.

LANSTAR A local area network utilizing a Northern Telecom PBX and standard twisted pair wire.

LATA See Local Access and Transport Area.

Learning Bridge A bridge designed to link together like as well as unlike LANs offered by different vendors supporting the IEEE 802.2 protocol.

LEC See Local Exchange Carrier.

Line A facility that is switched at one end and terminates a voice or data device at the other end. The connection from a PBX to an end-user's telephone is a line.

Line Interface Module A component of an Ericsson PBX supporting up to 172 voice and 172 data ports.

Local Co-located with a computer.

Local Access and Transport Area One of 160 local calling regions throughout the United States. Known as LATA's, they define the areas of local telephone service that can be provided by a local exchange company (LEC).

Local Exchange Carrier The local telephone operating company.

Local Loop The facility between an end user and a local exchange company's (LEC) central office.

Logical Unit (LU) An SNA point of access for users.

Log in A string of characters, usually a user's name, that is required by a computer, LAN or data network before access to system capabilities is allowed. It is used in conjunction with a password. (*See* password.)

Longitudinal Redundancy Check An error checking method that uses a block check character at the end of the block of data.

LU 6.2 A logical unit permitting peer-to-peer conversations between two application programs.

Machine Cycle The total time including Instruction Time and Execution Time for a computer to fetch an instruction and execute that instruction.

Mainframe Computer A computer defined by its fast processing speed, large storage capabilities in the gigabyte range, and price which often exceeds $100,000.

Manufacturing Automated Protocol (MAP) A protocol based closely on the IEEE 802.4

standard for a token bus network. This protocol was developed to permit devices from different vendors to work together effectively in a manufacturing environment.

Memory Address Register The register holding the next sequential instruction to be executed.

Meridian Call by Call Service Selection A Northern Telecom feature permitting enhanced services on selected calls.

Meridian Customer Defined Networking A Northern Telecom plan for an open architecture consisting of Meridian Network Services, Meridian Network Control, and Meridian SuperNode.

Meridian Data Network System (MDNS) A highly fault-tolerant transport system using intelligent nodes for connecting different networks together utilizing Northern Telecom equipment.

Meridian Network Control Northern Telecom's tools for enhanced network management, diagnostics, planning and design, and customer support services.

Meridian Network Services The software component of Northern Telecom's Meridian Customer Defined Networking.

Meridian SuperNode Northern Telecom's private network node that can be added to existing corporate networks to increase processing power and network management capabilities.

Mesh Architecture A network topology in which a workstation is connected to every other workstation.

Microcomputer A computer generally designed to be a single user desktop model costing less than $10,000.

Minicomputer A computer designed as a departmental multiuser, multitasking unit with less processing power, storage capacity, and cost than a mainframe.

Modem An electronic piece of equipment used to convert digital signals from a computer or other data device to analog signals that can pass over standard telephone lines. Commonly, the frequency or phase of an analog wave is modulated, or some combination is used. Modems also convert the signals back to digital form. The name modem comes from **mo**dulate/**dem**odulate

Modulate Alter or change the characteristics of a wave. Waves are modulated for the purpose of adding information to them. The modulated waves can then be transmitted. At the destination, the waves are demodulated to recover the information.

Multilink Procedure (MLP) The Multilink procedure is part of Level 2 of the X.25 Recommendation. It defines transmission over one or more network links.

Multipoint Connection A configuration in which several terminals share a common communications link.

Multistation Access Unit IBM's term on its token ring network for a wire center.

NetBIOS The Basic Input Output System found on a ROM chip on an IBM PC or compatible.

Network Access Control Devices Security devices that do not allow users to call into a network. Instead, the device calls the user back. Used with data networks.

Network Addressable Units (NAU) An SNA network component that can either send or receive data.

Network Architecture (topology) The network design, which can take many forms including bus, star, ring, or mesh.

Network Control Program (NCP) A physical unit running on a communications controller that helps manage network management functions.

Network Interface Card A circuit card that is placed in a microcomputer's expansion slot and serves as the "brains" governing the communications between a workstation and the rest of the local area network.

Network Layer The layer of the OSI model primarily concerned with internetwork connectivity and the establishment of a virtual circuit.

Network Operations Protocol (NOP) Northern Telecom's layered set of standards based on the OSI model.

Non-blocking A design wherein no calls will be blocked. Refers to a PBX that has been designed with sufficient talkpaths and facilities so that calls will never be blocked.

Normal Response Mode (NRM) The normal operating state of the HDLC protocol. Primary and secondary stations are used with primary stations controlling activity on a link.

Odd Parity An error-checking method in which the number of one-bits transmitted in each byte is always an odd number.

Off-Net Call A call placed from a network extension to a telephone number not on the network.

On-Net Call A call placed from a network extension to another network extension.

Operating System A set of programs that control a computer's operation.

Optic Fiber Cable Cable developed from very thinly stretched plastic or glass that can carry data at speeds of 100 megabit/seconds.

Oscillation The regular, back and forth movement associated with waves.

Out-of-Band Signaling Signaling information is not sent through the same channel as user information. A separate channel is provided for signaling information.

PABX *See* Private Branch Exchange.

Packet The packet is the vessel that carries information from location to location in a packet-switched network. It is analogous to an envelope.

Packet Assembler/Disassembler (PAD) A device used to packetize and reassemble messages in a packet-switched network.

Packet Network A network where information is formed into packets for transmission. Commonly used for data transmission.

Packetizing The process of breaking a message into packets. Used with packet-switched networks.

PAM *See* Pulse Amplitude Modulation.

Parallel Processing A technique used on some supercomputers in which several processing units divide up the work to increase the computer's speed.

Parity Bit A bit that is used for error checking rather than data. A one-bit or zero-bit is placed in this position to ensure that a byte reflects that appropriate number of one-bits for odd parity or even parity transmission.

Parity Check An error checking method in which the number of one-bits received in a byte is counted and then verified against an odd or even parity-checking standard.

Password A string of characters required by a computer, LAN or data network before access to system capabilities is allowed. The character string is usually a nonsense word or collection of characters. Some samples are: "yorban," "fql%e7," and "x13tgrt@." Passwords are intended to be uncommon strings difficult for would-be intruders to guess. They constitute secret codes known only to a user and the system. It is used in conjunction with a log in. (*See* log in.)

PBX   *See* Private Branch Exchange.

PCM   *See* Pulse Code Modulation.

Phased Implementation   An implementation procedure where new equipment is installed in steps, or phases, allowing users time to adjust to it. Old equipment is usually removed in phases, also. (Compare with changeout.)

Physical Layer   The layer of the OSI model responsible for activating, maintaining, and deactivating the physical link betwen a computer and its transmission media.

Physical Unit (PU)   The actual SNA network physical device or communication link.

Physical Unit Control Point   A subset of SSCP functions enabling a node to establish communication with other nodes.

Pipes   A method of transferring information from one file to another or one volume to another volume.

Pitch   The highness or lowness of a sound.

Point-to-Point Connection   A direct connection between each terminal and its computer.

Ports   The interface points where equipment or facilities can "plug in" to a PBX. Ports are located on circuit packs and are usually dedicated to a particular type of connection.

Presentation Layer   The layer of the OSI model concerned primarily with representing information.

PRI   *See* Primary Rate Interface.

Primary Rate Interface   The ISDN interface for access to a multiple user-switching device, such as a PBX. The interface is defined as 23 bearer channels and 1 data channel (23B + D, North American standard) or 30 bearer channels and one data channel (30B + D, European standard).

Primary Station   Under the HDLC protocol, a primary station controls transmission activity on a link. It is responsible for managing secondary stations and for performing link level error recovery. There can be only one primary station per link. (*See* also Secondary Station.)

Private Branch Exchange.   A telephone-switching system located on a customer's premises. Also called Private Automatic Branch Exchange (PABX).

Private Network   A network—dedicated to use by a private group of people. Contrast with public network. Public Network—A network available to the public at large. The most common public network is the Public Switched Network (PSN).

Proc   A procedure. Specifically, one of hundreds of procedures within a System 85 PBX that controls particular aspects of the switch's operation. Procs are used, for example, to add new users, change the locations of telephones, and activate trunks.

Progressive Control   A switching technique wherein calls are setup progressively. Each dialed digit generates pulses that directly control the movement of the switching equipment.

Protected Mode   A mode in which key registers and memory are protected for each program running so that a computer can be multitasking.

Protocol   A set of rules used by computers allowing communication and information transfer.

Protocol Converter   Devices that translate data conforming to one protocol to another protocol that is required by the computer receiving the information.

Public-Switched Network   The public network used to carry voice, data, and image traffic. Local exchange companies and common carriers combine to provide service to all members of the public.

Public Volumes   Volumes which contain information everyone has access to read but usually no access to delete or change.

Pulse Amplitude Modulation   An analog-sampling technique where the amplitude of a sample is replaced by an approximation of the signal's true amplitude. 256 amplitudes are used over the normal voice range (300 – 3,300 Hz).

Pulse Code Modulation   A sampling technique that converts an analog sample to one of 256 8-bit digital codes. Used frequently in digital PBXs and in T-1 carrier transmission systems.

Queue   A holding area. PBXs use a queue for calls waiting to be completed to an Automatic Call Distribution group. Queues are usually set up so that the first call to come in is the first call to go out.

Record Locking   The locking of a particular record while one user makes changes in it so that these changes will not be erased by a second user accessing the same record at the exact same time.

Registered Jack   The point in the Public Switched Network where registered telephone equipment can be plugged in. Commonly referred to as a wall jack.

Remote   Located too far from a computer to allow direct connection.

Remote Bridge   A bridge connecting a computer at a remote site with a network enabling a remote user to use the network's resources.

Response Time   The time required by a data system to accept user input and send a response during an interactive session.

Router   A device operating at the network-layer level of the OSI model designed to connect networks utilizing different protocols.

Sampling   The slicing of an analog voice signal into "pieces" or "samples" that may be passed through a time division multiplexed talkpath.

Satellite Hop   The transmission of a signal from an earth station to a satellite and back to an earth station. A call may make two or more hops before it reaches its final destination.

Secondary Station   Under the HDLC protocol, a secondary station is every station on a link that is not the primary station. Secondary stations cannot transmit without permission from a primary station. (*See also* Primary Station.)

Service Point   Locations for the management of non-SNA resources.

Session   A logical state between two NAUs that permits data transmission between them.

Session Layer   The layer of the OSI model concerned primarily with the organization and synchronization of conversations.

Session Protocol Unit   A data unit containing key information including connection ID, token selections, and parameters required for connection and quality of service.

Sidetone   The feedback from the mouthpiece to the earpiece of a telephone that causes a person to hear a portion of his or her own words. Sidetone allows a telephone conversation to more closely approximate the sound of a face-to-face conversation.

Simple Mail Transfer Protocol   A protocol for providing electronic mail that can travel in both directions under TCP/IP.

Simplex   A transmission mode in which information only flows in one direction.

Sine Wave   A wave form based on oscillation between minimum and maximum values.

Smart Terminal A terminal that has a number of features including memory and the ability to format its screen a number of different ways.

SOPHO-NET A packet-switched open system-wide area network used to link together disparate networks.

Space Division Switching A technique for routing calls used by first generation and early second generation PBXs. One call is carred over each physical path. The maximum, simultaneous call capacity is determined by the number of physical paths in the PBX.

Split An Automatic Call Distribution (ACD) group. In PBXs that support multiple ACD groups, each group is called a split.

Star Topology A network architecture featuring a central computer with cabling to all workstations radiating from this centralized unit.

Station Management The component of voice network management that deals with adding, changing, and removing peripheral telephone equipment.

Stored Program Control A technique for controlling the activities of an electronic switch by means of a program stored in the switch's memory. A very flexible and powerful method of control, it is used in all modern electronic switches.

Subarea The NCP, SSCP, and all network resources physically linked to them are known as a subarea.

Supercomputer The fastest of today's mainframe computers that can execute billions of instructions/second.

Switch Equipment that switches calls. Typically, central office equipment or a PBX.

Switched Circuit Network A network that establishes a physical connection for each call. Facilities are dedicated to the call until it is completed. Contrast with virtual circuit network.

Switching Switching is the ability to connect a call from an originating point to one of many destinations. Calls may be voice, data or image. There may be hundreds, thousands or even millions of destinations.

Synchronous A family of data communications protocols that send a block of data at a time. Typical of medium to high-speed data communications environments.

Synchronous Data Link Control (SDLC) A data link bit-oriented protocol commonly found on IBM computers.

Synchronous Terminal A terminal designed to send and receive the synchronous data transmission common to IBM's larger computers.

Synchronous Transmission Information transmitted in synchronous form which means that the sending and receiving computers must synchronize their clocks prior to transmission.

System Equipment combined in an integrated fashion. (*See* data system and voice system.)

System Fault Tolerance An attempt to build in a measure of redundancy in a network so that if one or more components of the network fail, the network as a whole will retain its integrity and be able to continue operations.

System Network Architecture (SNA) A set of layered protocols designed to facilitate data communications in an IBM computer system environment.

System Service Control Point (SSCP) An NAU that provides the services necessary to manage a network or portion of a network.

T-1 A transmission technique for carrying information digitally through the PSN. A facility with a bandwidth of 1.544 megabits/second is divided into 24 channels of 64 kilobits/second each.

Talkpath The route established for a call through a PBX.

Tandem In a network, it is the ability to pass a call from node to node until it reaches its final destination. Connections are set up as the call is routed so that when the destination is reached, a circuit from end to end is established allowing voice or data traffic to flow.

Task A specific job to be done. Businesses stay in business by performing tasks that are valued by society. Many tasks that used to be done by people are being performed by computers and other intelligent machines.

TDM *See* Time Division Multiplexing.

Terminal A keyboard and display screen used to access a host computer. Terminals usually have no computing capabilities of their own and depend entirely on the host to perform useful work (i.e., they are stupid).

Termination Block The point in the Public Switched Network separating Customer Premises Equipment (CPE), such as a PBX, from the network.

Time Division Multiplexing The division of a facility into timeslots to create multiple channels within a facility. It is a call-routing technique used by late second generation, third and fourth generation PBX's.

Time Division Switching *See* Time Division Multiplexing.

Time-sharing The processing of multiple calls simultaneously by using a "round robin" approach. One stage of each call is completed, then the next stage and the next until all stages are complete.

Timeslot The short intervals of time that the bandwidth of a facility is dedicated to a call. Associated with time division multiplexing.

Toll Office Switching center for the toll-switched network. Owned and operated by a toll vendor such as AT&T, MCI or Sprint.

Traffic A measure of the number and duration of calls within a system.

Traffic Management The component of voice network management that deals with the gathering and analysis of traffic data.

Transmission Control Protocol/Internet Protocol (TCP/IP) A protocol developed for the Department of Defense to provide inter-connectivity for heterogeneous computer systems.

Transport Layer The layer of the OSI model concerned with the transport of data including the monitoring of quality of service found in a transport connection.

Transport protocol data unit Data encapsulated in a packet containing user information and other control information.

Trough The lowest point of a wave; its maximum negative value.

Trunk A facility that is switched at both ends. Typically, a trunk connects a central office to a PBX.

Twisted Pair Cabling consisting of pairs of copper wires (two to several hundred) twisted together to minimize the effects of electronic interference and then placed together in a group. Often associated with PBXs because it is widely used to connect telephones to the common equipment. It is also used extensively in the Public Switched Network and is used in some local area networks.

User A person who uses a system.

Vertical Redundancy Check (VRC) An error-checking method that checks one bit at the end of each byte of information sent.

Video Display Terminal A generic term for a computer terminal that is also often called a cathode ray tube or CRT.

VINES A network operating system developed by Banyan.

Virtual Circuit A logical path between two locations as opposed to a physical path. Intelligent switching equipment is used to set up and route calls between the originating and terminating end. Local area networks and packet-switched networks use virtual circuits.

Virtual Circuit Network A network where it appears to end users that they have a dedicated facility while, in fact, many calls share it. Commonly used in packet networks. Contrast with switched circuit network.

Virtual Terminal Protocol A protocol for resolving differences between terminals by defining a generic type of terminal.

Voice systems Telephone systems and voice networks.

Volts The unit of measure of the strength of an electrical signal. It equals the power of a signal divided by the amount of current.

Volumes Partitioned portions of a hard disk dedicated to certain network users.

WATS *See* Wide Area Telephone Service.

Wide Area Telephone Service A service offered by common carriers allowing for reduced rates for calls within a calling area. The calling area may be part or all of the United States.

Wire Center A hub generally with bypass circuitry to which several different workstations can be connected.

Word Processor A software program used for creating and editing letters, documents, reports, etc. A computer may be dedicated to performing word processing. It is generally a minicomputer capable of supporting multiple users simultaneously.

Word Size The number of bits a processor can handle at one time.

Workstation Usually, a high performance microcomputer. They are often specialized for scientific or engineering tasks.

Xerox Network Systems (XNS) A protocol developed by Xerox for an office automation environment.

XMODEM An error-checking protocol commonly used with microcomputers in which a 128 byte block of information is sent along with a checksum.

Zero Insertion *See* Bit Stuffing.

Zone Unit Message A method of charging for telephone calls within a local calling area found in large metropolitan areas.

ZUM *See* Zone Unit Message.

Index

A

800 Service, 38
Add-on Data Module (ADM), 262
Advanced Program-to-Program Communications (APPC), 162
American Standard Code for Information Interchange (ASCII), 84–85
Amplitude, 8, 10
Amplitude modulation, 14
Analog private lines, 40
Analog transmission, 7
Analog wave theory, 8
AppleTalk, 141–142
Architecture, PBX, 53
Area code, 33
ARPANET, 203
Asynchronous Balanced Response Mode (ABRM), 208
Asynchronous Data Unit (ADU), 242
Asynchronous Response Mode (ARM), 208
Asynchronous transmission, 83
AT&T PBXs, 235–259
Attenuation, 13
Audix, 253–254
Authorization codes, 343
Automatic call distribution, 249
Automatic Facilities Test System (AFACT), 288, 291

B

B Channel, 224
Bandwidth, 15, 71
Baseband coaxial cabling, 114, 116
Basic Rate Interface (BRI), 223–224
Baudot code, 84
Bearer channel, 223, 224
Bel, 12
Bell operating companies, 31
Binary Synchronous Protocol (BSC), 177–179
Bit-robbed signalling, 43
Blocking architecture, 60
Blocking versus nonblocking architecture, 60
Bridge, 149–150
Broadband coaxial cabling, 116–117
Byte, 80

C

Carrier Sense Multiple Access with Collision Detection (CSMA/CD), 121–122
Carrier wave, 14
CCITT, 219
Cellular telephone service, 46
Central office, 31
Central Processing Unit (CPU), 97–98
Centralized System Management (CSM), 251–252
Changeout, 328

Channel Service Unit (CSU), 43
Clear channel signalling, 43
Coax Eliminator & Switching System (CEES), 274
Coaxial cable, 71
Comite Consultatif International Telegraphique et Telephonique (CCITT), 219
Common carriers, 31
Common equipment, 52
Components, voice/data network, 313–325
CompuServe, 215
Computerized Branch Exchange (CBX, See PBX), 49
Coordinate switches, 25
Crossbar switches, 25
Customer Premises Equipment (CPE), 52
Cyclic redundancy checking, 89–90
Cypress (Rolm), 284

D

D Channel, 224
Data channel, 223, 224
Data gathering, 306
Data Network Interface (DNI), 286–287
Data voice network:
  components, 313–325
  design, 304
  implementation, 327–329
dB loss, 13
Decibel, 11
Decibel loss, 13
Definity, 236, 242–245
Design, voice/data network, 304
Development, PBX, 50
Digital Communications Protocol (DCP), 248
Digital Network Architecture (DNA), 164–165
Digital private lines, 42
Digital Service Unit (DSU), 43
Digital Subscriber Line (DSL), 222, 224
Digital Trunk Interface, (DTI), 263
Dimension, 235
Disk server, 111
Distributed Communication System, (DCS), 257–258
Distributed intelligence, 239
Dual tone multi frequency (DTMF), 7

E

Earth station, 75
Electric signals, 13
Electronic blackboard, 45
Electronic Data Interchange (EDI), 188
Electronic information flow, 309
Electronic switches, 28
Electronic tandem network, 257
Encryption, 343–344
End office, 31
Enhanced trunking, 277
Entry Point (SNA), 292
Ericcson PBXs, 294–298
Ethernet, 122–123
ETN, 257
Exchange, 21
Extended Binary Coded Decimal Interchange Code (EBCDIC), 85–86
Extended Digital Subscriber Line (EDSL), 222, 224

F

Facilities, private network, 70
Facsimile, 45
Facsimile machine, 45, 189–190
Fax (*see* Facsimile), 45
Features, PBX, 63
Fiber Distributed Data Interface (FDDI), 137–141
Fiber optic cable, 71, 117
File server, 112
File Transfer Access & Management (FTAM), 187
"Fingerpointing," 338
FM radio, 14
Focal point (SNA), 292
Foreign Exchange Service (FX), 40
Frame, 208, 226
Freeze-motion video, 44
Frequency, 8
Frequency division multiplexing, 16
Frequency modulation, 14
Front end processors, 102
Full-motion video, 44

G

General Electric Information Services Co (GEISCO), 214
Government Open Systems Interconnect Profile (GOSIP), 196–197
Grade of service, 61
Group Switch Module (GSM), 294
Guard band, 15

H

Harris PBXs, 298–302
Hertz, 8

High-level Data Link Control (HDLC), 181–182, 207
Hop, satellite, 76

I

I. series recommendations, 219
IBM PC, 105
IEEE 802.3, 120–121
Image service, 44
Implementation, voice/data network, 327–329
In-WATS, 38
Information flow, 308
Intelligent network processor, 287–288
Inter-connect companies, 31
Interactive Terminal Interface (ITI), 211
Integrated Services Digital Network (ISDN), 217
    data link layer, 226–228
    network terminology, 219
    layer 1, 223–226
    layer 2, 226–228
    layer 3, 228–231
    network layer, 228–231
    physical layer, 223–226

K

Kermit, 90–91
Key equipment, 50

L

LANSTAR, 266–271
Link Access Protocol-B (LAP-B), 227
Link Access Protocol-D (LAP-D), 227
Leave word calling, 252–253
Line Interface Module (LIM), 294
Local Access and Transport Area (LATA), 33
Local exchange company, 31
Longitudinal redundancy checking, 89

M

Macintosh computer, 105–106, 166–167
Main/satellite/tributary networks, 257
Manufacturing Automation Protocol (MAP), 126
Meridian data networking system, 275–277
Meridian network control, 279
Meridian network services, 277
Meridian super node, 279–281
Message center/directory, 254
Messaging, 252–256
Metropolitan area networks, 141
Microwave, 71

Modems, 40
Modulation, 14
Multi-channel Data System (MCDS), 264
Multi-point circuit, 41
Multilink Procedure (MLP), 207
Multiple frame acknowledged information transfer, 227–228

N

Network:
    access control devices, 343
    administration, 341
    architecture, 117–120
    components, 313–325
    design, 304
    distributed communication system, 257–258
    electronic tandem, 257
    implementation, 327–329
    interface card, 113
    main/satellite/tributary, 257
    maintenance, 341–342
Network management functions, 341–344
Network management products, 344–351
Network management:
    centralized, 340
    decentralized, 340
    numbering plan, 33
    optimization, 342
    packet, 30
    private, 30
    public, 30
    security, 342–344
    security equipment, 351–352
    switched circuit, 30
    telephone, 30
    virtual circuit, 30
Nodes, private network, 68
Non-blocking architecture, 60
Normal Response Mode (NRM), 208
Number Plan Area (NPA), 33

O

Off-net calling, 324–325
On-net calling, 324–325
Open System Interconnect Model (OSI), 172–174
Operational flow, 307–309
OS/2, 135–136
Oscillation, 8
Out-WATS, 38

P

Packet Assembler/Disassembler (PAD), 202
Packet network, 200

Packetizing, 201
Panel switches, 24
Phased implementation, 328
Pitch, 8
Point-to-point circuit, 41
Primary Rate Interface (PRI), 223–226
Primary stations, 207
Private Automatic Branch Exchange (PABX, see PBX), 49
Private Branch Exchange (PBX), 49
   architecture, 53
   development, 50
   evolution, 56
   features, 63
   first generation, 56
   fourth generation, 63
   second generation, 57
   third generation, 59
Private Lines, 40
Private network, 68, 323–325
   facilities, 70
   nodes, 68
Protocol converters, 102
PS/2 (IBM), 105–106
Public Switch Network, 30, 50
Pulse Amplitude Modulation (PAM), 58
Pulse Code Modulation (PCM), 59

R

Regional Bell operating companies, 31
Remote access, 68
Rolm PBX, 283–294
Rotary dialing, 5
Routers, 150–151

S

Sample, 58
Satellite facilities, 74
Satellite hop, 76
Secondary stations, 207
Security, 342–344
Service point (SNA), 292
Sidetone, 5
Signalling System 7 (SS7), 231–232
Signalling:
   bit-robbed, 43
   clear channel, 43
Sine wave, 9
Slow-motion video, 44
Sound waves, 9
Space division switching, 57
Station Message Detail Recording (SMDR), 249–251

Steering digits, 22
Step-by-step switches, 22
Stored program control, 28
Strowger, Almon, 23
Supercomputer, 96
Switches:
   coordinate, 25
   crossbar, 25
   electromechanical, 22
   electronic, 28
   panel, 24
   progressive control, 22
   step-by-step, 22
Switching techniques, 19
Synchronous Data Link Control (SDLC), 180–181
Synchronous transmission, 83
System 25, 236, 238–240
System 75, 236, 238–240
System 85, 236–238
System fault tolerance, 142–144
System Network Architecture (SNA), 153–161

T

T-1 lines, 42
Teleconferencing, video, 44
Telenet, 213
Telephone and Telegraph Consultative Committee of the International Telecommunication Union, 329
Telephone networks, 30
Telephones, operation, 2
Terminals, 91–95
Time Division:
   multiplexing (TDM), 17, 57
   switching, 57
Time-sharing, 29
Timeslot, 17
Token bus networks, 124–126
Token-ring network (IBM), 128–133
Tone dialing, 6
Transmission Control Protocol/Internet Protocol (TCP), 190–194
Twisted pair, 54, 71, 113–114
Tymnet, 214

U

Unacknowledged information transfer, 227–228
Unified messaging, 254–256
Universal mailbox, 254

V

Vertical redundancy checking, 87, 89
Video:
   freeze-motion, 44
   full-motion, 44
   slow-motion, 44
Video teleconferencing, 43
Video transmission, 44
Virtual circuit, 200
Virtual Terminal (VT), 189
Voice/data network:
   components, 313–325
   design, 304
   implementation, 327–329
Volts, 13

W

Waves, 8
Wide Area Telecommunications Service (WATS), 37

X

X.1, 207
X.3, 204
X.21, 204–205
X.25, 203–204
X.28, 204
X.29, 205
X.75, 204
X.400, 188
X.500, 187
Xerox Network Systems (XNS), 196
XMODEM, 90